LIGHT–MATTER INTERACTION

T0177749

Light–Matter Interaction

Physics and Engineering at the Nanoscale

Second edition

John Weiner

Université Paul Sabatier, Toulouse France and IFSC Universidade des São Paulo, São Carlos, SP Brazil

Frederico Nunes

Federal University of Pernambuco, Recife, Brazil

OXFORD

UNIVERSITY PRESS

OXFORD
UNIVERSITY PRESS

Great Clarendon Street, Oxford, OX2 6DP,
United Kingdom

Oxford University Press is a department of the University of Oxford.
It furthers the University's objective of excellence in research, scholarship,
and education by publishing worldwide. Oxford is a registered trade mark of
Oxford University Press in the UK and in certain other countries

First Edition published in 2013
Second Edition published in 2017

Impression: 1

Published in the United States of America by Oxford University Press
198 Madison Avenue, New York, NY 10016, United States of America

British Library Cataloguing in Publication Data
Data available

Library of Congress Control Number: 2016945264

ISBN 978–0–19–879666–4 (hbk.)
ISBN 978–0–19–879667–1 (pbk.)

Printed and bound by
CPI Group (UK) Ltd, Croydon, CR0 4YY

Preface

Light–matter interaction pervades the disciplines of optical and atomic physics, condensed matter physics, electrical engineering, molecular biology, and medicine with frequency and length scales extending over many orders of magnitude. Deep earth and sea communications use frequencies of a few tens of Hz, and X-ray imaging requires sources oscillating at hundreds of petaHz (10^{15} s^{-1}). Length scales range from thousands of kilometres to a few hundred picometres. Although the present book makes no pretence to offer an exhaustive treatise on this vast subject, it does aim to provide advanced undergraduates, graduate students, and researchers from diverse disciplines, the principal tools required to understand and contribute to rapidly advancing developments in light–matter interaction centred at *optical* frequencies and length scales, from a few hundred nanometres to a few hundredths of a nanometre. Classical electrodynamics, with an emphasis on the *macroscopic* expression of Maxwell's equations, physical optics, and quantum mechanics provide their own perspectives and physical interpretations at these length scales. Circuit theory and waveguide theory from electrical engineering furnish useful analogies and often offer important insights into the nature of these interactions. A principal aim of this book is to deploy this arsenal of powerful tools so as to render the subject in forms not likely to be encountered in standard physics or engineering courses, while not straying too far into eccentricity.

This book builds on an earlier one, *Light–Matter Interaction, Physics and Engineering at the Nanoscale*, that I wrote with Frederico Nunes. Much of the material in Chapters 2, 4–8, and 12 remains essentially unchanged from this earlier work, although the organisation has been improved and some minor mistakes corrected. Chapters 1 and 3 have been expanded somewhat, and chapters 9–11 are entirely new. The motivation behind writing this book was to include the subject matter of current research interest, such as metamaterials and light forces on atomic and nanoscale objects. The chapter on momentum in fields and matter grew out of the considerations on light forces, and I was surprised to find how subtle and slippery this subject is. It is my hope that the chapter will provide a deeper understanding of how momentum transport can affect the nature of forces and force distributions on ponderable objects. The subject of momentum flux between fields and matter is as important as the more frequently treated (via Poynting's theorem) subject of energy flux.

After a historical synopsis of the major milestones in the human understanding of light and matter in Chapter 1, the subject begins in earnest with a review of conventional electrodynamics in Chapter 2, *Elements of Classical Electrodynamics*. The intent is to reacquaint the reader with electric and magnetic force fields and their interactions with ponderable media through Maxwell's equations and accompanying force laws, such as the common Lorentz force law. We emphasise here, macroscopic quantities of permittivity and permeability, and through the constitutive relations, polarisation and magnetisation fields. Dipole radiation, space-propagating and surface-propagating wave solutions to Maxwell's equations are all fundamental to understanding energy and momentum transport around, and through, atomic-scale and nanoscale structured materials. The chapter ends with a development of plane wave propagation in homogenous media and at dielectric and metallic surfaces.

Chapter 3, *Physical Optics of Plane Waves*, introduces the phasor representation, as well as the first mention of the expressions for energy density and energy flux. These key notions will recur continually throughout the book in various contexts. The second half of the chapter treats reflection and transmission, total internal reflection, the Fresnel coefficients, real material interfaces, and plane-wave behaviour in a lossy, conductive medium at high frequency.

Chapter 4, *Energy flow in Polarisable Matter*, covers the time evolution of energy flux when electromagnetic waves propagate through media with electric polarisation. We point out analogies between the behaviour of classical fields in bulk matter with the energy dynamics of reactive and dissipative electronic circuits. In the section on polarisation and polarisability, it is shown how the macroscopic electric, polarisation, and displacement fields can be related to microscopic atomic and molecular properties through the Clausius–Mossotti equation that expresses the dielectric constant of a material (a macroscopic property) in terms of the microscopic polarisability of the constituent atoms or molecules.

Chapter 5, *The Classical Charged Oscillator and the Dipole Antenna*, is next presented for its own intrinsic and practical interest as well as an application of the foregoing principles. It is shown how a 'real' antenna can be built up from an array of oscillating charges and how an array of macro-antennas can be used to concentrate the spatial direction of emission or reception. The treatment here is fundamental with a fairly conventional engineering perspective, but it lays the groundwork for a thorough understanding of atomic, molecular, and nanoscale dipole emitters and absorbers.

Chapter 6, *Black-body Radiation*, reviews the Rayleigh–Jeans and Planck distributions. The presentation shows how any radiation law must be the product of mode counting and the distribution of energy per mode. It is shown that the key to avoiding the 'ultraviolet catastrophe', and to obtaining agreement with experimental measurement, is to use the Planck distribution. This chapter also provides some necessary background material and context for the discussion of dipole emitters interacting with hyperbolic metamaterials.

Chapter 7, *Surface Waves*, is devoted to a fairly extensive discussion of waves at the interface between dielectrics and metals, because they play such an important role in 'plasmonic' structures and devices. In fact, this propagation can be expressed in terms of circuit and waveguide theory, familiar to electrical engineers. At the opening of the twentieth century, surface waves were thought to be the means by which radio transmission was carried beyond the earth's curvature, and the importance of this subject motivated the extensive analysis that appears in Arnold Sommerfeld's celebrated *Series of Lectures on Theoretical Physics, Volume VI*; especially chapter VI, *Problems of Radio*. Although the ionosphere was found to be responsible for long-distance radio transmission,

Sommerfeld's analysis laid the groundwork for understanding atomic and molecular emission near surfaces and the importance of anisotropic metamaterials for reflection and transmission.

Chapter 8, *Transmission Lines and Waveguides*, establishes the correspondence between classical electromagnetics and circuit properties such as capacitance, inductance, and impedance. Rectangular and cylindrical geometries are discussed at length because of their importance in conventional microwave-scale waveguides as well as in nano-fabricated light-guiding devices. TM and TE waveguide modes (as distinct from TM and TE polarisation) are discussed in detail. The chapter ends with a presentation of how waveguide modal analysis and impedance matching can be used to guide the design of nanoscale optical devices.

Chapter 9 introduces the notions of 'left-handed materials', negative-index metamaterials, and waveguides, and how they may be used to tailor light flow. The field of metamaterials develops new directions and applications with the appearance of each monthly, or even bi-weekly, issue of the principal research journals. To try to present the 'latest and greatest' in this chapter would be futile, so the emphasis is rather on the basic physics, and especially, transmission and reflection in periodic stacked layers. This geometry is the simplest implementation of fabricated anisotropic materials with engineered properties of transmission and reflection.

Chapter 10 examines the meaning of momentum in electromagnetic fields and how that momentum interacts with ponderable media. Energy conservation in electromagnetics enters by way of Poynting's theorem, and the Poynting vector expresses energy power flux (Watts per m^2 in SI units). The Einstein thought experiment establishes the need for a similar conservation principle for momentum transmission between fields and matter. The question of field momentum is crucial to a thorough understanding of light forces (which must be equivalent to the time rate of change of momentum as it passes between field and object), such as the radiation pressure force and the dipole-gradient force. We examine the Abraham-Minkowski controversy on the 'correct' way to express optical momentum inside ponderable matter and discuss, in some detail, the key experiments whose motivation was to resolve the controversy. The experiments, at least at this writing, have only managed to send the conflicting analyses in new directions. The chapter ends with an extended discussion of light momentum on a point dipole (standing in for a two-level atom) and summarises important articles cited and referenced in Chapter 10. This discussion is a natural lead-in to the next chapter on atom-optical forces, optical cooling, and trapping.

Chapter 11 presents the simplest and most intuitive approach to atom-light-field interactions: the atom as a damped harmonic oscillator with spontaneous emission as the damping agent. The next step is the semiclassical two-level atom, initially introduced as a point dipole (but now with two internal states) at the end of Chapter 10. The semiclassical two-level atom sets the stage for establishing light forces at the atomic level: the dipole-gradient force and the radiation pressure force. Finally, the optical Bloch equations are introduced, which facilitates the presentation of the last section on atom Doppler cooling.

Chapter 12, *Radiation in Classical and Quantal Atoms*, introduces light–matter inter-action at the atomic scale (a few hundred picometres) and at interaction energies less than, or comparable to, the chemical bond. Under these conditions the subject can be very well understood through a semiclassical approach in which the light field is treated classically and the atom quantally. We therefore retain the *classical* electrodynamics treatment while presenting a very simple *quantum* atomic structure with dipole tran-sitions among atomic and molecular internal states. We take a physically intuitive, wave mechanical approach to the quantum description in order to bring out the analogies be-tween classical light waves, quantum matter waves, classical dipole radiation, and atomic radiative emission.

A number of Appendices have been included that provide supplementary discus-sion of the analytical tools used to develop the physics and engineering of light–matter interaction. Appendix A lists numerical values of important fundamental constants and dimensions of electromagnetic quantities. Appendix B is a brief discussion of systems of units in electricity and magnetism. Although the Système International (SI) has now been almost universally adopted, it is still worthwhile to understand how this system is related to others; what quantities and units can be chosen for 'convenience' and what are the universal constraints that all systems must respect. Students should not be deterred from studying earlier articles and texts simply because of an unfamiliar system of units. Appendix C is a brief review of vector calculus that readers have probably already seen, but who might find a little refresher discussion useful. Appendix D discusses how the important differential operations of vector calculus can be recast in different coordinate systems. Although the Cartesian system is usually the most familiar, spherical and cy-lindrical coordinates are practically indispensable for frequently encountered problems. Much of the book deals with harmonically oscillating fields, and Appendix E is a suc-cinct review of the quite useful phasor representation of these fields. Finally, Appendices F, G, and H present the properties of the special functions, Laguerre, Legendre, and Hermite, respectively, that are so commonly encountered in electrodynamics and quan-tum mechanics. These Appendices are an integral part of the book, not just some 'boiler plate' nailed on at the end. Readers are strongly encouraged to pay as much attention to them as they do to the Chapters.

Most of the material in this book is not new nor original with the authors. Excellent texts and treatises on classical electrodynamics, physical optics, circuit theory, wave-guide and transmission line engineering, atomic physics, and spectroscopy are readily available. The real aim of this book is take the useful elements from these disciplines and to organise them into a course of study applicable to light–matter interaction at the nano-scale and the atomic scale. To the extent, for example, that waveguide mode analysis and sound design practice in microwave propagation inform the nature of light transmission around and through fabricated nanostructures, they are relevant to the purposes of this book. Rugged, reliable coherent light sources in the optical and near-infrared regime, together with modern fabrication technologies at the nanoscale, have opened a new area of light–matter interaction to be explored. This exploration is far from complete, but the present book is intended to serve as a point of entry and a useful account of some of the principal features of this new terrain.

Fundamental Constants and Symbols

m_u	atomic mass constant	1.660539×10^{-27} kg
m_p	proton mass	1.672622×10^{-27} kg
m_e	electron mass	9.109382×10^{-31} kg
g	acceleration of gravity	9.80665 m s^{-2}
G	gravitation constant	6.674287×10^{-11} m^3 kg^{-1} s^{-2}
r_e	classical electron radius	2.817941×10^{-15} m
F	force	N
G	momentum	kg m s^{-1}
ε_0	vacuum permittivity	8.854187×10^{-12} F m^{-1}
μ_0	vacuum permeability	12.566370×10^{-7} N A^{-2}
$\varepsilon, \varepsilon_r$	dielectric constant	unitless
μ, μ_r	relative permeability	unitless
h	Planck constant	6.626070×10^{-34} J s
$\hbar$	Planck constant/2π	1.054572×10^{-34} J s
c	vacuum speed of light	299792458 m s^{-1}
ν	frequency	s^{-1}
ω	angular frequency	s^{-1}
λ	wavelength	m
k	wave vector	m^{-1}
T	temperature	K
k_B	Boltzmann constant	1.380650×10^{-23} J K^{-1}
σ	Stefan–Boltzmann constant	5.670367×10^{-8} W m^{-2} K^{-4}
N_A	Avogadro constant	6.022142×10^{23} mol^{-1}
R	molar gas constant	8.314472 J mol^{-1} K^{-1}
μ_B	Bohr magneton	$927.400999 \times 10^{-26}$ J T^{-1}
e	electron charge	1.602177×10^{-19} C
q	electric charge	C
ρ	electric charge density	C m^{-3}
Φ	magnetic flux	Wb
E	electric field	V m^{-1}
B	magnetic induction field	Wb m^{-2}
D	electric displacement field	C m^{-2}
H	magnetic field strength	A m^{-1}
J	electric current density	A m^{-2}
P	electric polarisation field	C m^{-2}
M	magnetisation field	Wb m^{-2}
S	power flux density	W m^{-2}

χ, χ_e	electric susceptibility	unitless
χ_m	magnetic susceptibility	unitless
L	inductance	Wb A^{-1}
C	capacitance	F
$E, \mathscr{E}$	energy	J
P	power	W
I	electric current	C s^{-1}
R	electric resistance	Ω
V	voltage, potential	J C^{-1}
ρ	resistivity	$\Omega\,\text{m}$
σ, κ	conductivity	S m^{-1}
Z	impedance	Ω
Y	admittance	S
n	refractive index	unitless
η	Re[n]	unitless
κ	Im[n]	unitless
R	reflection coefficient	unitless
T	transmission coefficient	unitless

Contents

1

Historical Synopsis of Light–Matter Interaction

The phrase 'light–matter interaction' covers a vast realm of physical phenomena from classical to quantum electrodynamics, from black holes and neutron stars to mesoscopic plasmonics, and nanophotonics to subatomic quantum objects. The term 'interaction' implies that light and matter are distinct entities that influence one another through some intermediate agent. The history of scientific inquiry from the earliest times to the present day might be neatly summarised into three questions: what is the nature of light itself, of matter itself, and of the interaction agent? We now know from Einstein's celebrated equation $E = mc^2$ that light (E) and matter (m) are fundamentally manifestations of the same 'thing', related by a universal proportionality constant: the square of the speed of light in vacuum (c^2). Nevertheless, under ambient physical conditions normally found on earth, the distinction between light and matter makes sense; their interaction meaningful and worth studying.

1.1 Light and matter in antiquity

In the fifth century BC, Leucippus, a Greek philosopher from Miletus (now in Turkey), founded the school of *atomism* in which the universe is composed of immutable, indestructible, indivisible atoms and the space through which they move: the *void*. His best student was Democritus (460–370 BC) who elaborated the atomistic construct of the universe, attributing natural phenomena to the motion of atoms and the diversity of material objects to their shapes and interlocking structures. The most extensive account of the Leucippus-Democritus atomic theory appears in an extended epic poem, *De rerum natura* (*The Nature of Things*) by Lucretius, a Roman, who lived much later (99–55 BC).

A contemporary of Democritus, the Greek philosopher Empedocles (490–430 BC), proposed that the cosmos was composed of four elements: fire, air, water, and earth. Like the atomist school, these elements were immutable and the diversity of nature arose

Light-Matter Interaction. Second Edition. John Weiner and Frederico Nunes.
© John Weiner and Frederico Nunes 2017. Published 2017 by Oxford University Press.

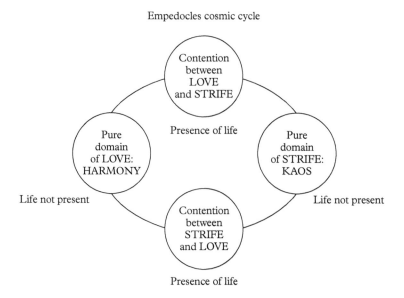

Figure 1.1 *The cosmic cycle of Empedocles. Creative Commons by-sa 3.0, Paolo Anghileri.*

from their combinations. The dynamics of the combinations are affected by two forces, repulsive and attractive, known as love and strife, respectively. Figure 1.1 illustrates Empedocles' scheme.

Empedocles is also credited with proposing the first theory of light. His idea was that light particles stream out of the eyes and contact material objects. Euclid (about 300 BC) assumed this flux moved in straight lines and used the idea to explain some optical phenomena in *Euclid's Optics*, a very influential early treatise on optics. *Euclid's Optics*, in turn, influenced Claudius Ptolemy (AD 90–168), a Roman citizen living in Egypt, whose writing on geocentric astronomy was considered definitive until the European Renaissance.

1.2 The Golden Age of Sciences in Islam

The Golden Age of Sciences in Islam was around the year AD 1000, at the time of Ibn-i-Sina (Avecenna), the last of the Islamic mediaevalists, and Ibn-al-Haitham (Alhazen, AD 965–1039), the first of the modernists. Alhazen enunciated that a ray of light, when passing through a medium, takes a path that is easier and 'quicker', anticipating Fermat's *principle of least time* by many centuries (see Section 1.3). Contrary to Empedocles and Ptolomy, Alhazen believed that the eye detected light from an external source. Figure 1.2 shows an illustration from the title page of a Latin translation of Alhazen's *Book of Optics*.

Figure 1.2 *Illustration from the title page of the 1572 edition of* Opticae Thesarus–*Latin translation of Alhazen's* Book of Optics. *The figure illustrates many of the properties and uses of light. Perspective, refraction, reflection, the rainbow, the periscope, and early photonic naval defences are represented. Figure in public domain.*

1.3 Light and matter in the European Renaissance

By the beginning of the seventeenth century, the certitude of received ideas was crumbling. Earth as the centre of the universe and Europe as the centre of the earth was cast into doubt. The Americas were discovered by European explorers between 1492 and 1504, the earth had been circumnavigated by 1522, and the Ptolemaic geocentric astronomical system had been effectively overthrown by the Copernican heliocentric revolution of 1543. The invention of the telescope in 1608 enabled Galileo to show that Jupiter's moons revolved around that planet, not the earth. This discovery of observational astronomy ran counter to Church dogma, but Pope Urban VIII was actually sympathetic to Galileo's scientific way of thinking. Unfortunately, Galileo published a 'dialogue' consisting of a conversation among three people: Salviati, a smart scientist who bore a striking resemblance to Galileo and who argued for the Copernican system; Sagredo, a sort of moderator in the dialogue who asked intelligent questions; and Simplicio, who tried to defend the conventional Aristotelian method of speculating by pure thought before eventually having recourse to cite the mysteries of the unknowable hand of God. Unfortunately, Simplicio's arguments uncomfortably paralleled those of Urban VIII himself. The Pope did not take kindly to being embarrassed and Galileo lost his protection from the Inquisition. Figure 1.3 shows the frontispiece and title page of the *Dialogue* published in 1632.

Into this fluid situation stepped René Descartes (1596–1650) with a new world view. Descartes proposed that the universe consisted only of matter and motion. Forces could only be propagated among massive bodies by actual contact, and therefore the apparent space between celestial bodies, the 'void' of Democritus, was actually filled with a kind of very fine-grained material medium or plenum. Light emission, reflection, refraction,

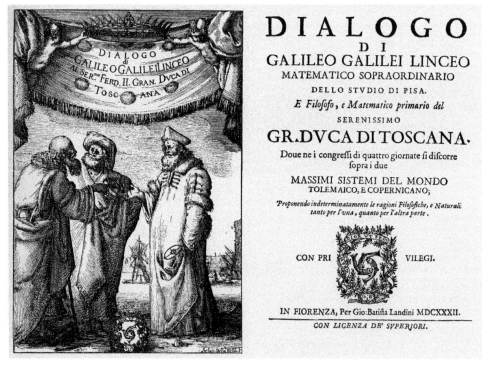

Figure 1.3 *Frontispiece and title page of the* Dialogue *published in Florence in 1632. Simplicio is on the left. Figure in public domain.*

and absorption, were all explained in terms of material flux. The notion of force 'fields' and action at a distance had no place in the Cartesian system of the universe. Descartes' interpretation of refraction, however, was severely challenged by Pierre de Fermat (1601–1665) who explained the deviation of light rays on the basis of the *principle of least time*. Applying this principle, Fermat derived that the sines of the incident and refracted angles are in constant ratio, essentially the equivalent of what we now commonly term 'Snell's law'. Descartes also derived this law, but his interpretation of light as particle flux required greater velocity in the denser medium, whereas the principle of least time imposed a slower velocity. Fermat's principle is in accord with the modern expression for the velocity of light, $v = c/n$ where n, the index of refraction, is unity in free space and greater than unity in material media.

The next significant observation was light 'diffraction', a term coined by Francesco Grimaldi (1618–1663) to describe the appearance of light beyond the geometrical shadow boundary defined by the supposed rectilinear motion of light-particle flux. Figure 1.4 shows a portrait of Grimaldi and his diagram for light appearing beyond the geometric limits of an angular opening. Diffraction was also observed by Robert Hooke (1635–1703) who conjectured that light was due to rapid vibratory motion of the very small particles of which ordinary matter is composed. Furthermore Hooke had the brilliant insight that light (still considered as a kind of matter flux) propagated outwards from the centre of each tiny vibrating centre in circular figures and that light 'rays' were trajectories at right angles to these circular figures. This view of light propagation laid the foundation for the construction of wave fronts from which Hooke was able to explain refraction. He also tried to interpret colours in terms of refraction, but his colour theory was challenged by Isaac Newton (1642–1727) who correctly interpreted colour

 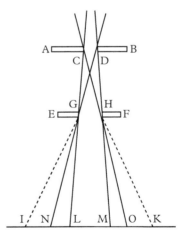

Figure 1.4 *Grimaldi and his diffraction diagram that shows that the rays* I, K *extend beyond the angular limits* N, O *defined by the two apertures. Figures in public domain.*

as an intrinsic property of light and not a distortion of it due to refraction. Although Hooke took the first steps toward a wave theory of light, it was Christiaan Huygens (1629–1695) who put it on a firmer foundation by expressing refraction in terms of the principle that *each element of a wavefront may be regarded as the centre of a secondary disturbance giving rise to spherical waves. The wavefront at any later time is the envelope of all such secondary wavelets.* This principle was later refined and extended by the French engineer Augustin-Jean Fresnel (1788–1827) to establish modern wave optics, based on the Huygens-Fresnel principle. It successfully explains light intensity modulations due to diffraction.

Throughout antiquity and into the seventeenth century, however, light and matter were not considered to be intrinsically different. Light was simply a manifestation of matter, either in linear flux, or in vibratory motion. Huygens also discussed the phenomenon of double diffraction in Iceland crystal and interpreted it in terms of the propagation of ordinary and 'extraordinary' waves within the crystal. These studies were later considered by Newton and led to the discovery of light polarisation. Newton believed that longitudinal, compressive waves could never account for polarisation and his arguments essentially laid to rest the wave interpretation of light until it was revived by Thomas Young (1773–1829) who demonstrated interference in the celebrated *Young's double slit experiment.* Figure 1.5 shows the classic diagram of the two-slit experiment that provided powerful evidence of the wave nature of light.

Meanwhile, the atom theory of matter was being transformed from an antique speculative philosophical proposition to a working scientific hypothesis that was refined, enlarged, and tested by quantitative experiments. In 1643, Evangelista Torricelli (1608–1647) established that the 'air' of everyday experience exerted pressure and was therefore a tangible, if rarified, material composed of particles in constant motion.

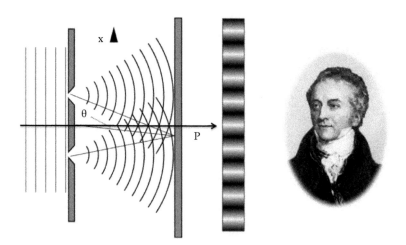

Figure 1.5 *Thomas Young and a diagram of the two-slit experiment. Notice that a bright band always appears on the centre-line. Left figure created under Creative Commons by-sa 3.0; right figure in public domain.*

In England, Newton published the *Philosophiae Naturalis Principia Mathematica* in 1687, laying out his laws of motion and theory of gravitation supported by precise, quantitative celestial mechanics. Although Newton did not like to admit it, his gravitational force law did not require an intervening plenum between bodies with mass. Contrary to Descartes, action-at-a-distance was tacitly, if not overtly, admitted.

Robert Boyle (1627–1691) established one of the fundamental gas laws—that the pressure was inversely proportional to volume at fixed temperature—and helped the science of chemistry emerge from obscurantist alchemy by publishing a truly scientific treatise: *The Sceptical Chymist*. His distinction between heterogeneous mixtures and homogenous compounds laid the groundwork for the truly revolutionary advances of the eighteenth century.

1.4   The revolution accelerates

At the beginning of the eighteenth century the nature of light was still very much in doubt. The notion, commonly attributed to Newton, that light consisted of beams of particles, or 'corpuscles,' travelling in straight lines through homogenous media, held sway. But from 1801 to 1803, Thomas Young carried out well-designed diffraction experiments that put the wave theory of light back in the race. Although Young's experiments (especially the double-slit experiment published in 1807) dealt a body blow to the corpuscular theory, light waves were still considered to be vibratory motion in the longitudinal direction, along the direction of propagation. As such, they could not account for polarisation. In 1821, the French engineer Augustin-Jean Fresnel (1788–1827) showed that polarisation was consistent with the wave picture if the periodic vibration was *transverse* to the direction of propagation. This finding removed the principal remaining objection to the wave model of light. The dramatic *Arago spot* demonstration, illustrated in Figure 1.6, in which a bright central spot can be observed in the geometric shadow of an opaque circular screen, laid to rest the corpuscular model forever (or so it seemed at the time).

At the same time that Young and Fresnel were correctly characterising the wave nature of light, Michael Faraday (1791–1867) was carrying out experiments in electricity and magnetism. Faraday pictured magnetic influences acting on bodies not in direct contact as *lines of force*. These lines, originating and terminating in closed loops, were the beginnings of force *fields* acting on bodies through space with no actual physical contact. Together with Newton's theory of gravitation, these ideas elevated action-at-a-distance to serious consideration. The next step was the great unification of electric, magnetic, and optical phenomena effected by James Clerk Maxwell (1831–1879). In 1865, Maxwell published *A Dynamical Theory of the Electromagnetic Field* in which he set forth the proposition that light was in fact a transverse electromagnetic wave. In modern form, the four Maxwell equations and the Lorentz force law constitute a unified classical theory of electricity, magnetism, and light.

While Young, Fresnel, and Maxwell were putting light, electricity, and magnetism on a firm foundation, an understanding of ponderable matter and chemical interaction

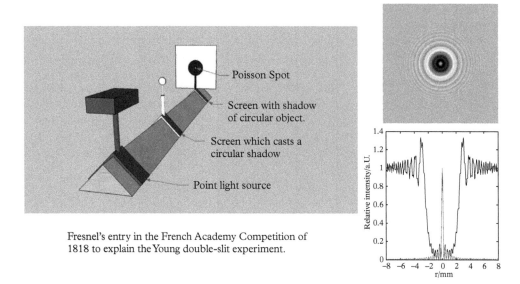

Poisson Spot

Screen with shadow of circular object.

Screen which casts a circular shadow

Point light source

Fresnel's entry in the French Academy Competition of 1818 to explain the Young double-slit experiment.

Figure 1.6 *The celebrated Arago or Poisson spot experiment diagrammed on the left with the simulated interference measurement and intensity trace through the centre on the right. All figures, Creative Commons by-sa 3.0, Thomas Reisinger.*

was also advancing rapidly. Building on Boyle's experiments with gases, Joseph Priestly (1733–1804) conducted extensive experiments on 'airs' and identified nitric (NO) and nitrous (N_2O) oxides, and oxygen (O_2), which he called 'dephlogisticated air'. This term refers to 'phlogiston', a substance that was thought to be contained within matter and expelled during combustion. Although now discredited by modern understanding of oxidation, Priestly used the principle of phlogiston to rationalise his observations. In France, Joseph Louis Gay-Lussac (1778–1850) annunciated another physical gas law, complementary to Boyle's law. Gay-Lussac found that for a given quantity of gas, the volume was directly proportional to the temperature at constant pressure. Back in England, John Dalton (1766–1844) ascertained that elements combine in simple number ratios, and he began to determine atomic masses. Dalton also determined that the pressure of a mixture of gases was equal to the sum of the pressures of the individual constituents. However, the major unifying advance in understanding chemical reactivity was carried out by Antoine Lavoisier (1743–1794), considered the father of modern chemistry. Lavoisier was the first to clearly annunciate that between reactants and products of a chemical reaction, mass was conserved. The recognition of mass conservation as a fundamental principle distinguished light, now recognised as an electromagnetic field wave, from ponderable matter. Finally, Dmitri Mendeleev (1834–1907) unified the growing list of elements into a rational order with the periodic table in 1869. Figure 1.7 shows Mendeleev's early periodic chart.

ОПЫТЪ СИСТЕМЫ ЭЛЕМЕНТОВЪ.

ОСНОВАННОЙ НА ИХЪ АТОМНОМЪ ВѢСѢ И ХИМИЧЕСКОМЪ СХОДСТВѢ.

```
                           Ti = 50   Zr = 90    ? = 180.
                            V = 51   Nb = 94   Ta = 182.
                           Cr = 52   Mo = 96    W = 186.
                           Mn = 55   Rh = 104,4 Pt = 197,4.
                           Fe = 56   Rn = 104,4 Ir = 198.
                        Ni = Co = 59  Pl = 106,6 O = 199.
         H = 1            Cu = 63,4  Ag = 108  Hg = 200.
              Be = 9,4 Mg = 24  Zn = 65,2  Cd = 112
              B = 11   Al = 27,4  ? = 68   Ur = 116  Au = 197?
              C = 12   Si = 28    ? = 70   Sn = 118
              N = 14   P = 31   As = 75   Sb = 122  Bi = 210?
              O = 16   S = 32   Se = 79,4  Te = 128?
              F = 19   Cl = 35,6 Br = 80   I = 127
         Li = 7 Na = 23  K = 39  Rb = 85,4  Cs = 133  Tl = 204.
                       Ca = 40  Sr = 87,6  Ba = 137  Pb = 207.
                        ? = 45  Ce = 92
                       ?Er = 56  La = 94
                       ?Yt = 60  Di = 95
                       ?In = 75,6 Th = 118?
```

Д. Менделѣевъ

Dmitri Mendeleev

Figure 1.7 *Dmitri Mendeleev and his early periodic chart of the elements. Figures in the public domain.*

As the nineteenth century drew to a close, there were grounds for great satisfaction. Although the inner structure of the atomic elements was not known, their chemical identity and relationships to each other were firmly established. The periodic table not only rationalised known elements but predicted those not yet discovered, leading Oliver Lodge to write in 1888: 'The whole subject of electrical radiation seems to be working itself out splendidly'. There was only the somewhat perplexing null result of the Michelson–Morley experiment in 1887 that failed to measure the velocity of the 'luminiferous aether' through which light was supposed to travel, and the failure of the equipartition of energy principle to account for the spectral distribution of black-body radiation. But these were considered minor blemishes on a near perfect masterwork of human understanding.

1.5   One scientific revolution spawns another

The null result of the Michelson–Morley experiment and its subsequent refinements were finally explained by the theory of special relativity proposed in 1905 by Albert Einstein (1879–1955). The principal elements of the Michelson–Morley experiments are depicted in Figure 1.8. Special relativity showed that the speed of light in a vacuum was not just a property of the electromagnetic wave but a universal constant independent of the inertial reference frame in which it is measured. It follows that any experiment attempting to determine the luminiferous aether wind, with respect to the earth's motion, must yield a null result. Furthermore, in a subsequent paper published later in the same

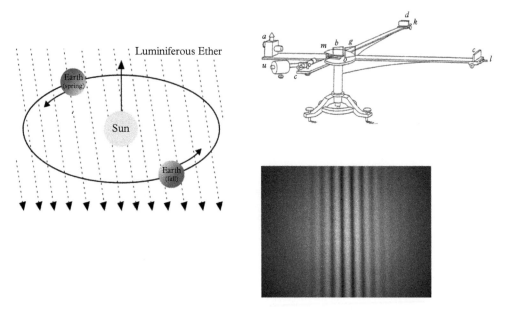

Figure 1.8 *The Michelson–Morley experiment. The diagram on the left shows how the speed of light, travelling through a 'luminiferous aether', should show a Doppler shift with respect to the motion of the earth in its orbit around the sun. As the earth moves toward (away) from the direction of propagation of light from a distant source, the shift should be blue (red). The schematic on the right shows Michelson's interferometer setup that was designed with sufficient resolution such that the shift in the interference pattern would be measurable. Left figure, © cc by-sa 3.0, Cronholm144. Right top figure, public domain. Right bottom figure, Creative Commons by-sa 2.5, Alain Le Rille.*

year, Einstein showed that special relativity implied the equivalence of matter and energy through the famous formula, $m = E/c^2$.

The 'new philosophy' originating with Descartes and culminating in the achievements of Maxwell and Mendeleev appeared to establish a fundamental distinction between light and matter. Within a few decades of their master works, however, special relativity announced a fundamental equivalence between them. Furthermore, in a subsequent paper of the same year, Einstein proposed the quantisation of radiation that re-established some of the old Newtonian corpuscular properties of light. Energy quanta in the form of $E = h\nu$, where ν is the frequency of light and h is Planck's constant, also removed the 'ultraviolet catastrophe' from the Rayleigh-Jeans formulation of black-body radiation. Finally, the quantum mechanics established by Max Born, Erwin Schrödinger, Werner Heisenberg, and Paul Dirac in the first decades of the twentieth century explained the periodic table of the chemical elements in terms of the inner structure of atoms. Quantum mechanics in its non-relativistic form is quite adequate for an understanding of matter outside the atomic nucleus. Together with Dirac's initial formulation of a quantised version of electromagnetism, the second revolution in physics was almost complete. In 1916, Einstein was able to generalise his theory of relativity and apply it to gravity.

The result was a new geometric conception of 'space-time' relevant to cosmological energy and length scales. Of the four known classes of forces—gravity, electromagnetism, the strong, and the weak, nuclear forces—only two have been unified into a consistent theoretical framework: electromagnetism and the weak force. To date, a quantum theory of gravity does not exist and current efforts to unify the force classes into a 'theory of everything' using string theory or M-theory continue. However, for the purposes of this book, classical electrodynamics, essentially as formulated by the Maxwell's equations and the Lorentz force law, together with non-relativistic quantum mechanics, is perfectly adequate to describe light–matter interactions.

1.6 Summary

In this chapter we have briefly examined notions of light and matter from antiquity to the present day, and the major developments are illustrated in Figure 1.9. The first

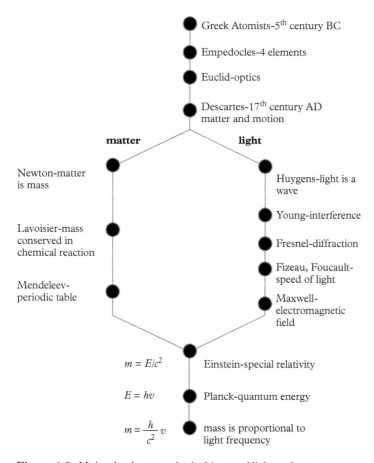

Figure 1.9 *Major developments in the history of light and matter.*

ideas were embedded in speculative philosophical 'theories of everything' without much thought to testing them through experimentation. Evidence of a more modern, scientific approach was found in Alhazen's *Book of Optics*, a treatise very influential in the West during the European Renaissance. Galileo, Grimaldi, Hooke, Huygens and Newton established major milestones in the seventeenth century on the route to understanding the nature of light and its interaction with matter. Newton's conception of the corpuscular character of light held sway for most of the eighteenth century until the theory came under serious attack at the beginning of the nineteenth century from Fresnel and Young, who demonstrated convincing experimental evidence that light was really a wave. The proposition that light was a *transverse* wave removed the last defensive rampart of the Newtonian corpuscular school, and James Clerk Maxwell produced the crowning achievement with the *Theory of the Electromagnetic Field* in 1865. In the meantime, theories of matter evolved from a philosophical, speculative atomism of antiquity, through the Cartesian 'plenum' to Boyle's and Priestly's experiments with gases that lent credibility to integer combining numbers (volumes). Dalton produced evidence for chemical combinations of whole numbers and began to determine atomic masses, while the unifying achievement of Mendeleev's periodic table of the elements lent predictive power to the atomic theory of matter. Black-body radiation and the Michelson–Morley experiment showed that the triumphalism of the late nineteenth century was premature, and that quantal atomic structure, special relativity, and the quantisation of the radiation fields were required to explain not only those two experiments but also the entire body of atomic and molecular spectroscopy. The bifurcation of the antique world into matter and light now seems to be returning to a more unified view of the equivalence of light and matter.

1.7   Further reading

1. E. T. Whittaker, *A History of the Theories of Aether and Electricity: From the Age of Descartes to the Close of the Nineteenth Century*, BiblioLIfe, Charleston, South Carolina (1910).

2. F. Wilczek, *The Lightness of Being*, Basic Books, New York (2008).

3. S. Hawking and L. Mlodinow, *The Grand Design*, Bantam Books, New York (2010).

4. J. Gribbin, *The Scientists*, Random House, New York (2004).

5. D. C. Lindberg, *The Beginnings of Western Science*, University of Chicago Press, Chicago (2007).

6. C. A. Pickover, *Archimedes to Hawking*, Oxford University Press, Oxford (2008).

2
Elements of Classical Electrodynamics

2.1 Introduction

Before we can discuss the *interaction* of light and matter we have to establish a working vocabulary for the vector fields of classical electrodynamics and the equations that relate to them. These equations, Maxwell's equations, together with the Lorentz force law (or some other force law), serve as the postulates of that theory. A class of solutions to these equations describes the propagation of plane electromagnetic waves, and we will explore this propagation in vacuum, in dielectrics, and in good conductors. We will also study polarisation fields in bulk matter, the relation between the macroscopic polarisation field and the microscopic atomic polarisability, and finally develop further the central role played by the radiating dipole in the construction of antennas and antenna arrays. Later, in Chapter 6, we will also describe how standing waves are established in a cavity and how to calculate the energy density by counting modes. The energy density is then applied to the problem of classical black-body radiation theory, and we will describe how it leads to a fundamental contradiction between the predictions of classical theory and experimental measurement.

2.2 Relations among classical field quantities

Since virtually all students now learn electricity and magnetism with the SI (Système International) or rationalized MKS (metres, kilograms, seconds) family of units, we adopt it here. A discussion of units and fundamental quantities in electromagnetism is presented in Appendix B. There we will see that although SI is now the most common choice, other unit systems have their advantages, and in any case, all expressions can be readily transposed from one system to another.

The present choice of SI means that we write Coulomb's force law between two electric point charges q, q' separated by a distance r as

$$\mathbf{F} = \frac{1}{4\pi\varepsilon_0}\left(\frac{qq'}{r^3}\right)\mathbf{r} \tag{2.1}$$

Light-Matter Interaction. Second Edition. John Weiner and Frederico Nunes.

and Ampère's force law (force per unit length) of magnetic induction between two infinitely long wires carrying electric currents I, I', separated by a distance r as

$$\frac{d\,|\mathbf{F}|}{dl} = \frac{\mu_0}{2\pi}\left(\frac{II'}{r}\right) \tag{2.2}$$

where ε_0 and μ_0 are called the *permittivity of free space* and the *permeability of free space*, respectively. In this unit system, the permeability of free space is *defined* as

$$\frac{\mu_0}{4\pi} \equiv 10^{-7} \tag{2.3}$$

and the numerical value of the permittivity of free space is fixed by the condition that

$$\frac{1}{\varepsilon_0\mu_0} = c^2 \tag{2.4}$$

where c is the speed of light in vacuum. Therefore we must have

$$\frac{1}{4\pi\varepsilon_0} = 10^{-7}c^2 \tag{2.5}$$

The time-independent vector fields issuing from these force expressions are the electric field, $\mathbf{E}$, and the magnetic induction field, $\mathbf{B}$. The E-field is the Coulomb force per unit charge,

$$\mathbf{E} = \frac{1}{4\pi\varepsilon_0}\frac{q}{r^3}\mathbf{r} \tag{2.6}$$

and the magnetic induction B-field is given by the Biot–Savart law,

$$\mathbf{B} = \frac{\mu_0}{4\pi}\int\frac{\mathbf{I}\times\mathbf{r}}{r^3}dl = \frac{\mu_0}{4\pi}I\int\frac{d\mathbf{l}\times\mathbf{r}}{r^3} \tag{2.7}$$

where I is the current running in a wire and $d\mathbf{l}$ is an element of the wire length. The Lorentz force law succinctly summarises the effect of the E- and B-fields on a charged particle moving with velocity $\mathbf{v}$,

$$\mathbf{F} = q(\mathbf{E} + \mathbf{v}\times\mathbf{B}) \tag{2.8}$$

In addition to these two fields, the displacement field $\mathbf{D}$ and magnetic field $\mathbf{H}$ are needed to describe the modification of force fields in dielectric or conductive matter. In linear, isotropic materials, these two additional fields are linked to **E** and **B** by the 'constitutive relations'

$$D = \varepsilon E \tag{2.9}$$

$$H = \frac{1}{\mu} B \tag{2.10}$$

where ε and μ are the permittivity and permeability of the material, respectively. These material parameters are related to those of free space by

$$\varepsilon = \varepsilon_0 \varepsilon_r \tag{2.11}$$

$$\mu = \mu_0 \mu_r \tag{2.12}$$

where ε_r and μ_r are the *relative* (and unitless) permittivity and permeability, respectively. The relative permittivity ε_r is often called the *dielectric constant,* but the relative permeability rarely finds application and does not have a common alternative name. Unfortunately, many authors use ε for the dielectric constant, so there is danger of confusion between permittivity (with MKS units of $C^2/N \cdot m^2$) and the unitless dielectric constant. However, one can always discern from the context what is meant by the symbol ε. There is also some difference of opinion among authors as to the most appropriate nomenclature for the B-field and H-field. Here we follow conventional usage and call the B-field the 'magnetic induction field', and the H-field, the 'magnetic field'.

2.3   Classical fields in matter

In free space, **E,B,D,** and **H** are related by

$$D = \varepsilon_0 E \tag{2.13}$$

$$B = \mu_0 H \tag{2.14}$$

When these force fields act on a material medium, however, the **D**- and **H**-fields take on added terms. Matter consists of positively charged core nuclei surrounded by distributions of negatively charged electrons. If the core nuclei are arranged according to some symmetric spatial extension the material is crystalline, and if not, the material may be a glassy solid, a liquid, or a gas. The electric charge distribution may be bound to the nuclei or delocalised throughout the crystal structure. In any case, if **E** and **B** are present, the Lorentz forces acting on the electric charge distribution will clearly modify it. These modifications can be characterised by the introduction of two new fields: the polarisation

P and the magnetisation **M**. In the presence of matter, the displacement and magnetic fields now are expressed as

$$\mathbf{D} = \varepsilon_0 \mathbf{E} + \mathbf{P} \tag{2.15}$$
$$\mathbf{B} = \mu_0 \mathbf{H} + \mathbf{M} \tag{2.16}$$

For E- and B-fields that are not too strong, the P- and M-fields themselves are proportional to **E** and **H**:

$$\mathbf{P} = \varepsilon_0 \chi_e \mathbf{E} \tag{2.17}$$
$$\mathbf{M} = \mu_0 \chi_m \mathbf{H} \tag{2.18}$$

where χ_e, χ_m are electric and magnetic *susceptibility*. Equations 2.15 and 2.16 then become

$$\mathbf{D} = \varepsilon_0 (1 + \chi_e) \mathbf{E} \tag{2.19}$$
$$\mathbf{B} = \mu_0 (1 + \chi_m) \mathbf{H} \tag{2.20}$$

and from Equations 2.9 and 2.10 we see that the relative permittivity and permeability can be written in terms of the corresponding susceptibilities as

$$\varepsilon_r = 1 + \chi_e \tag{2.21}$$
$$\mu_r = 1 + \chi_m \tag{2.22}$$

The relative permittivity and permeability are unitless but may be complex; the imaginary parts reflecting absorptive loss:

$$\varepsilon_r = \varepsilon' + i\varepsilon'' = 1 + \chi_e' + i\chi_e'' \tag{2.23}$$
$$\mu_r = \mu' + i\mu'' = 1 + \chi_m' + i\chi_m'' \tag{2.24}$$

2.4   Maxwell's equations

In the foregoing, we have presented the fields of classical electrodynamics, **D, E, B, H**, and the fields induced in materials, **P** and **M**, which represent the response of matter to electric and magnetic forces. The equations governing the spatial and temporal behaviour of these fields are Maxwell's equations,

$$\nabla \cdot \mathbf{D} = \rho_{\text{free}} \tag{2.25}$$

$$\nabla \cdot \mathbf{B} = 0 \tag{2.26}$$

$$\nabla \times \mathbf{E} = -\frac{\partial \mathbf{B}}{\partial t} \tag{2.27}$$

$$\nabla \times \mathbf{H} = \mathbf{J}_{\text{free}} + \frac{\partial \mathbf{D}}{dt} \tag{2.28}$$

The first two equations make statements about field sources. The first states that the source of the displacement field $\mathbf{D}$ is the 'free' electric charge density, ρ_{free}. In fact, the total charge density is composed of two terms: the free charge density and the 'bound' charge, ρ_{bound}:

$$\rho = \rho_{\text{free}} + \rho_{\text{bound}} \tag{2.29}$$

The bound charge density is defined as the negative divergence of the polarisation field. In Chapter 10, Section 10.4.3, we will see what motivates that definition:

$$\rho_{\text{bound}} = -\nabla \cdot \mathbf{P} \tag{2.30}$$

so that from Equations 2.15, 2.25, and 2.30 we see that

$$\nabla \cdot \mathbf{E} = \frac{\rho}{\varepsilon_0} \tag{2.31}$$

The second source equation, Equation 2.26, states that the magnetic induction field $\mathbf{B}$ does *not* originate from a magnetic charge density, ρ_m. In fact, Equation 2.26 implies that magnetic source 'charges' do not exist: magnetic monopoles have never been found in nature (although Dirac's quantum electrodynamics implies that they must exist somewhere or have existed at some time). The second pair of equations, termed Faraday's law and the Maxwell–Ampère law, respectively, describe the spatial and temporal behaviour of the fields. Equation 2.28, Ampère's law, introduces another field, the charge current density, $\mathbf{J}_{\text{free}}$. Just as Equation 2.29 expresses the total charge density as the sum of the free and bound charge densities, so the total current density $\mathbf{J}$ is composed of the sum of free and bound current densities:

$$\mathbf{J} = \mathbf{J}_{\text{free}} + \mathbf{J}_{\text{bound}} \tag{2.32}$$

The bound current density is defined by two contributions: one from the polarisation field and the other from the magnetisation field:

$$\mathbf{J}_{\text{bound}} = \frac{\partial \mathbf{P}}{\partial t} + \frac{1}{\mu_0} \nabla \times \mathbf{M} \tag{2.33}$$

and from Equation 2.28

$$\mathbf{J}_{\text{free}} = -\frac{\partial \mathbf{D}}{\partial t} + \nabla \times \mathbf{H} \tag{2.34}$$

We see how $\mathbf{D}$ and $\mathbf{H}$ play analogous roles for $\mathbf{J}_{\text{free}}$ as $\mathbf{P}$ and $\mathbf{M}$ play for $\mathbf{J}_{\text{bound}}$. The terms $\partial \mathbf{P}/\partial t$ and $\partial \mathbf{D}/\partial t$ are called the polarisation current and the displacement current, respectively.

2.4.1   Charge-current continuity

If we take the divergence of both sides of the Maxwell–Ampère law (Equation 2.28) and use Equation 2.25, we find

$$\nabla \cdot \mathbf{J}_{\text{free}} + \frac{\partial \rho_{\text{free}}}{\partial t} = 0 \tag{2.35}$$

This equation is called the charge-current continuity equation and states that the free current leaving or entering a closed surface must be equal to the time rate of free charge in that volume. The continuity condition is important because it implies that current and charge sources of fields $\mathbf{D}$ and $\mathbf{H}$ cannot be independently specified. They are constrained by charge-current conservation.

2.5   Static fields, potentials, and energy

2.5.1   Electric field energy

The E-field resulting from a single point charge, Equation 2.6, can be written as

$$\mathbf{E} = \frac{1}{4\pi\varepsilon_0} \frac{q}{r^2} \hat{\mathbf{r}} \tag{2.36}$$

where r is the distance from the charge and the radial unit vector $\hat{\mathbf{r}} = \mathbf{r}/r$. Written in this way, the familiar $1/r^2$ radial fall-off of the field is made explicit. Any vector field is specified by the divergence and the curl of the field, and in the case of Equation 2.36, the divergence is

$$\nabla \cdot \mathbf{E} = \frac{1}{\varepsilon_0} \rho(\mathbf{r}) \tag{2.37}$$

where $\rho(\mathbf{r})$ is the charge density at the position $\mathbf{r}$.

It can be easily shown by the application of Stokes' theorem (see Appendix C for a discussion of the differential and integral operations common in vector fields) to Equation 2.36 that the curl of $\mathbf{E}$ is always null:

$$\nabla \times \mathbf{E} = 0 \tag{2.38}$$

By the principal of vector field superposition, this property is true not just for the E-field of a single point charge but for any spatial distribution of point charges. The E-field is said to be *irrotational* and therefore can be set equal to the gradient of a scalar function, say,

$$\mathbf{E} = -\nabla V \tag{2.39}$$

This characterisation of the E-field as the gradient of a scalar *potential* V is justified by the vector calculus identity that the curl of the gradient of any scalar function is null:

$$\nabla \times \mathbf{E} = -\nabla \times \nabla V = 0 \tag{2.40}$$

The negative sign on the right-hand side of Equation 2.39 is a convention. A unit analysis shows that V has units of energy per charge, and the work done to bring n charges from infinity to a given spatial configuration is

$$\mathscr{E}_{\text{elec}} = \frac{1}{2} \sum_{i=1}^{n} q_i V(\mathbf{r}_i) \tag{2.41}$$

The factor of 1/2 on the right-hand side of Equation 2.41 avoids double counting of mutual pairwise interactions among charges. This expression can be generalised to a smooth charge density distribution $\rho(\mathbf{r})$,

$$\mathscr{E}_{\text{elec}} = \frac{1}{2} \int \rho V \, d\tau \tag{2.42}$$

and taking into account Equations 2.37, and 2.39, the energy required to assemble the charge distribution can be expressed in terms of the resulting E-field,

$$\mathscr{E}_{\text{elec}} = \frac{\varepsilon_0}{2} \int E^2 \, d\tau \tag{2.43}$$

where the integration is over all space. Equation 2.43 can simply be interpreted as the energy of the E-field.

2.5.2   Magnetic field energy

We know from Equation 2.26 that the divergence of the B-field is null,

$$\nabla \cdot \mathbf{B} = 0 \tag{2.44}$$

but we find from the Biot–Savart law describing the magnetic field issuing from current flowing in a wire, Equation 2.7, that the curl of the B-field is given by

$$\nabla \times \mathbf{B} = \mu_0 \mathbf{J} \tag{2.45}$$

where J is the distribution of the source current densities giving rise to **B**. Just as we introduced the scalar potential $V(\mathbf{r})$ to characterise the curl-less E-field, so we can introduce
a vector potential **A** to characterise the divergence-less B-field,

$$\mathbf{B} = \nabla \times \mathbf{A} \tag{2.46}$$

and setting the divergence of the vector potential to zero, we find the standard expression
for **A** is

$$\mathbf{A}(\mathbf{r}) = \frac{\mu_0}{4\pi} \int \frac{\mathbf{J}(\mathbf{r}')}{\mathsf{r}} \, d\tau' \tag{2.47}$$

where $\mathsf{r} = |\mathbf{r}-\mathbf{r}'|$, the distance between the point **r** where **A** is evaluated and $\mathbf{r}'$ the position
of the current density **J**. Just as it is possible to write the energy of some distribution of
electric charge as an integral over the product of the charge density distribution $\rho(\mathbf{r})$
and the scalar potential $V(\mathbf{r})$ (Equation 2.42), we can write the magnetic energy as the
integral over a scalar product of the current density distribution $\mathbf{J}(\mathbf{r})$ and the vector
potential $\mathbf{A}(\mathbf{r})$.

$$\mathscr{E}_{\text{mag}} = \frac{1}{2} \int \mathbf{J} \cdot \mathbf{A} \, d\tau \tag{2.48}$$

By recognising that **J** is proportional to the curl of **B** from Equation 2.46, substituting
into Equation 2.48, and integrating by parts, we can write the magnetic energy entirely
in terms of the B-field in an expression analogous to Equation 2.43

$$\mathscr{E}_{\text{mag}} = \frac{1}{2\mu_0} \int B^2 \, d\tau \tag{2.49}$$

where the integral is taken over all space.

2.5.3   Poynting's theorem

From Equations 2.43 and 2.49 it appears plausible that the energy of an electromagnetic
field is the sum of the energy of the constituent parts,

$$\mathscr{E}_{\text{em}} = \frac{1}{2} \int \left(\varepsilon_0 E^2 + \frac{1}{\mu_0} B^2 \right) d\tau \tag{2.50}$$

and this result can, in fact, be obtained directly from one of the basic postulates of
classical electrodynamics, the Lorentz force law,

$$\mathbf{F} = q(\mathbf{E} + \mathbf{v} \times \mathbf{B}) \tag{2.51}$$

The force **F** acting on charged particles set in motion with velocity **v** does work on the particles at the rate

$$\frac{d\mathscr{E}_{\text{mech}}}{dt} = q\mathbf{E} \cdot \mathbf{v} \tag{2.52}$$

where $\mathscr{E}_{\text{mech}}$ is the mechanical energy of the particle system. Of course, the B-field, always acting at right angles to the direction of motion, can do no work on the particles. But $q = \int \rho d\tau$ and $\mathbf{v} = \mathbf{J}_{\text{free}}/\rho$, so

$$\frac{d\mathscr{E}_{\text{mech}}}{dt} = \int \mathbf{E} \cdot \mathbf{J}_{\text{free}} \, d\tau \tag{2.53}$$

Now we can use the Maxwell–Ampère law, Equation 2.28, to eliminate $\mathbf{J}_{\text{free}}$ in Equation 2.53. The dot product in the integrand becomes

$$\mathbf{E} \cdot \mathbf{J}_{\text{free}} = \mathbf{E} \cdot (\nabla \times \mathbf{H}) - \mathbf{E} \cdot \frac{\partial \mathbf{D}}{\partial t} \tag{2.54}$$

The next step is to invoke a vector field identity and the Faraday law, Equation 2.27, to rewrite the first term on the right of Equation 2.54. The vector field identity is

$$\nabla \cdot (\mathbf{E} \times \mathbf{H}) = \mathbf{H} \cdot (\nabla \times \mathbf{E}) - \mathbf{E} \cdot (\nabla \times \mathbf{H}) \tag{2.55}$$

and Faraday's law is

$$\nabla \times \mathbf{E} = -\frac{\partial \mathbf{B}}{\partial t}$$

so

$$\mathbf{E} \cdot (\nabla \times \mathbf{H}) = -\mathbf{H} \cdot \frac{\partial \mathbf{B}}{\partial t} - \nabla \cdot (\mathbf{E} \times \mathbf{H}) \tag{2.56}$$

and

$$\mathbf{E} \cdot \mathbf{J}_{\text{free}} = -\mathbf{H} \cdot \frac{\partial \mathbf{B}}{\partial t} - \mathbf{E} \cdot \frac{\partial \mathbf{D}}{\partial t} - \nabla \cdot (\mathbf{E} \times \mathbf{H}) \tag{2.57}$$

Now we substitute the $\mathbf{E} \cdot \mathbf{J}_{\text{free}}$ expression back into the integrand in Equation 2.53, and use Stokes' theorem to convert the volume integral of the $\nabla \cdot (\mathbf{E} \cdot \mathbf{H})$ term to a surface integral with integrand $\mathbf{E} \cdot \mathbf{H}$. The resulting expression for the time rate of work done by the electromagnetic field on the charged particles is

$$\frac{d\mathscr{E}_{\text{mech}}}{dt} = -\int \left(\mathbf{H} \cdot \frac{\partial \mathbf{B}}{\partial t} + \mathbf{E} \cdot \frac{\partial \mathbf{D}}{\partial t} \right) d\tau - \oint_{\sigma} (\mathbf{E} \times \mathbf{H}) \cdot d\mathbf{a} \tag{2.58}$$

where the integral of the second term on the right is taken over the surface σ bounding the volume integral in the first term. The two terms in the volume integral are identified with the electric and magnetic parts of the electromagnetic field energy. In a free-space volume, or in any medium with negligible polarisation, the decrease in the field energy, as work is being done on the system of particles, can be written as

$$-\frac{d\mathcal{E}_{em}}{dt} = -\frac{d}{dt}\int \frac{1}{2}\left(\frac{1}{\mu_0}B^2 + \varepsilon_0 E^2\right)d\tau \tag{2.59}$$

confirming that the total electromagnetic field energy can be identified as the sum of the separate static magnetic and electric terms as in Equation 2.50. We can now write Equation 2.58 as an energy conservation expression,

$$\frac{d\mathcal{E}_{mech}}{dt} + \frac{d\mathcal{E}_{em}}{dt} = -\oint \mathbf{E}\times\mathbf{H}\cdot d\mathbf{a} \tag{2.60}$$

where the term on the right is interpreted as an integral over the energy flux (energy per unit area per unit time) flowing across the bounding surface. This cross-product expression is termed the *Poynting vector*, $\mathbf{S}$:

$$\mathbf{S} = (\mathbf{E}\times\mathbf{H}) \tag{2.61}$$

Notice that energy increasing within the volume is equal to a negatively signed energy flux *into* the volume across the boundary surface. The negative sign on the Poynting vector is simply a consequence of the sign convention on the vector surface differential $d\mathbf{a}$, positive pointing *outwards*.

2.6   Three examples of problem solving in electrostatics

The following three examples illustrate many of the useful analytical tools needed to solve problems in electrostatics. They are graded in increasing difficulty, and understanding every line is not essential in a first reading (especially the third example), but the use of cylindrical and spherical coordinates for the appropriately symmetric problem, and the consequent separation of variables, needs to be thoroughly understood by any serious student of light–matter interaction. The same strategy will apply when we consider the structure of the hydrogen atom in Chapter 12. Gauss's law plays a pivotal role in many of the applications we will study in the following chapters. We write out in detail here the solutions to three problems; the results of which will be useful in subsequent topics. The integral form of Gauss's law is also developed at the end of this section.

Example 1

Estimate the electron oscillation frequency in a hydrogen atom using the differential form of Gauss's law.

Solution 1

To solve this problem, let us consider the following model for the hydrogen atom. We assume that the electron is represented by a cloud of charge density ρ, symmetrically distributed in a sphere of radius R centred at the origin $x = 0$ where a positively charged proton is located:

$$\rho = \frac{e}{V} = \frac{e}{4/3\pi R^2} \tag{2.62}$$

The force of attraction between the proton and the electron charge distribution is given by Coulomb's law:

$$\mathbf{F} = -\frac{e^2}{4\pi\varepsilon_0 r^2}\hat{r} \tag{2.63}$$

Under the action of a constant external E-field, the cloud will be displaced from its equilibrium position. Let us call x the displacement distance between charge centres at which a new balance of forces will occur. Figure 2.1 illustrates this displacement. We seek to calculate the restoring force on the electron charge density using Gauss's law. This restoring force will be exerted by the charge fraction δq located at the surface of the sphere with radius x. The fraction of charge outside the volume $V_x = 4/3\pi x^3$ produces no force on the proton. Taking ρ as the electron charge density, we have

$$\delta q = (\rho)(V_x) = -e\left(\frac{x}{R}\right)^3 . \tag{2.64}$$

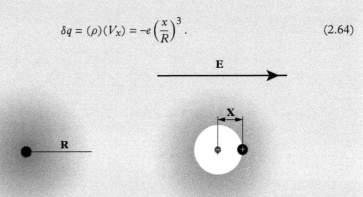

Figure 2.1 *Displacement of the electron charge cloud under the influence of a constant external E-field.*

continued

Example 1 *continued*

The magnitude of the effective Coulomb force between the proton and the displaced electron charge cloud is now

$$|\mathbf{F}| = \frac{e\delta q}{4\pi\varepsilon_0 x^2} = -\frac{e^2}{4\pi\varepsilon_0 x^2}\frac{x^3}{R^3} = -Kx \tag{2.65}$$

which shows that small displacements result in a linear restoring force with the force constant K:

$$K = \frac{e^2}{4\pi\varepsilon_0 R^3} \tag{2.66}$$

Remembering that the frequency of oscillation is given by

$$\omega = \sqrt{\frac{K}{m}} = \sqrt{\frac{e^2}{4\pi\varepsilon_0 R^3 m_e}} \tag{2.67}$$

and taking $R = 0.529^{-10}$ m as the radius of the hydrogen atom in the ground state, and $m_e = 9.109 \times 10^{-31}$ kg, the rest mass of the electron, we calculate the charge frequency of oscillation $\omega_e = 4.13 \times 10^{16}$ s^{-1}.

Example 2

Calculate the E-field of a three-dimensional static electric dipole from the potential.

Solution 2

In Section 2.8 we will study the oscillating dipole and will emphasize the time dependence of the field and radiation. Here we study the static dipole and illustrate the usefulness of calculating the E-field by first writing down the scalar potential, then taking its divergence.

A dipole is a distribution of two equal but opposite charges, separated by a distance a as shown in Figure 2.2. The dipole is a vector, which by convention points from the negative to the positive charge. In Figure 2.2 the dipole p is oriented along z, $p = qa\hat{e}_z$. The calculation of the E-field is carried out through the electric potential using $\mathbf{E} = -\nabla V(x,y,z)$. Since the dipole is oriented along the z-axis and has cylindrical symmetry around this axis, the dipole field is invariant with rotation about the azimuthal angle φ. From Figure 2.2, at point r we have for the potential,

$$V(x,y,z) = \frac{q}{4\pi\varepsilon_0}\left(\frac{1}{r_1} - \frac{1}{r_2}\right) \tag{2.68}$$

Example 2 *continued*

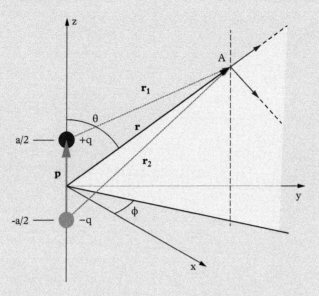

Figure 2.2 *Electric dipole in a system of polar and cartesian coordinates.*

Since we know from the law of cosines,

$$r_1^2 = r^2 + \left(\frac{a}{2}\right)^2 - ar\cos\theta \qquad \text{and} \qquad r_2^2 = r^2 + \left(\frac{a}{2}\right)^2 + ar\cos\theta \qquad (2.69)$$

Equation 2.68 takes the form

$$V(r,\theta) = \frac{q}{4\pi\varepsilon_0}\left[\frac{1}{\left(r^2 + \left(\frac{a}{2}\right)^2 - ar\cos\theta\right)^{1/2}} - \frac{1}{\left(r^2 + \left(\frac{a}{2}\right)^2 + ar\cos\theta\right)^{1/2}}\right] \qquad (2.70)$$

Taking advantage of the dipole cylindrical symmetry, we calculate the electric field by applying the gradient operation to the potential in polar coordinates,

$$\mathbf{E}(r,\theta) = E_r\hat{\boldsymbol{\varepsilon}}_r + E_\theta\hat{\boldsymbol{\varepsilon}}_\theta \qquad (2.71)$$

with

$$E_r = -\frac{\partial V}{\partial r} \qquad \text{and} \qquad E_\theta = -\frac{1}{r}\frac{\partial V}{\partial \theta} \qquad (2.72)$$

continued

Example 2 *continued*

Carrying out the derivative operations on Equation 2.70 we find

$$E_r = \frac{q}{4\pi\varepsilon_0}\left[\frac{r - \frac{a}{2}\cos\theta}{\left(r^2 + \left(\frac{a}{2}\right)^2 - ar\cos\theta\right)^{3/2}} - \frac{r + \frac{a}{2}\cos\theta}{\left(r^2 + \left(\frac{a}{2}\right)^2 + ar\cos\theta\right)^{3/2}}\right] \tag{2.73}$$

and

$$E_\theta = \frac{q}{4\pi\varepsilon_0}\left[\frac{\frac{a}{2}\sin\theta}{\left(r^2 + \left(\frac{a}{2}\right)^2 - ar\cos\theta\right)^{3/2}} + \frac{\frac{a}{2}\sin\theta}{\left(r^2 + \left(\frac{a}{2}\right)^2 + ar\cos\theta\right)^{3/2}}\right] \tag{2.74}$$

In many practical problems, the dipole potential and field are of interest when $r \gg a$: the region of space termed the *far field*. Under these conditions we can write

$$\left[r^2 + \left(\frac{a}{2}\right)^2 \pm ar\cos\theta\right]^n = r^{2n}\left[1 + \left(\frac{a}{2r}\right)^2 \pm \left(\frac{a}{r}\right)\cos\theta\right]^n$$

$$\simeq r^{2n}\left[1 + n\left[\left(\frac{a}{2r}\right)^2 \pm \left(\frac{a}{r}\right)\cos\theta\right]\right] \tag{2.75}$$

and therefore write the dipole potential Equation 2.68 in the far field as

$$V(r,\theta) = \frac{q}{4\pi\varepsilon_0}\frac{1}{r}\left\{\left[1 - \frac{1}{2}\left[\left(\frac{a}{2r}\right)^2 - \left(\frac{a}{r}\right)\cos\theta\right]\right] - \left[1 - \frac{1}{2}\left[\left(\frac{a}{2r}\right)^2 + \left(\frac{a}{r}\right)\cos\theta\right]\right]\right\} \tag{2.76}$$

which, after simplification, yields

$$V(r,\theta) \simeq \frac{qa\cos\theta}{4\pi\varepsilon_0 r^2} \tag{2.77}$$

and the far-field dipole E-field

$$\mathbf{E}(r,\theta) = \frac{2qa\cos\theta}{4\pi\varepsilon_0 r^3}\hat{\boldsymbol{\varepsilon}}_r + \frac{qa\sin\theta}{4\pi\varepsilon_0 r^3}\hat{\boldsymbol{\varepsilon}}_\theta \tag{2.78}$$

We can write the dipole potential and E-field, Equations 2.77 and 2.78, in vector form as

$$V(r,\theta) = \frac{\boldsymbol{p}\cdot\hat{\boldsymbol{\varepsilon}}_r}{4\pi\varepsilon_0 r^2} \tag{2.79}$$

and

$$\mathbf{E}(r,\theta) = \frac{(2\boldsymbol{p}\cdot\hat{\boldsymbol{\varepsilon}}_r)\hat{\boldsymbol{\varepsilon}}_r - (\boldsymbol{p}\cdot\hat{\boldsymbol{\varepsilon}}_\theta)\hat{\boldsymbol{\varepsilon}}_\theta}{4\pi\varepsilon_0 r^3} \tag{2.80}$$

where $\hat{\boldsymbol{\varepsilon}}_r$, $\hat{\boldsymbol{\varepsilon}}_\theta$ are unit vectors in polar coordinates. As discussed at some length in Appendix D, the dipole vector E-field can be expressed either in polar or cartesian coordinates. The unit

Example 2 *continued*

vectors of the E-field in polar coordinates are $\hat{\boldsymbol{e}}_r, \hat{\boldsymbol{e}}_\theta, \hat{\boldsymbol{e}}_\varphi$, and the dipole $\boldsymbol{p}$ in Figure 2.3 can be written in terms of the polar components as

$$\boldsymbol{p} = \left(\boldsymbol{p} \cdot \hat{\boldsymbol{e}}_r\right)\hat{\boldsymbol{e}}_r + \left(\boldsymbol{p} \cdot \hat{\boldsymbol{e}}_\theta\right)\hat{\boldsymbol{e}}_\theta + \left(\boldsymbol{p} \cdot \hat{\boldsymbol{e}}_\varphi\right)\hat{\boldsymbol{e}}_\varphi \tag{2.81}$$

The relations between the unit vectors in polar coordinates and cartesian coordinates are given by

$$\hat{\boldsymbol{e}}_r = \hat{\boldsymbol{e}}_x \sin\theta \cos\varphi + \hat{\boldsymbol{e}}_y \sin\theta \sin\varphi + \hat{\boldsymbol{e}}_z \cos\theta \tag{2.82}$$

$$\hat{\boldsymbol{e}}_\theta = \hat{\boldsymbol{e}}_x \cos\theta \cos\varphi + \hat{\boldsymbol{e}}_y \cos\theta \sin\varphi - \hat{\boldsymbol{e}}_z \sin\theta \tag{2.83}$$

$$\hat{\boldsymbol{e}}_\varphi = -\hat{\boldsymbol{e}}_x \sin\varphi + \hat{\boldsymbol{e}}_y \cos\varphi \tag{2.84}$$

By multiplying the first of these relations by $\cos\theta$ and the second by $-\sin\theta$ we can easily show that

$$\cos\theta\hat{\boldsymbol{e}}_r - \sin\theta\hat{\boldsymbol{e}}_\theta = \hat{\boldsymbol{e}}_z \tag{2.85}$$

In Figure 2.3 we see that in an r, θ plane, carrying out the dot products in Equation 2.81 with the help of Equations 2.82, 2.83, and 2.84 results in

$$\boldsymbol{p} = p(\cos\theta\hat{\boldsymbol{e}}_r - \sin\theta\hat{\boldsymbol{e}}_\theta) \tag{2.86}$$

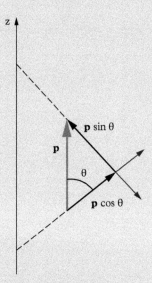

Figure 2.3 *Relation between cartesian and polar coordinates for dipole oriented along z-axis.*

continued

Example 2 *continued*

and therefore

$$(p \cdot \hat{\varepsilon}_\theta)\hat{\varepsilon}_\theta = -p\sin\theta\hat{\varepsilon}_\theta = p - (p\cos\theta)\hat{\varepsilon}_r = p - (p \cdot \hat{\varepsilon}_r)\hat{\varepsilon}_r \qquad (2.87)$$

Substituting this last expression into Equation 2.78 results in

$$\mathbf{E}(r,\theta) = \frac{(3p \cdot \hat{\varepsilon}_r)\hat{\varepsilon}_r - p}{4\pi\varepsilon_0 r^3} \qquad (2.88)$$

Example 3

Analyse the polarisation of a dielectric sphere of radius R immersed in a uniform electric field with permittivity ε.

Solution 3

Consider a dielectric sphere placed in a uniform electric field $\mathbf{E}$ as shown in Figure 2.4. We analyse the problem in polar coordinates. The potential along the z direction is given by

$$V = -\mathbf{E} \cdot \hat{z} = -E_0 r \cos\theta \qquad (2.89)$$

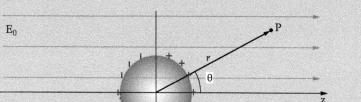

Figure 2.4 *Dielectric sphere immersed in a uniform electric field oriented along the z direction.*

Our first goal is to determine the potential within, and external to, the sphere. To find the potential we solve Laplace's equation,

$$\nabla^2 V = 0 \qquad (2.90)$$

Example 3 *continued*

in both regions and then join the solutions at the boundary. We show in Appendix D, Equation D.37 how to express the Laplacian operator in polar coordinates. An equivalent, slightly different form is

$$\nabla \cdot \nabla V = \frac{1}{r}\frac{\partial^2}{\partial r^2}(rV) + \frac{1}{r^2 \sin\theta} \cdot \frac{\partial}{\partial\theta}\left(\sin\theta\frac{\partial V}{\partial\theta}\right) + \frac{1}{r^2 \sin^2\theta}\frac{\partial^2 V}{\partial\varphi^2} \tag{2.91}$$

The problem of the dielectric sphere subject to a constant electric field along the z-axis is obviously cylindrically symmetric. Before attacking this problem directly, we examine the form of the solutions to Laplace's equation for spherical symmetry. Assuming for the moment a spherically symmetric potential V, we write V as a product of a radial function $R(r)/r$, a polar angular function $\Theta(\theta)$, and an azimuthal angular function $\Phi(\varphi)$:

$$V(r,\theta) = \frac{R(r)}{r}\,\Theta(\theta)\,\Phi(\varphi) \tag{2.92}$$

Now we substitute this form back into Equation 2.91 and arrange all terms with r and θ dependence on the left-hand side, and all terms with φ dependence on the right-hand side. After some algebra, the result is

$$\frac{r^2 \sin^2\theta}{R(r)}\frac{\partial^2 R(r)}{\partial r^2} + \frac{\sin\theta}{\Theta(\theta)}\left[\cos\theta\frac{\partial\Theta(\theta)}{\partial\theta} + \sin\theta\frac{\partial^2\Theta(\theta)}{\partial\theta^2(\theta)}\right] = -\frac{1}{\Phi(\varphi)}\frac{\partial^2\Phi}{\partial\varphi^2} \tag{2.93}$$

Since r, θ, φ are independent variables, the only way that Equation 2.93 can be true is for each side to be equal to a constant, called the 'separation constant'. We can choose the form of this constant to be anything we want, so we choose it to be m^2. From the right-hand side of Equation 2.93 we then have

$$\frac{1}{\Phi(\varphi)}\frac{\partial^2\Phi}{\partial\varphi^2} = -m^2 \tag{2.94}$$

and we can see by simple inspection, and can verify by substitution, that the solutions are

$$\Phi(\varphi) = e^{-im\varphi} \tag{2.95}$$

In order for Φ to be a single-valued function in φ (modulo 2π), m must be an integer, $m = 0, \pm1, \pm2, \ldots$. After division by $\sin^2\theta$, the left-hand side of Equation 2.93, also equal to m^2, can be rearranged to

$$\frac{r^2}{R(r)}\frac{\partial^2 R}{\partial r^2} + \frac{1}{\sin\theta\Theta(\theta)}\left[\cos\theta\frac{\partial\Theta(\theta)}{\partial\theta} + \sin\theta\frac{\partial^2\Theta(\theta)}{\partial\theta^2}\right] - \frac{m^2}{\sin^2\theta} = 0 \tag{2.96}$$

The first term in Equation 2.96 depends only on r and the next two only on θ. Therefore, they can also be separated and set equal to a separation constant, the form of which we choose to be $l(l+1)$:

continued

Example 3 *continued*

$$\frac{r^2}{R(r)}\frac{d^2R}{dr^2} = l(l+1) \tag{2.97}$$

$$-\left\{\frac{1}{(\sin\theta)\Theta(\theta)}\left[\cos\theta\frac{d\Theta(\theta)}{d\theta} + \sin\theta\frac{d^2\Theta(\theta)}{d\theta^2}\right] - \frac{m^2}{\sin^2\theta}\right\} = l(l+1) \tag{2.98}$$

Equation 2.98 can be made more concise by recognising that the term in square brackets is an expanded differential of a product of functions. Multiplying both sides by Θ we have

$$\frac{1}{\sin\theta}\frac{d}{d\theta}\left(\sin\theta\frac{d\Theta}{d\theta}\right) - \frac{m^2}{\sin^2\theta}\Theta = -l(l+1)\Theta \tag{2.99}$$

The radial and angular differential equations can finally be rearranged to

$$\frac{d^2R}{dr^2} - \frac{l(l+1)}{r^2} = 0 \tag{2.100}$$

$$\frac{d^2\Theta}{d\theta^2} + \cot\theta\frac{d\Theta}{d\theta} + \left[l(l+1) - \frac{m^2}{\sin^2\theta}\right]\Theta = 0 \tag{2.101}$$

We have already observed that m must be zero or an integer in order to maintain Φ as a single-valued function. In this field-matching problem of a dielectric sphere immersed in a constant electric field, oriented along z, the E-field will show no φ dependence, and we can set $m = 0$. The resulting expression is called Legendre's equation:

$$\frac{d^2\Theta}{d\theta^2} + \cot\theta\frac{d\Theta}{d\theta} + [l(l+1)]\,\Theta = 0 \tag{2.102}$$

Often, the independent variable of the Legendre equation is taken to be $x = \cos\theta$ rather than θ itself. Then Equation 2.102 takes the form

$$\frac{d}{dx}\left[\left(1-x^2\right)\frac{d\Theta}{dx}\right] + l(l+1)\Theta = 0 \tag{2.103}$$

The physically admissible solutions to Equation 2.103 are a family of polynomials in $\cos\theta$ called the Legendre polynomials. They are labelled by the index l which must only assume the positive integer values $l = 0, 1, 2, \ldots$. The first few Legendre polynomials are listed in Table 2.1 The *Rodrigues' formula* is a concise expression of any desired member l of the Legendre polynomials

$$P_l(x) = \frac{1}{2^l l!}\frac{d^l}{dx^l}\left(x^2 - 1\right)^l \tag{2.104}$$

The solution to the separated radial equation, Equation 2.100, is

$$R(r) = Ar^l + Br^{-(l-1)} \tag{2.105}$$

Example 3 *continued*

where A and B are, as yet, undetermined constants. The fact that Equation 2.105 really is a solution can be verified by direct substitution.

Table 2.1 *Legendre polynomials*

Index l	Function	Polynomial
0	$P_0(x)$	1
1	$P_1(x)$	x
2	$P_2(x)$	$1/2\left(3x^2 - 1\right)$
3	$P_3(x)$	$1/2\left(5x^3 - 3x\right)$
4	$P_4(x)$	$1/8\left(35x^4 - 30x^2 + 3\right)$

Now we are in a position to write the general solution, Equation 2.92, to Maxwell's equation governing the response of a dielectric sphere immersed in a constant electric field:

$$\frac{V(r,\theta)}{r} = U(r,\theta,\varphi) = \sum_{l=0}^{\infty}\left[Ar^l + Br^{-(l+1)}\right]P_l(\cos\theta)e^{-im\varphi} \tag{2.106}$$

Because the problem has cylindrical symmetry, and the solution is invariant about the z-axis of the applied E-field, the $m = 0$ solution of $\Phi(\varphi) = e^{-im\varphi}$ is the appropriate choice, and $\Phi(\varphi) = 1$. Furthermore, we must fit the solution to the physical boundary conditions at the origin, at infinity, and at the interface on the surface of the sphere. Since the potential must remain finite as $r \to 0$, it is clear that we should choose $B = 0$ for r inside r_0, the radius of the sphere. In this region

$$U_{\text{in}}(r,\theta) = \sum_{l=0}^{\infty}A_l r^l P_l(\cos\theta) \tag{2.107}$$

Outside the sphere, in the limit $r \to \infty$ we see from Equation 2.89 that the asymptotic solution must take on the form $-E_0\cos\theta$. In order to fit this result, $A_1 = -E_0$, with all other $A_l = 0$. The solution outside the sphere must therefore have the form

$$U_{\text{out}}(r,\theta) = -E_0 r\cos\theta + \sum_{l=0}^{\infty}B_l r^{-(l+1)}P_l(\cos\theta) \tag{2.108}$$

In order to determine the remaining parameters A_l and B_l we apply *continuity conditions* at the sphere surface. The solutions must match there, and in order to match smoothly, the first derivatives must match as well. Therefore, we have at $r = r_0$

$$\sum_{l=0}^{\infty}A_l r_0^l P_l(\cos\theta) = -E_0 r_0\cos\theta + \sum_{l=0}^{\infty}B_l r^{-(l+1)}P_l(\cos\theta) \tag{2.109}$$

continued

Example 3 *continued*

for the first continuity condition. For the second condition, we remember that

$$\mathbf{E}_r = -\frac{\partial U(r,\theta)}{\partial \theta} \tag{2.110}$$

and that components of the D-field normal to the boundary must be continuous. Then we have for the second continuity condition

$$-\varepsilon_{in}\sum_{l=0}^{\infty} A_l l r_0^{(l-1)} P_l(\cos\theta) = -\varepsilon_{out}\left[-E_0 r_0\cos\theta + \sum_{l=0}^{\infty} B_l r_0^{-(l+1)} P_l(\cos\theta)\right] \tag{2.111}$$

where ε_{in}, ε_{out} are the permittivities inside and outside the sphere. Next, we will eliminate all terms with summations in Equations 2.109 and 2.111 except one, by using the orthogonality and normalisation properties of the Legendre polynomials:

$$\int_{-1}^{+1} P_l(\cos\theta) P_m(\cos\theta)\, d(\cos\theta) = \frac{2}{2m+1}\delta(ml) \tag{2.112}$$

The result is

$$A_m r_0^m\left(\frac{2}{2m+1}\right) = -\frac{2}{3}E_0 r_0 + B_m r_0^{-(m+1)}\left(\frac{2}{2m+1}\right) \tag{2.113}$$

and

$$B_m r_0^{-(m+1)}\left(\frac{2}{2m+1}\right)\left[1 + \frac{\varepsilon_{out}}{\varepsilon_{in}}\left(\frac{m+1}{m}\right)\right] = \frac{2}{3}E_0 r_0\left[1 - \frac{\varepsilon_{out}}{\varepsilon_{in}}\cdot\frac{1}{m}\right] \tag{2.114}$$

or,

$$B_m r_0^{-(m+1)}\left(\frac{2}{2m+1}\right) = \frac{2}{3}E_0 r_0\frac{\left[1 - \frac{\varepsilon_{out}}{\varepsilon_{in}}\cdot\frac{1}{m}\right]}{\left[1 + \frac{\varepsilon_{out}}{\varepsilon_{in}}\left(\frac{m+1}{m}\right)\right]} \tag{2.115}$$

Substituting Equation 2.115 into the right-hand side of Equation 2.113 results in the following expression involving A_m and r_0:

$$A_m r_0^m\left(\frac{2}{2m+1}\right) = -\frac{2}{3}E_0 r_0 + \frac{2}{3}E_0 r_0\frac{\left[1 - \frac{\varepsilon_{out}}{\varepsilon_{in}}\cdot\frac{1}{m}\right]}{\left[1 + \frac{\varepsilon_{out}}{\varepsilon_{in}}\left(\frac{m+1}{m}\right)\right]} \tag{2.116}$$

The continuity condition should hold for any choice of r_0, the radius of the sphere, and we see that if we choose $m = 1$, then the expression for A_m will be independent of r_0. Then we find

Example 3 *continued*

$$A_1 = -E_0 \left[\frac{3\varepsilon_{\text{out}}}{\varepsilon_{\text{in}} + 2\varepsilon_{\text{out}}} \right] \tag{2.117}$$

and from Equation 2.115

$$B_1 = r_0^3 E_0 \left[\frac{\varepsilon_{\text{in}} - \varepsilon_{\text{out}}}{\varepsilon_{\text{in}} + 2\varepsilon_{\text{out}}} \right] \tag{2.118}$$

Now we can substitute A_1 and B_1 back into Equations 2.108 and 2.107 to obtain the expressions for the scalar potential $U(r, \theta)$ outside and inside the dielectric sphere:

$$U(r, \theta)_{\text{out}} = -E_0 r \cos \theta + E_0 \frac{r_0^3}{r^2} \cos \theta \left[\frac{\varepsilon_{\text{in}} - \varepsilon_{\text{out}}}{\varepsilon_{\text{in}} + 2\varepsilon_{\text{out}}} \right] \qquad \text{outside potential} \tag{2.119}$$

$$U(r, \theta)_{\text{in}} = -E_0 r \cos \theta \left[\frac{3\varepsilon_{\text{out}}}{\varepsilon_{\text{in}} + 2\varepsilon_{\text{out}}} \right] \qquad \text{inside potential} \tag{2.120}$$

From Equations 2.119 and 2.120 we can easily calculate the E-field outside and inside the sphere, which was the initial goal we set out to reach:

$$\mathbf{E}_{\text{in}}(r, \theta) = \left[\frac{3\varepsilon_{\text{out}}}{\varepsilon_{\text{in}} + 2\varepsilon_{\text{out}}} \right] E_0 \hat{e}_z \tag{2.121}$$

$$\mathbf{E}_{\text{out}}(r, \theta) = E_0 \hat{z} + E_0 \left[\frac{\varepsilon_{\text{in}} - \varepsilon_{\text{out}}}{\varepsilon_{\text{in}} + 2\varepsilon_{\text{out}}} \right] \frac{r_0^3}{r^3} (2 \cos \theta \, \hat{e}_r + \sin \theta \, \hat{e}_\theta) \tag{2.122}$$

2.6.1 Integral form of Gauss's law

Maxwell's equations are usually summarised in their differential form, but they can be cast in an integral form, often more useful for practical calculation. The integral form of Equation 2.31 is found by applying Gauss's theorem of vector calculus (see Appendix C),

$$\int_V \nabla \cdot \mathbf{K} \, d\tau = \int_S \mathbf{K} \cdot d\sigma \tag{2.123}$$

This theorem states that the integral of the divergence of a vector field $\mathbf{K}$ over a volume is equal to the integral of the field itself over a surface enclosing the volume. Applying this theorem to the differential form of Gauss's law we have

$$\int_V \nabla \cdot \mathbf{D} \, d\tau = \varepsilon_0 \int_V \nabla \cdot \mathbf{E} \, d\tau = q = \varepsilon_0 \int_S \mathbf{E} \cdot d\sigma \tag{2.124}$$

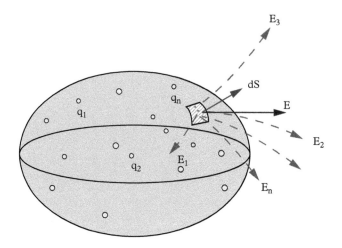

Figure 2.5 *Charges* $q_1, q_2, q_3, \dots$ *are point sources on which E-field lines originate. The integral form of Gauss's law states that the surface integral of E over the volume enclosing these charges is equal to the total charge divided by the permittivity.*

where q is the total charge enclosed by the volume, $q = \int_V \rho \, d\tau$. Note that, strictly speaking, ρ in the integrand includes bound as well as free charge density. Figure 2.5 illustrates the physical significance of the integral form of Gauss's law.

2.7   Dynamic fields and potentials

2.7.1   Maxwell's equations revisited

In Section 2.4 we wrote down Maxwell's equations

$$\nabla \cdot \mathbf{E} = \frac{\rho_{\text{free}}}{\varepsilon_0} \qquad\qquad \nabla \times \mathbf{E} = -\frac{\partial \mathbf{B}}{\partial t} \qquad \text{Faraday's law}$$

$$\nabla \cdot \mathbf{B} = 0 \qquad\qquad \nabla \times \mathbf{H} = \frac{\partial \mathbf{D}}{\partial t} + \mathbf{J} \qquad \text{Maxwell–Ampère law}$$

In the case of static fields, the E-field is *irrotational*, $\nabla \times \mathbf{E} = 0$, and the B-field is 'divergenceless', $\nabla \cdot \mathbf{B} = 0$. Therefore, the E-field can be specified by the gradient of a scalar function, the scalar potential V, and the B-field is determined by the curl of a vector function, the vector potential $\mathbf{A}$:

$$\mathbf{E} = -\nabla V \qquad\qquad\qquad\qquad \mathbf{B} = \nabla \times \mathbf{A} \qquad (2.125)$$

In the case of fields changing in time, the E-field is no longer irrotational because

$$\frac{\partial \mathbf{B}}{\partial t} = \nabla \times \frac{\partial \mathbf{A}}{\partial t}$$

and substituting into Faraday's law we have

$$\nabla \times \mathbf{E} = -\nabla \times \frac{\partial \mathbf{A}}{\partial t}$$

or

$$\nabla \times \left(\mathbf{E} + \frac{\partial \mathbf{A}}{\partial t} \right) = 0 \tag{2.126}$$

The quantity $\mathbf{E} + \partial \mathbf{A}/\partial t$ is itself irrotational and can therefore be written as the gradient of a scalar function:

$$\mathbf{E} + \frac{\partial \mathbf{A}}{\partial t} = -\nabla V$$

or

$$\mathbf{E} = -\nabla V - \frac{\partial \mathbf{A}}{\partial t} \tag{2.127}$$

Then, from Maxwell's E-field divergence equation,

$$\nabla \cdot \mathbf{E} = \nabla \cdot \left[-\nabla V - \frac{\partial \mathbf{A}}{\partial t} \right] = \frac{\rho_{\text{free}}}{\varepsilon_0}$$

or

$$\nabla^2 V + \frac{\partial}{\partial t} (\nabla \cdot \mathbf{A}) = -\frac{\rho_{\text{free}}}{\varepsilon_0} \tag{2.128}$$

Substituting $\mathbf{B} = \nabla \times \mathbf{A}$ into the left-hand side of the Maxwell–Ampère law and taking the time derivative of Equation 2.127 we find

$$\nabla \times (\nabla \times \mathbf{A}) = \mu_0 \mathbf{J} - \mu_0 \varepsilon_0 \nabla \left(\frac{\partial V}{\partial t} \right) - \mu_0 \varepsilon_0 \frac{\partial^2 \mathbf{A}}{\partial t^2}$$

and using the vector field identity $\nabla \times (\nabla \times \mathbf{A}) = \nabla(\nabla \cdot \mathbf{A}) - \nabla^2 \mathbf{A}$, we can simplify to

$$\left(\nabla^2 \mathbf{A} - \mu_0 \varepsilon_0 \frac{\partial^2 \mathbf{A}}{\partial t^2} \right) - \nabla \left(\nabla \cdot \mathbf{A} + \mu_0 \varepsilon_0 \frac{\partial V}{\partial t} \right) = -\mu_0 \mathbf{J} \tag{2.129}$$

Equations 2.128 and 2.129 are just a reformulation of the Maxwell relations in terms of the time-dependent potentials rather than fields.

2.7.2 Lorentz gauge

The vector potential $\mathbf{A}$ can be adjusted to a certain extent since the B-field only depends on the curl of $\mathbf{A}$, not $\mathbf{A}$ itself. In particular, since $\nabla \times (\nabla f) = 0$, where f is some arbitrary scalar function, we can add the gradient of a scalar function to the vector potential without changing its curl. The choice of what scalar function to use is called the choice of *gauge*, and one particularly useful choice of scalar function is called the Lorentz gauge. It is defined by

$$f = \nabla \cdot \mathbf{A} = -\mu_0 \varepsilon_0 \frac{\partial V}{\partial t} \qquad (2.130)$$

Substituting Equation 2.130 into Equation 2.129 we see that it simplifies to

$$\nabla^2 \cdot \mathbf{A} - \mu_0 \varepsilon_0 \frac{\partial^2 \mathbf{A}}{\partial t^2} = -\mu_0 \mathbf{J} \qquad (2.131)$$

and Equation 2.128 becomes

$$\nabla^2 V - \mu_0 \varepsilon_0 \frac{\partial^2 V}{\partial t^2} = -\frac{\rho_{\text{free}}}{\varepsilon_0} \qquad (2.132)$$

The strategy then, given the source terms $\mathbf{J}$ and ρ, is to solve the two inhomogeneous wave equations, Equations 2.131 and 2.132, for $\mathbf{A}$ and V, and then use Equations 2.125 and 2.127 to find the fields $\mathbf{B}$ and $\mathbf{E}$.

2.7.3 Retarded potentials

The points in space where the potentials are evaluated may be very far from the location of the sources. Therefore, as the sources $\mathbf{J}(\mathbf{r}', t)$, $\rho(\mathbf{r}', t)$ vary with time, the changes in the potentials at $V(\mathbf{r}, t)$, $\mathbf{A}(\mathbf{r}, t)$ will only take effect at a later time t, governed by the distance from the source position to the points in space where the potentials are evaluated, and the speed of light. The time that the changes actually take place at the sources is called the 'retarded time' t_r, and the time at which these changes take effect at the potential points is just denoted by t. The relation between the two is $t_r = t - \ell/c$, where ℓ is the distance from the source point to the field point, $\ell = |\mathbf{r} - \mathbf{r}'|$. The spatial relations between coordinate origin, source position, and field position are shown in Figure 2.6. The formal solutions for distributed charge and current density sources, $\rho(\mathbf{r}', t_r)$, $\mathbf{J}(\mathbf{r}', t_r)$ are

$$V(\mathbf{r}, t) = \frac{1}{4\pi \varepsilon_0} \int \frac{\rho(\mathbf{r}', t_r)}{\ell} d\tau' \qquad (2.133)$$

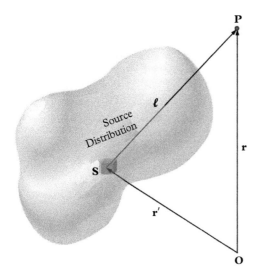

Figure 2.6 *Relations between coordinate origin O, position r′ of differential volume dτ′ in distributed source, and distance ℓ = |r − r′| between differential source volume dτ′, located at point S, and field point P.*

and

$$A(r, t) = \frac{\mu_0}{4\pi} \int \frac{J(r', t_r)}{\ell} d\tau' \qquad (2.134)$$

Using these formal solutions for V and $\mathbf{A}$, we can obtain formal solutions to the fields from Equations 2.127 and 2.125:

$$E(r, t) = \frac{1}{4\pi \varepsilon_0} \int \left[\frac{\rho(r', t_r)}{\ell^2} \hat{\ell} + \frac{\dot{\rho}(r', t_r)}{c\ell} \hat{\ell} - \frac{\dot{J}(r', t_r)}{c^2 \ell} \right] d\tau' \qquad (2.135)$$

and

$$B(r, t) = \frac{\mu_0}{4\pi} \int \left[\frac{J(r', t_r)}{\ell^2} + \frac{\dot{J}(r', t_r)}{c\ell} \right] \times \hat{\ell} d\tau' \qquad (2.136)$$

In Equations 2.135 and 2.136, $\dot{J}$ denotes the time derivative of the source current density, and thus we write

$$\frac{\partial A}{\partial t} = \frac{\mu_0}{4\pi} \int \frac{\dot{J}}{\ell} d\tau' \qquad (2.137)$$

which is used in the third term on the right-hand side of Equation 2.135.

2.8   Dipole radiation

In this section, we use the results for the time-dependent retarded potentials developed in Section 2.7 to calculate the E-field and B-field solutions for a harmonically oscillating dipole source. We take the dipole as two charges $+q$ and $-q$ placed at the ends of very short conducting wire of length a. The charges oscillate along the wire with a frequency ω such that the time dependence of the charge at any point along the wire is $q(t) = q_0 \cos(\omega t)$. The electric dipole, aligned along the z-axis, is then given by

$$\mathbf{p}(t) = q(t)a\hat{z} = q_0 a \cos(\omega t)\hat{z} = p_0 \cos(\omega t)\hat{z} \qquad (2.138)$$

The current $\mathbf{i}$ associated with this dipole is just

$$\mathbf{i}(t) = \frac{dq}{dt}\hat{z} = -q_0\omega \sin \omega t \hat{z} \qquad (2.139)$$

Substituting the current into Equation 2.134, the vector potential for the harmonic oscillator is

$$\mathbf{A}(r,t) = \frac{\mu_0}{4\pi} \int_{a/2}^{a/2} \frac{-q_0\omega \sin\left[\omega\left(t-\frac{r}{c}\right)\right]\hat{z}}{r} dz \qquad (2.140)$$

But since the length a of the oscillator is short compared to the wavelength, we can simply replace the integral with the integrand multiplied by a:

$$\mathbf{A}(r,t) \simeq -\frac{\mu_0 p_0 \omega}{4\pi r} \sin\left[\omega\left(t-\frac{r}{c}\right)\right]\hat{z} \qquad (2.141)$$

We can write down the retarded scalar potential for some point $\mathbf{r}$ at time t, $V(\mathbf{r},t)$ by using Equation 2.133 and Figure 2.7:

$$V(\mathbf{r},t) = \frac{1}{4\pi\varepsilon_0}\left[\frac{q_0 \cos\left[\omega(t-r_1/c)\right]}{r_1} - \frac{q_0 \cos\left[\omega(t-r_2/c)\right]}{r_2}\right] \qquad (2.142)$$

The denominators r_1, r_2 can be written in terms of $\mathbf{r}$ and the angle θ as

$$r_{1,2} = \left[r^2 \mp ra\cos\theta + \left(\frac{a}{2}\right)^2\right]^{1/2} \qquad (2.143)$$

Now, if the point where the field is to be evaluated is sufficiently far from the dipole such that $|\mathbf{r}| \gg a$, and the length of the dipole is sufficiently short such that the wavelength of light $\lambda = \omega/c \gg a$, then it can be easily shown, by expanding $1/r_{1,2}$ in a Taylor series, that

$$V(r,\theta,t) \simeq \frac{p_0\cos\theta}{4\pi\varepsilon_0}\left[-\frac{\omega}{rc}\sin\left[\omega\left(t-\frac{r}{c}\right)\right] + \frac{1}{r^2}\cos\left[\omega\left(t-\frac{r}{c}\right)\right]\right] \qquad (2.144)$$

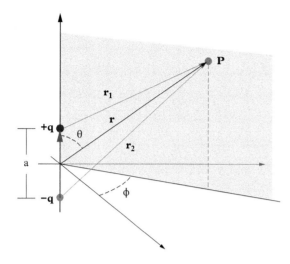

Figure 2.7 *Electric dipole oscillator aligned along z-axis. The maximum length of the dipole is a and the coordinates of field point p, with respect to the dipole midpoint, are indicated in spherical coordinates r, θ, φ.*

In order to calculate the time-varying E-field, we need to know $\mathbf{A}(\mathbf{r}, t)$ as well as $V(r, \theta, t)$ because

$$\mathbf{E} = -\nabla V - \frac{\partial \mathbf{A}}{\partial t} \tag{2.145}$$

Using Equation 2.141 for the vector potential of a dipole aligned along the z-axis,

$$\frac{\partial \mathbf{A}}{\partial t} = -\frac{\mu_0 p_0 \omega^2}{4\pi r} \cos\left[\omega\left(t - \frac{r}{c}\right)\right] \hat{\mathbf{z}} \tag{2.146}$$

expressed in spherical coordinates as

$$\frac{\partial \mathbf{A}}{\partial t} = -\frac{\mu_0 p_0 \omega^2}{4\pi r} \cos\left[\omega\left(t - \frac{r}{c}\right)\right] (\cos\theta\hat{\mathbf{r}} - \sin\theta\hat{\boldsymbol{\theta}}) \tag{2.147}$$

Substituting Equations 2.144 and 2.147 into 2.145 and writing the gradient operator ∇ in spherical coordinates (see Appendix D),

$$\nabla = \frac{\partial}{\partial r}\hat{\mathbf{r}} + \frac{1}{r}\frac{\partial V}{\partial \theta}\hat{\boldsymbol{\theta}}$$

we find the radial component of the E-field

$$\mathbf{E}_r = \frac{2p_0 \cos\theta}{4\pi\varepsilon_0} \left\{\frac{1}{r^3}\cos\left[\omega\left(t - \frac{r}{c}\right)\right] - \frac{\omega}{r^2 c}\sin\left[\omega\left(t - \frac{r}{c}\right)\right]\right\} \hat{\mathbf{r}} \tag{2.148}$$

and the angular component

$$\mathbf{E}_\theta = \frac{p_0 \sin\theta}{4\pi\varepsilon_0}\left\{\frac{1}{r^3}\cos\left[\omega\left(t-\frac{r}{c}\right)\right]-\frac{\omega}{r^2 c}\sin\left[\omega\left(t-\frac{r}{c}\right)\right]\right.$$

$$\left.-\frac{\omega^2}{rc^2}\cos\left[\omega\left(t-\frac{r}{c}\right)\right]\right\}\hat{\theta}\qquad(2.149)$$

Then, using $\mathbf{B} = \nabla \times \mathbf{A}$ we find, for the oscillator B-field,

$$\mathbf{B}_\varphi = -\frac{\mu_0 p_0 \sin\theta}{4\pi}\left\{\frac{\omega}{r^2}\sin\left[\omega\left(t-\frac{r}{c}\right)\right]+\frac{\omega^2}{rc}\cos\left[\omega\left(t-\frac{r}{c}\right)\right]\right\}\hat{\varphi}\qquad(2.150)$$

2.8.1   Far field

In the 'far field', where $r \gg \lambda$, the only surviving E-field term is

$$\mathbf{E}_\theta = -\frac{p_0 \sin\theta}{4\pi\varepsilon_0}\frac{\omega^2}{rc^2}\cos\left[\omega\left(t-\frac{r}{c}\right)\right]\hat{\theta}\qquad(2.151)$$

and the surviving B-field term is

$$\mathbf{B}_\varphi = -\frac{\mu_0 p_0 \sin\theta}{4\pi}\frac{\omega^2}{rc}\cos\left[\omega\left(t-\frac{r}{c}\right)\right]\hat{\varphi}\qquad(2.152)$$

which can be rewritten, using $\mu_0\varepsilon_0 = 1/c^2$, as

$$\mathbf{B}_\varphi = -\frac{p_0 \sin\theta}{4\pi\varepsilon_0}\frac{\omega^2}{rc^3}\cos\left[\omega\left(t-\frac{r}{c}\right)\right]\hat{\varphi}\qquad(2.153)$$

Comparing Equations 2.151 and 2.153, we see that the field amplitudes are related simply by a factor of c,

$$\mathbf{B}_\varphi = \frac{1}{c}\mathbf{E}_\theta\qquad(2.154)$$

that the $\mathbf{E}_\theta$ and $\mathbf{B}_\varphi$ components are in phase, and that they are orthogonal to each other and to the radial propagation direction $\hat{r}$. The components E_θ and B_φ in the far field as $r \to \infty$ become the plane waves discussed in Section 3.1.

The Poynting vector describes the energy flux emitted into the far-field region by the oscillating dipole:

$$\mathbf{S} = \frac{1}{\mu_0}(\mathbf{E}\times\mathbf{B}) = \frac{\mu_0 p_0^2 \omega^4}{16\pi^2 c}\left(\frac{\sin^2\theta}{r^2}\right)\cos^2\left[\omega\left(t-\frac{r}{c}\right)\right]\hat{r}\qquad(2.155)$$

The optical-cycle-averaged energy flux $\langle \mathbf{S} \rangle$ (energy per unit area per unit time) is

$$\langle \mathbf{S} \rangle = \frac{\mu_0 p_0^2 \omega^4}{32\pi^2 c} \left(\frac{\sin^2 \theta}{r^2} \right) \tag{2.156}$$

and the total power W (energy per unit time) emitted by the oscillator is calculated by integrating the energy flux over all space

$$W = \int \langle \mathbf{S} \rangle \cdot d\mathbf{a} = \frac{\mu_0 p_0^2 \omega^4}{32\pi^2 c} \int \frac{\sin^2 \theta}{r^2} r^2 \sin\theta \, d\theta \, d\varphi = \frac{\mu_0 p_0^2 \omega^4}{12\pi c} \tag{2.157}$$

2.8.2   Quasi-stationary regime

In the regime where $r \lesssim \lambda$, or equivalently where $\omega r/c \lesssim 2\pi$, we can drop the retardation term in the time arguments of Equations 2.148, 2.149, and 2.150, and rewrite them as

$$\mathbf{E}_r = \frac{2p_0 \cos\theta}{4\pi \varepsilon_0 r^3} \left[\cos\omega t - \left(\frac{\omega r}{c} \right) \sin\omega t \right] \hat{r} \tag{2.158}$$

$$\mathbf{E}_\theta = \frac{p_0 \sin\theta}{4\pi \varepsilon_0 r^3} \left[\cos\omega t - \left(\frac{\omega r}{c} \right) \sin\omega t - \left(\frac{\omega r}{c} \right)^2 \cos\omega t \right] \hat{\theta} \tag{2.159}$$

$$\mathbf{B}_\varphi = -\frac{p_0 \sin\theta}{4\pi \varepsilon_0 c r^3} \left[\left(\frac{\omega r}{c} \right) \sin\omega t + \left(\frac{\omega r}{c} \right)^2 \cos\omega t \right] \hat{\varphi} \tag{2.160}$$

We see that, in the quasi-stationary regime, the surviving E-field and B-field terms from an oscillator source (the first terms on the right in Equations 2.158–2.160) are in quadrature, not in phase as in far-field radiation. This regime corresponds to conventional engineering circuit response where the product of frequencies ω (radio and microwave) and circuit size r are well within the $\omega r/c \lesssim 2\pi$ criterion.

Note that, when the quasi-stationary regime goes to the truly stationary, $\omega \to 0$, Equations 2.158–2.160 go to their electrostatic dipole limits,

$$\mathbf{E}_r = \frac{2p_0 \cos\theta}{4\pi \varepsilon_0 r^3} \qquad \mathbf{E}_\theta = \frac{p_0 \sin\theta}{4\pi \varepsilon_0 r^3} \qquad \mathbf{B}_\varphi = 0$$

2.9   Light propagation in dielectric and conducting media

So far we have assumed that light propagates either through a vacuum or through a gas so dilute that we need consider only the isolated field–particle interaction. Here we study the propagation of light through a continuous dielectric (non-conducting) medium and near the surface of a good conductor. Interaction of light with such media permits us to re-examine the important quantities of polarisation, magnetisation, susceptibility, index of refraction, extinction coefficient, and absorption coefficient. We shall see later how the polarisation field can be usefully regarded as a density of transition dipoles induced in the dielectric by the oscillating light field. Formally, as the definitions Equations 2.17 and 2.18 show, the magnetisation $\mathbf{M}$ of materials is on an equal footing with the polarisation $\mathbf{P}$, but in practice, electric polarisation is encountered much more commonly. We will restrict the present discussion, therefore, to the response of materials to the incident electric field component of a plane wave. We begin by recalling the definition of material polarisation $\mathbf{P}$ with respect to an applied electric field $\mathbf{E}$ as

$$\mathbf{P} = \varepsilon_0 \chi_e \mathbf{E} \tag{2.161}$$

where χ_e is the linear electric *susceptibility*: an intrinsic property of the medium responding to the light field. Recall the relation between the electric field $\mathbf{E}$, the polarisation $\mathbf{P}$, and the displacement field $\mathbf{D}$ in a material medium. In the SI system of units the relation is

$$\mathbf{D} = \varepsilon_0 \mathbf{E} + \mathbf{P} \tag{2.162}$$

Furthermore, for isotropic materials, in all systems of units, the so-called 'constitutive relation' between the displacement field $\mathbf{D}$ and the imposed electric field $\mathbf{E}$, is written

$$\mathbf{D} = \varepsilon \mathbf{E} \tag{2.163}$$

with ε being referred to as the *permittivity* of the material. Therefore,

$$\mathbf{D} = \varepsilon_0 (1 + \chi)\mathbf{E} \tag{2.164}$$

and

$$\varepsilon = \varepsilon_0 (1 + \chi) \tag{2.165}$$

The susceptibility χ is often a strong function of frequency ω around resonances and can be spatially anisotropic. It is a complex quantity, having a real dispersive component χ' and an imaginary absorptive component χ''

$$\chi = \chi' + i\chi'' \tag{2.166}$$

Real and imaginary parts of the permittivity are related to the susceptibility by

$$\varepsilon = \varepsilon_0 (1 + \chi') + i\varepsilon_0 \chi'' \tag{2.167}$$

with

$$\varepsilon' = \varepsilon_0 (1 + \chi') \tag{2.168}$$
$$\varepsilon'' = \varepsilon_0 \chi'' \tag{2.169}$$

A number of familiar expressions in free space become modified in a dielectric medium. For example:

$$\left(\frac{kc}{\omega}\right)^2 = 1 \quad ; \text{free space}$$

$$\left(\frac{kc}{\omega}\right)^2 = 1 + \chi \; ; \text{dielectric}$$

In a dielectric medium kc/ω becomes a complex quantity that is conventionally expressed as

$$\frac{kc}{\omega} = \eta + i\kappa \tag{2.170}$$

where η is the *refractive index* and κ is the *extinction coefficient* of the dielectric medium. The relations between the refractive index, the extinction coefficient, and the two components of the susceptibility are

$$\eta^2 - \kappa^2 = 1 + \chi' \tag{2.171}$$

$$2\eta\kappa = \chi'' \tag{2.172}$$

Note that, in a transparent lossless dielectric medium,

$$\eta^2 - \kappa^2 = 1 + \chi' = \frac{\varepsilon'}{\varepsilon_0} \tag{2.173}$$

But if the medium has an absorptive component

$$2\eta\kappa = \frac{\varepsilon''}{\varepsilon_0} \tag{2.174}$$

The unitless terms $\varepsilon'/\varepsilon_0$ and $\varepsilon''/\varepsilon_0$ are called the real and imaginary parts of the relative permittivity or the dielectric constant. They are denoted as

$$\varepsilon_r' = \frac{\varepsilon'}{\varepsilon_0} \tag{2.175}$$

$$\varepsilon_r'' = \frac{\varepsilon''}{\varepsilon_0} \tag{2.176}$$

The subscript r emphasizes *relative* permittivity. Unfortunately, in the scientific literature the dielectric constant is often just written $\varepsilon = \varepsilon' + i\varepsilon''$ with the same notation as the permittivity, and one must decipher from the context whether the permittivity (units of $C^2/J \cdot m$) or relative permittivity–dielectric constant (unitless) is intended. For example, it is clear from expressions involving the refractive index, such as

$$n = \eta + i\kappa = \sqrt{\varepsilon} \tag{2.177}$$

or a propagation parameter

$$k = k_0 n = k\sqrt{\varepsilon} \tag{2.178}$$

that ε is the dielectric constant, whereas in an expression involving a constitutive relation, such as

$$\mathbf{D} = \varepsilon \mathbf{E} \tag{2.179}$$

the ε factor denotes a material permittivity. Note, however, that in the expression

$$\mathbf{D} = \varepsilon \varepsilon_0 \mathbf{E} \tag{2.180}$$

The factor ε is the unitless dielectric constant (relative permittivity) and ε_0 is the vacuum permittivity.

In a dielectric medium the travelling wave solutions of Maxwell's equation become

$$\mathbf{E} = \mathbf{E}_0 e^{i(kz-\omega t)} \longrightarrow \mathbf{E}_0 e^{[i\omega\left(\frac{\eta z}{c}-t\right)-\omega\frac{\kappa}{c}z]} \tag{2.181}$$

the relation between magnetic and electric field amplitudes is

$$\mathbf{B}_0 = \sqrt{\varepsilon_0 \mu_0}\, \mathbf{E}_0 \longrightarrow \mathbf{B}_0 = \sqrt{\varepsilon_0 \mu_0}\, (\eta + i\kappa)\, \mathbf{E}_0$$

and the period-averaged field energy density is

$$\bar{\rho}_\omega = \frac{1}{2}\varepsilon_0\, |\mathbf{E}|^2 \longrightarrow \bar{\rho}_\omega = \frac{1}{2}\varepsilon_0 \eta^2\, |\mathbf{E}|^2 \tag{2.182}$$

Now, the light-beam intensity in a dielectric medium is attenuated exponentially by absorption:

$$\bar{I}_\omega = \frac{1}{2}\varepsilon_0 c\, |\mathbf{E}|^2 \longrightarrow \bar{I}_\omega = \frac{1}{2}\varepsilon_0 \eta^2\, |\mathbf{E}|^2 \left(\frac{c}{\eta}\right) = \frac{1}{2}\varepsilon_0 c\eta E_0^2 e^{-2\frac{\omega\kappa}{c}z} = \bar{I}_0 e^{-Kz} \tag{2.183}$$

where

$$\bar{I}_0 = \frac{1}{2}\varepsilon_0 c\eta E_0^2 \tag{2.184}$$

is the intensity at the point where the light beam enters the medium, and

$$K = 2\frac{\omega\kappa}{c} = \frac{\omega}{\eta c}\chi'' \tag{2.185}$$

is termed the *absorption coefficient*. Note that the energy flux $\bar{I}_\omega$ in the dielectric medium is still the product of the energy density

$$\bar{\rho}_\omega = \frac{1}{2}\varepsilon_0\eta^2 |\mathbf{E}|^2 \tag{2.186}$$

and the speed of propagation c/η. Note also that, although light propagating through a dielectric maintains the same frequency as in vacuum, the wavelength contracts as

$$\lambda = \frac{c/\eta}{\nu} \tag{2.187}$$

2.10 Summary

In this chapter we have passed in review many of the ideas, concepts, and quantities already familiar to the reader as a warm-up exercise. We first discussed the basic force fields E and B and the constitutive relations that join them to D and H. We then introduced Maxwell's equations, the notion of bound charge, bound current, and the charge-current continuity relation. We then introduced the expressions for field potential, field energy, and energy transport by Poynting's theorem. We then tried to concretise these ideas with three illustrative examples of increasing complexity. The examples are useful in themselves but they also showcase certain techniques and approaches to solving problems commonly encountered in electrostatics. The focus then passed to time-dependent 'dynamic' solutions to Maxwell's equations where the vector potential and the Lorentz gauge were introduced. The reader then got a first taste of dipole radiation in the near field, the far field, and the 'quasi-stationary regime', before the chapter concluded with some useful expressions for light propagation in dielectric media.

2.11 Exercises

1. A plane wave propagates in free space with an E-field and B-field given by Equations 3.14 and 3.16. Calculate the Poynting vector $\mathbf{S}$ associated with this wave.

2. As shown in Figure 2.8, a plane wave of wavelength 620 nm in air is incident from the left on a lossless, non-magnetic dielectric slab with dielectric constant $\varepsilon' = 2.25$. Calculate the reflection and transmission coefficients R, T, the reflected power, and the transmitted power assuming an E-field amplitude of 1 V/m. Assume the dielectric slab to be infinitely thick.

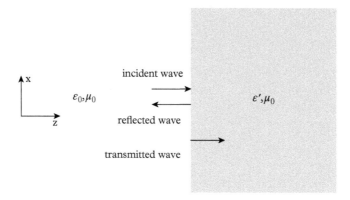

Figure 2.8 *Plane wave propagating from left to right, reflected at the interface.*

3. Now reconsider the dielectric slab in Figure 2.8 as a slightly lossy dielectric with an index of refraction corresponding to SiO at $\lambda_0 = 620\,\text{nm}$. At this wavelength $n = \eta + i\kappa = 1.969 + i0.01175$. Calculate the reflection and transmission coefficients, the reflected power, and the transmitted power for this case.

4. Silicon nitride, Si_3N_4, is a very low loss dielectric material commonly used in micro- and nanofabrication. The index of refraction of silicon nitride in the visible region of the spectrum is 2.05. Calculate the ratio of the impedance of a plane wave propagating in Si_3N_4 to that of a plane wave propagating in free space.

5. The bulk plasma frequency of silver (Ag) metal is, $\omega_p = 1.3 \times 10^{16}\,\text{s}^{-1}$. A plane wave with an amplitude of 1 V/m, impinging on a silver surface, penetrates to a skin depth of about 5 nm. Calculate the wavelength of light and the current density J_c induced in the Ag near the surface.

6. Consider a hydrogen atom subject to a *constant*, uniform electric field. Suppose the hydrogen atom consists of a positive point charge with a proton mass, sitting at the origin. A negatively charged uniform electron charge density cloud represents the electron. The volume integral of the electron charge density is equal to the proton charge. Because of the external E-field, the electron charge cloud is displaced from the proton. Use Coulomb's law to calculate the restoring force between the charge centres. Use this restoring force in a harmonic oscillator model to estimate a 'natural' hydrogen atom frequency.

7. With the electron in the ground state of the hydrogen atom, the total energy of the system is known to be $-13.6\,\text{eV}$.[1] The total energy is the sum of the kinetic energy of the system in the ground state and the potential energy. If the kinetic energy of the electron is 13.6 eV, what is the potential energy (due to the Coulomb attractive force between the electron and proton)? Use your answer and the Planck relation $\mathscr{E} = \hbar\omega$

---

[1] The unit of energy 'eV' is the electron volt and is the energy acquired by an electron after passing through a potential difference of 1 V. One eV is equivalent to $1.602 \times 10^{-19}\,\text{J}$.

to calculate the corresponding frequency ν. Compare this frequency to the harmonic oscillator frequency you calculated in Exercise 6.. Remember: the angular frequency and cyclic frequency are related by $\omega = 2\pi\nu$.

8. Redo Exercise 6 supposing that the applied E-field is harmonically oscillating with an angular frequency ω. (a) Estimate the average value of the oscillating dipole established between the positive and negative charge centres of the atom. (b) Calculate the rate of energy (in Watts) emitted by the radiating dipole.

9. The ionosphere can be considered a charged-particle plasma gas. Calculate the plasma resonance frequency ω_p. Assume that the density of free electrons is 10^5 cm^{-3}. As we shall see in Chapter 7, according to the dispersion diagram for the charged-particle-plasma-gas model, waves can propagate into the plasma above ω_p but are reflected below it. In radio transmission, above what frequency are we broadcasting to the solar system (and beyond)?

2.12 Further reading

1. D. J. Griffiths, *Introduction to Electrodynamics*, 3rd edition, Pearson Addison Wesley, Prentice Hall (1999).

2. J.-P. Pérez, R. Carles, and R. Fleckinger, *Électromagnétism Fondements et applications*, 3ème édition, Masson (1997).

3. M. Born and E. Wolf, *Principles of Optics*, 6th edition, Pergamon Press (1980).

4. H. A. Haus, *Waves and Fields in Optoelectronics*, Prentice Hall (1984).

5. D. M. Pozar, *Microwave Engineering*, 3rd edition, John Wiley & Sons (2005).

6. M. Mansuripur, *Field, Force, Energy and Momentum in Classical Electrodynamics*, Bentham e-Books (2011).

7. F. Nunes, T. Vasconcelos, M. Bezerra, and J. Weiner, *Electromagnetic energy density in dispersive and dissipative media. Journal of the Optical Society of America B*, vol 28, pp. 1544–52 (2011).

3

Physical Optics of Plane Waves

3.1 Plane electromagnetic waves

3.1.1 Phasor form of the vector fields

Up to now we have tacitly assumed that the vector force fields of electricity and magnetism are real. However, for most of the following chapters we shall be studying electromagnetic waves of various sorts, propagating through dielectric and conductive media. The sources of these waves will be posited to be dipoles harmonically oscillating at the angular frequency $\omega = 2\pi\nu$ and removed spatially to negative infinity. Therefore, the time dependence of the fields will always be set to a factor of $e^{-i\omega t}$, and the spatial oscillation will always have an associated phase, $e^{i\varphi}$. The general form of the vector fields can then be expressed as

$$\mathbf{A}(\mathbf{r}, t) = \mathbf{A}(\mathbf{r})e^{i\varphi}e^{-i\omega t} \tag{3.1}$$

The use of $\mathbf{A}$ here should not be confused with the vector potential. Here, it is just a generic vector field. If the spatial dependence of the wave is sinusoidal as well, then the spatial phase φ can be written as

$$\varphi = \mathbf{k} \cdot \mathbf{r} + \delta \tag{3.2}$$

where $\mathbf{k}$ is called the propagation vector, the magnitude of which is related to the wavelength λ by

$$k = \frac{2\pi}{\lambda} \tag{3.3}$$

and δ is just the reference phase at the wave origin, usually set to zero. The direction of the propagation vector is given by

$$\mathbf{k} = k_x\hat{\mathbf{x}} + k_y\hat{\mathbf{y}} + k_z\hat{\mathbf{z}} \tag{3.4}$$

Light-Matter Interaction. Second Edition. John Weiner and Frederico Nunes.
© John Weiner and Frederico Nunes 2017. Published 2017 by Oxford University Press.

The form[1], always expressed as that of the general vector field, Equation 3.1, is then

$$\mathbf{A}(\mathbf{r}, t) = \mathbf{A}(\mathbf{r})e^{i(k_x x + k_y y + k_z z)} e^{-i\omega t} \tag{3.5}$$

The vector form of the amplitude $\mathbf{A}(\mathbf{r})$ reflects the fact that in most instances the amplitudes components A_x, A_y, A_z are not equal, and the amplitude is *polarised* linearly, circularly, or radially. In the subsequent discussion we will assume the simplest case of linear polarisation with electromagnetic vector fields aligned along some cartesian axis. The direction of polarisation is described by vector $\hat{\mathbf{e}}(\mathbf{r})$ of unit length so that we can write as

$$A(x, y, z, t) = \underbrace{\hat{\mathbf{e}}(\mathbf{r}) \cdot \mathbf{A}(\mathbf{r})e^{i(k_x x + k_y y + k_z z)} e^{-i\omega t}}_{\text{phasor}} \tag{3.6}$$

and the complex spatial part of the field is called a *phasor*. Unless otherwise explicitly stated, we will assume from here forward that the usual fields of electromagnetism, E,B,D,H,P,M,J, and S are in the phasor form.

3.1.2 Decoupling Maxwell's curl equations

We saw in Section 2.8.1 that the far-field spherical waves morphed asymptotically into local plane waves. We can also simply consider that the dipole source is removed to $-\infty$ and posit the plane wave form as possible solutions to Maxwell's equations. The two curl equations, Equations 2.27 and 2.28, are at the heart of classical electrodynamics. They are two coupled first-order differential equations, but they can be uncoupled to form two second-order differential equations that admit propagating electromagnetic wave solutions. The standard way to carry out the uncoupling is to apply the curl operation to Equations 2.27 and 2.28, invoke a vector calculus identity, and posit source-free conditions. The curl operations yield

$$\nabla \times (\nabla \times \mathbf{E}) = \nabla (\nabla \cdot \mathbf{E}) - \nabla^2 \mathbf{E} = \nabla \times \left(-\frac{\partial \mathbf{B}}{\partial t} \right) \tag{3.7}$$

$$= -\frac{\partial}{\partial t} (\nabla \times \mathbf{B}) = -\mu_0 \varepsilon_0 \frac{\partial^2 \mathbf{E}}{\partial t^2} \tag{3.8}$$

[1] The reader should be mindful that physics literature and engineering literature use different conventions to describe time-harmonic phasor fields. The form $A = A_0 e^{i(k \cdot r - \omega t)}$ is the physics convention. The form $A = A_0 e^{-j(\beta \cdot r - \omega t)}$ is the engineering convention. An important consequence of this choice is that complex quantities $\tilde{a}$ must be expressed as $\tilde{a} = a + i\alpha$ with the physics convention and $\tilde{a} = a - j\alpha$ with the engineering convention. Thus, a complex propagation parameter in the physics convention would be written $\tilde{k} = k + i\kappa$, while engineers would write $\tilde{\beta} = \alpha - j\gamma$. Most of the circuit, transmission line, and waveguide literature is written by engineers.

In source-free regions $\nabla \cdot \mathbf{E} = 0$ so

$$\nabla^2 \mathbf{E} = \mu_0 \varepsilon_0 \frac{\partial^2 \mathbf{E}}{\partial t^2} = \frac{1}{c^2} \frac{\partial^2 \mathbf{E}}{\partial t^2} \tag{3.9}$$

Similarly,

$$\nabla \times (\nabla \times \mathbf{B}) = \nabla(\nabla \cdot \mathbf{B}) - \nabla^2 \mathbf{B} = \nabla \times \left(\mu_0 \varepsilon_0 \frac{\partial \mathbf{E}}{\partial t} \right) \tag{3.10}$$

$$= \mu_0 \varepsilon_0 \frac{\partial}{\partial t} (\nabla \times \mathbf{E}) = -\mu_0 \varepsilon_0 \frac{\partial^2 \mathbf{B}}{\partial t^2} \tag{3.11}$$

According to Equation 2.26, the divergence of the magnetic induction field is always null, $\nabla \cdot \mathbf{B} = 0$, so,

$$\nabla^2 \mathbf{B} = \mu_0 \varepsilon_0 \frac{\nabla^2 \mathbf{B}}{\partial t^2} = \frac{1}{c^2} \frac{\nabla^2 \mathbf{B}}{\partial t^2} \tag{3.12}$$

Equations 3.9 and 3.12 clearly have solutions of the form of plane waves propagating with velocity c. We shall generally be concerned with light propagating in free space, or in dielectric and conductive materials. We start by describing the properties of the simplest form of light propagation: the plane electromagnetic wave composed of two fields, electric and magnetic, oscillating at a single frequency ω. Figure 3.1 shows the essential

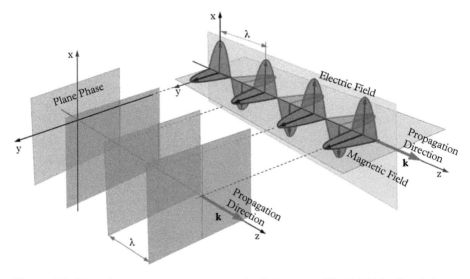

Figure 3.1 *Plane electromagnetic wave propagating in free space. The E-field is aligned along x, the B-field along y, and the wave vector k is oriented along z.*

features of the plane light wave. The electric field **E** constituent of the electromagnetic field, propagating in free space as a plane wave in direction **k**, is given by

$$\mathbf{E} = \hat{\mathbf{e}}_{\mathrm{E}} \tilde{E}_0 e^{i(\mathbf{k} \cdot \mathbf{r} - \omega t)} \tag{3.13}$$

In particular, with propagation in the z direction, the wave has the form

$$\mathbf{E} = \hat{\mathbf{e}}_{\mathrm{E}} \tilde{E}_0 e^{i(k_z z - \omega t)} \tag{3.14}$$

The complex field amplitude is $\tilde{E}_0$, the polarisation direction $\hat{\mathbf{e}}_{\mathrm{E}}$, and the frequency ω. The frequency, wavelength λ, and velocity v of the constant phase front are related by

$$\omega = \frac{2\pi}{\lambda} v = kv \tag{3.15}$$

In free space v is the speed of light c, and when in other homogeneous materials, is given by

$$v = \frac{c}{n}$$

where n is the *index of refraction* of the material. The index of refraction in free space is unity, and in matter, always greater. In materials with negligible absorption, n is a real number greater than unity, but in absorbing 'lossy' materials, n is complex with a positive imaginary term. As light of a single frequency propagates through materials of different indices of refraction, ω remains constant, but the phase velocity changes as c/n and the wavelength shortens by λ/n. It is important not to confuse the 'angular' frequency ω with the frequency of optical cycles ν. The relation is always

$$\omega = 2\pi \nu$$

The magnetic induction field **B**, the other constituent of the plane electromagnetic wave propagating in free space in the z direction, is given by

$$\mathbf{B} = \hat{\mathbf{e}}_{\mathrm{B}} \tilde{B}_0 e^{i(k_z z - \omega t)} \tag{3.16}$$

with complex amplitude $\tilde{B}_0$ and polarisation direction $\hat{\mathbf{e}}_{\mathrm{B}}$. The reason for denoting the field amplitudes complex is to express a possible phase relation between them. In fact, the divergence and curl relations in Maxwell's equations ensure that the E- and B-fields are orthogonal to the direction of propagation and that the amplitudes are in phase. The E- and **B**-fields combine to form the electromagnetic (E-M) field. They are mutually orthogonal and orthogonal to the direction of propagation:

$$\hat{\mathbf{e}}_{\mathrm{E}} \cdot \hat{\mathbf{e}}_{\mathrm{B}} = 0 \quad \text{and} \quad \hat{\mathbf{e}}_{\mathrm{E}} \times \hat{\mathbf{e}}_{\mathrm{B}} = \hat{\mathbf{e}}_{\mathrm{k}}$$

where $\hat{e}_k$ is the unit vector of the propagation parameter k. The relative amplitude between magnetic induction and electric fields in vacuum is given by,

$$B_0 = \frac{1}{c} E_0 = \sqrt{\varepsilon_0 \mu_0} E_0 \qquad (3.17)$$

and for a travelling wave through free space and homogeneous matter, the time oscillation of the **E**- and **B**-fields is always in phase.

3.1.3   Plane wave field relations from Maxwell's equations

For plane waves with E-field and H-field in the $x - y$ plane, propagating along z in a dielectric medium with permittivity ε and permeability μ, and oscillating at frequency ω, we have, from the curl relations of Faraday's law and the Maxwell–Ampère law, Equations 2.27 and 2.28;

$$\frac{\partial E_x}{\partial z} = -\mu \frac{\partial H_y}{\partial t} = i\omega\mu H_y \qquad (3.18)$$

$$\frac{\partial E_y}{\partial z} = \mu \frac{\partial H_x}{\partial t} = -i\omega\mu H_x \qquad (3.19)$$

and

$$\frac{\partial H_y}{\partial z} = -\varepsilon \frac{\partial E_x}{\partial t} = i\omega\varepsilon E_x \qquad (3.20)$$

$$\frac{\partial H_x}{\partial z} = \varepsilon \frac{\partial E_y}{\partial t} = -i\omega\varepsilon E_y \qquad (3.21)$$

Equations 3.18 and 3.20, relating E_x and H_y, are coupled, as are Equations 3.19 and 3.21, relating E_y and H_x. The first of these pairs can be decoupled by differentiating Equation 3.18 by z and Equation 3.20 by t. The order of differentiation is interchangeable, which results in

$$\frac{\partial^2 E_x}{\partial z^2} - \mu\varepsilon \frac{\partial^2 E_x}{\partial t^2} = 0 \qquad (3.22)$$

Taking into account the harmonic time variation at frequency ω, we have

$$\frac{\partial^2 E_x}{\partial z^2} + \mu\varepsilon\omega^2 E_x^2 = 0 \qquad (3.23)$$

The product of the material parameters is $\mu\varepsilon = 1/v^2$, where v is the phase velocity of the plane wave in the medium. The plane wave propagation parameter is $k = \omega/v$ so

Equation 3.23 can be written finally as the one-dimensional wave equation for the E_x component of the plane wave:

$$\frac{\partial^2 E_x}{\partial z^2} + k^2 E_x^2 = 0 \tag{3.24}$$

Clearly, analogous equations can be written for all the field components in the relations Equations 3.18–3.21. Furthermore, substitution of the plane wave E-field solution, Equation 3.14, into the second-order wave equation, Equation 3.9, results in

$$\nabla^2 \mathbf{E} + k^2 \mathbf{E} = 0 \tag{3.25}$$

with $k = \omega/c = \omega\sqrt{\mu_0 \varepsilon_0}$ when the waves are propagating in free space. Equation 3.25 is clearly valid for each of the Cartesian components and can be written simply in terms of the wave amplitude,

$$\nabla^2 E + k^2 E = 0 \tag{3.26}$$

and as such it is called the Helmholtz scalar wave equation. The analogous equation is obviously valid for the plane-wave H-field, and the relation between the E-field and H-field plane-wave amplitudes is specified by Equation 3.19:

$$B = \frac{E}{c} \quad \text{and therefore} \quad E = \sqrt{\frac{\mu_0}{\varepsilon_0}} H \tag{3.27}$$

The proportionality factor between the amplitudes, $\sqrt{\mu_0/\varepsilon_0}$, has units of impedance and is variously called the *impedance of free space*, or *intrinsic impedance*, or *wave impedance* and is denoted by Z_0:

$$Z_0 = \sqrt{\frac{\mu_0}{\varepsilon_0}} = 376.73 \quad \text{Ohms} \tag{3.28}$$

In lossless media μ and ε are real numbers, but in lossy dielectrics or conductive materials such as metals or semiconductors, the permittivity becomes complex. We examine the interaction of plane waves and lossy or conductive media in the next section.

3.1.4 Plane waves in a lossy, conductive medium (low frequency limit)

In this section we examine the propagation of plane waves in a lossy conductor. The constitutive relations for permeability, permittivity, and conductivity, μ, ε, σ, respectively, characterise the medium. As usual, we do not consider 'magnetic' materials so the permeability remains μ_0 even inside the conductor, but the permittivity becomes complex, $\varepsilon = \varepsilon_0(\varepsilon' + i\varepsilon'')$. Here we assume that σ is real, which is perfectly adequate for RF and microwave frequencies. Later in Section 3.2.4 we shall see that at optical frequencies the conductivity must also be generalised to a complex quantity.

We start again with Faraday's law and the Maxwell–Ampère law including both displacement and charge current in a medium characterised by μ, ε, σ. Assume further that Ohm's law obtains so that $\mathbf{J} = \sigma\mathbf{E}$:

$$\nabla \times \mathbf{E} = -\frac{\partial \mathbf{B}}{\partial t} = -\mu\frac{\partial \mathbf{H}}{\partial t}$$

$$\nabla \times \mathbf{H} = \frac{\partial \mathbf{D}}{\partial t} + \mathbf{J} = \varepsilon\frac{\partial \mathbf{E}}{\partial t} + \sigma\mathbf{E}$$

Let us posit plane solutions with the E-field aligned along x, B-field aligned along y, and propagating along z:

$$\begin{aligned} E_x &= E_0 e^{i(k_z z - \omega t)} \\ B_y &= B_0 e^{i(k_z z - \omega t)} \end{aligned}$$
(3.29)

The single frequency, harmonic time dependence results in

$$\nabla \times \mathbf{E} = i\mu\omega\mathbf{H} \tag{3.30}$$
$$\nabla \times \mathbf{H} = (-i\varepsilon\omega + \sigma)\mathbf{E} \tag{3.31}$$

Take the curl of both sides of Equation 3.30 and apply the standard 'curl-curl' identity

$$\nabla \times \nabla \times \mathbf{E} = i\mu\omega\nabla \times \mathbf{H} \tag{3.32}$$
$$\nabla(\nabla \cdot \mathbf{E}) - \nabla^2\mathbf{E} = i\mu\omega\nabla \times \mathbf{H} \tag{3.33}$$
$$-\nabla^2\mathbf{E} = i\mu\omega\nabla \times \mathbf{H} \tag{3.34}$$

Because, as usual, we assume a source-free environment where the plane wave is propagating, $\nabla \cdot \mathbf{E}$ evaluates to zero in Equation 3.33. Now substituting Equation 3.31 into the right-hand side of Equation 3.34 results in

$$\nabla^2\mathbf{E} + \mu\varepsilon\omega^2\left(1 + i\frac{\sigma}{\omega\varepsilon}\right)\mathbf{E} = 0 \tag{3.35}$$

This is clearly a 'Helmholtz-like' wave equation, and since we have posited the E-field along x,

$$\frac{\partial^2 E_x}{\partial z^2} + \mu\varepsilon\omega^2\left(1 + i\frac{\sigma}{\omega\varepsilon}\right)E_x = 0 \tag{3.36}$$

and the H-field along y

$$\frac{\partial^2 H_y}{\partial z^2} + \mu\varepsilon\omega^2\left(1 + i\frac{\sigma}{\omega\varepsilon}\right)H_y = 0 \tag{3.37}$$

The k of Equation 3.26, assumed real, generalises to a complex expression

$$k = \sqrt{\mu\varepsilon}\omega \left(1 + i\frac{\sigma}{\omega\varepsilon}\right)^{1/2} \qquad (3.38)$$

For non-magnetic lossy materials we can write $\mu_r = 1$ and $\varepsilon = \varepsilon_0\varepsilon_r = \varepsilon_0(\varepsilon' + \varepsilon'')$. Then:

$$k = \frac{2\pi}{\lambda_0} \left[\varepsilon' + i\left(\varepsilon'' + \frac{\sigma}{\omega\varepsilon_0}\right)\right]^{1/2} \qquad (3.39)$$

3.1.4.1 Plane waves in a lossy dielectric (but poor conductor)

If $\sigma/\omega\varepsilon_0 \ll \varepsilon''$ then the material is considered a poor conductor or good insulator and the propagation parameter becomes

$$k \rightarrow \frac{2\pi}{\lambda_0}(\varepsilon' + i\varepsilon'')^{1/2} \qquad (3.40)$$

A plane wave propagating through this medium reflects dissipative loss due to ε'':

$$E_x = E_0 e^{i\left[2\pi/\lambda_0 (\varepsilon' + i\varepsilon'')^{1/2} z - \omega t\right]} \qquad (3.41)$$

In many cases $|\varepsilon''/\varepsilon'| \ll 1$ in which case we can write

$$k \simeq \frac{2\pi}{\lambda_0}\sqrt{\varepsilon'}\left(1 + i\frac{\varepsilon''}{2\varepsilon'}\right) \qquad (3.42)$$

and Equation 3.41 can be written as

$$E_x = E_0 e^{-(2\pi/\lambda_0)\frac{\varepsilon''}{2\sqrt{\varepsilon'}}z} e^{i[(2\pi/\lambda_0)\sqrt{\varepsilon'}z - \omega t]} \qquad (3.43)$$

The first exponential factor with the real argument damps the amplitude E_0 as the wave propagates along z. The second exponential factor represents the usual harmonic wave oscillation through a medium characterised as $\sqrt{\varepsilon'}$. The two factors evidently represent a damped travelling plane wave. As $\varepsilon'' \rightarrow 0$, the damping vanishes and the oscillatory term represents a plane wave propagating through a lossless dielectric with real refractive index $\eta = \sqrt{\varepsilon'}$. The wave impedance in a lossy medium is again determined from Faraday's law. We find in this case:

$$E_x = \sqrt{\frac{\mu_0}{\varepsilon_0(\varepsilon' + i\varepsilon'')}}H_y \qquad (3.44)$$

and

$$Z = \sqrt{\frac{\mu_0}{\varepsilon_0(\varepsilon' + i\varepsilon'')}} \tag{3.45}$$

In the limit of a lossless, non-magnetic dielectric, $\varepsilon'' \to 0$,

$$Z = \sqrt{\frac{\mu_0}{\varepsilon}} = Z_0 \frac{1}{\sqrt{\varepsilon'}} \tag{3.46}$$

3.1.4.2   *Plane waves in a good conductor (low frequency limit)*

As in Section 3.1.4, here we continue to treat σ as a real quantity. We shall see in Section 3.2.4 that although this assumption is adequate for radio and microwave frequencies, it breaks down at optical frequencies.

When $\sigma/\omega\varepsilon_0 \gg \varepsilon''$ we have a good conductor. The generalised propagation parameter, Equation 3.39 becomes, as $\varepsilon''\omega\varepsilon_0/\sigma \to 0$,

$$k \to \frac{2\pi}{\lambda_0}\left(\varepsilon' + i\frac{\sigma}{\omega\varepsilon_0}\right)^{1/2} \tag{3.47}$$

The good conductor (often a metal such as silver or gold) is then characterised by the real part of the permittivity and the conductivity of the metal at frequency ω. If the dissipative part ε'' of the permittivity is retained, then the imaginary terms in Equation 3.39 can be grouped together and labelled as a 'total' $\varepsilon''_{\text{tot}}$:

$$\varepsilon''_{\text{tot}} = \varepsilon'' + \frac{\sigma}{\omega\varepsilon_0} \tag{3.48}$$

and an 'equivalent conductivity' including both terms of the imaginary part of the permittivity can be defined as

$$\sigma_{\text{equiv}} = \omega\varepsilon_0\varepsilon''_{\text{tot}} \tag{3.49}$$

Returning to Equation 3.172, we see that if $\sigma/\omega\varepsilon_0 \gg \varepsilon'$ then a plane wave entering the conductor will by damped by the conductivity term:

$$k \simeq \frac{2\pi}{\lambda_0}\sqrt{i}\left(\frac{\sigma}{\omega\varepsilon_0}\right)^{1/2} \tag{3.50}$$

Using the identity $\sqrt{i} = (1 + i)/\sqrt{2}$,

$$k \simeq \frac{2\pi}{\lambda_0}(1 + i)\left(\frac{\sigma}{2\omega\varepsilon_0}\right)^{1/2} \tag{3.51}$$

The imaginary term damps the plane wave propagating into the conductor. The depth of penetration, the *skin depth* δ, is defined as inverse of this imaginary term

$$\delta = \frac{\lambda_0}{2\pi} \left(\frac{2\omega\varepsilon_0}{\sigma} \right)^{1/2} \tag{3.52}$$

As a specific example, take silver metal as a good conductor and $\lambda_0 = 514.5\,\text{nm}$, the 'green line' of an Argon-ion laser. At this wavelength, $\varepsilon' \simeq -15$ and $\varepsilon'' \simeq 0.3$. The conductivity of silver is $\sigma \simeq 6.3 \times 10^7\,\text{S/m}$. The ratio $\sigma/\omega\varepsilon_0 \simeq 2 \times 10^3$ is much greater than the real or imaginary parts of the dielectric constant. The expression for the skin depth is therefore valid, and for this example, $\delta \simeq 2.6 \times 10^{-9}\,\text{m}$. At optical frequencies an incident plane only penetrates a few nanometres below the surface of a good metal.

The impedance is again determined from Faraday's law, specifying the relation between the E- and H-fields in the conductor:

$$E_x = \sqrt{\frac{\mu_0}{\varepsilon_0}} \left[\frac{(1+i)}{\sqrt{2}} \right]^{-1} \sqrt{\frac{\omega\varepsilon_0}{\sigma}} H_y \tag{3.53}$$

and

$$Z = Z_0 e^{-i(\pi/2)} \sqrt{\frac{\omega\varepsilon_0}{\sigma}} \tag{3.54}$$

Note that the E-field lags the H-field in the conductor by a phase angle of 45 degrees.

3.1.5   Energy density and flux in harmonic phasor fields

The discussion of energy density and power in Section 2.5 must now be extended to account for the complex form of harmonic phasor fields. The expressions for the E-field energy and B-field energy (Equations 2.43 and 2.49) in the static case and in vacuum are

$$\mathscr{E}_{\text{elec}} = \frac{1}{2} \int \mathbf{E} \cdot \mathbf{D}\, d\tau$$

$$\mathscr{E}_{\text{mag}} = \frac{1}{2} \int \mathbf{H} \cdot \mathbf{B}\, d\tau$$

If the fields are time-harmonic phasors then the optical-cycle-averaged expressions for the electric and magnetic field energies in vacuum are

$$\mathscr{E}_{\text{elec}} = \frac{\varepsilon_0}{4} \int \mathbf{E} \cdot \mathbf{E}^*\, d\tau \tag{3.55}$$

$$\mathscr{E}_{\text{mag}} = \frac{\mu_0}{4} \int \mathbf{H} \cdot \mathbf{H}^*\, d\tau \tag{3.56}$$

From the relation between the amplitudes of the electric and magnetic fields

$$|\mathbf{H}| = \frac{1}{\mu_0 c}|\mathbf{E}| \tag{3.57}$$

It is clear that in vacuum

$$\mathcal{E}_{\text{mag}} = \mathcal{E}_{\text{elec}} \tag{3.58}$$

the magnetic field energy of the wave is equal to the electric field energy. Starting with an expression analogous to Equation 2.53 we can develop a power balance relation between work done on moving charges and the energy flow from plane-wave phasor fields. In this case we will include charges flowing in conductors $\mathbf{J}_c$ as well as freely moving charged particles $\mathbf{J}_{\text{free}}$ and write the total current as

$$\mathbf{J}_{\text{tot}} = \mathbf{J}_{\text{free}} + \mathbf{J}_c \tag{3.59}$$

and the time rate of doing work on, or by, these charges by the electromagnetic field is

$$\frac{d\mathcal{E}}{dt} = \mathbf{E} \cdot \mathbf{J}_{\text{tot}} \tag{3.60}$$

Now we use the Maxwell–Ampère law and Faraday's law in their forms appropriate for time-harmonic phasors. We generalise the permittivity to include the possibility of lossy dielectrics and conductors and write $\varepsilon = \varepsilon_0(\varepsilon' + i\varepsilon'')$, but do not consider the much less frequent case of magnetic materials. Therefore, we leave the permeability as μ_0 and write

$$\nabla \times \mathbf{H}^* = \varepsilon^* \frac{\partial \mathbf{E}^*}{\partial t} + \mathbf{J}_{\text{tot}}^* = i\omega\varepsilon^*\mathbf{E}^* + \mathbf{J}_{\text{tot}}^* \tag{3.61}$$

$$\nabla \times \mathbf{E} = -\mu_0 \frac{\partial \mathbf{H}}{\partial t} = i\omega\mu_0\mathbf{H} \tag{3.62}$$

Now we 'dot-multiply' the conjugate Faraday's law by $\mathbf{E}$ and the Maxwell–Ampère law by $\mathbf{H}^*$ and invoke the vector field identity

$$\nabla \cdot (\mathbf{E} \times \mathbf{H}^*) = (\nabla \times \mathbf{E}) \cdot \mathbf{H}^* - (\nabla \times \mathbf{H}^*) \cdot \mathbf{E} \tag{3.63}$$

Substituting the result of the dot-multiplications into Equation 3.63 we find

$$\nabla \cdot (\mathbf{E} \times \mathbf{H}^*) = i\omega\left(\mu_0|H|^2 - \varepsilon_0\varepsilon'|E|^2\right) - \omega\varepsilon_0\varepsilon''|E|^2 - \mathbf{J}_{\text{free}}^* \cdot \mathbf{E} - \sigma|E|^2 \tag{3.64}$$

Then applying Stokes' theorem to the term on the left, rearranging the terms on the right, and dividing all terms by 2,

$$-\frac{1}{2}\int \mathbf{J}_{\text{free}}^{*}\cdot\mathbf{E}\,dV =\frac{1}{2}\oint_{S}(\mathbf{E}\times\mathbf{H}^{*})\cdot d\mathbf{a}+\frac{1}{2}\int \omega\varepsilon_{0}\varepsilon''|E|^{2}\,dV+\frac{1}{2}\int \sigma|E|^{2}\,dV$$
$$+\frac{i\omega}{2}\left[\int (\varepsilon_{0}\varepsilon'|E|^{2}-\mu_{0}|H|^{2})\,dV\right] \tag{3.65}$$

This power balance equation requires careful interpretation. The term on the left can be considered a source term from which energy is emitted. The negative sign denotes power flowing away from the emitter through the surface S of the enclosing volume V. The first term on the right is immediately recognisable as the cycle-averaged power passing out of the volume enclosed by surface S. This term is the phasor version of Equations 2.58 and 2.60, and we can identify the integrand with the Poynting vector

$$\mathbf{S}=\mathbf{E}\times\mathbf{H}^{*} \tag{3.66}$$

The next two terms indicate dissipative loss: absorptive loss from ε'' and conductive loss from σ due to material present within the enclosed volume. The last term on the right describes energy 'stored' in the material. Taking into account Equation 3.57, we can write this last term as

$$\frac{d\mathcal{E}_{\text{stored}}}{dt}=\frac{i\omega}{2}\left[\int \chi'\varepsilon_{0}|E|^{2}\right]dV=-\frac{1}{2}\left[\int \mathbf{E}\cdot\frac{d\mathbf{P}}{dt}\right]dV \tag{3.67}$$

and we see that this term represents power flow into the polarisation of the material. Since polarisation is a density of dipoles, we can think of this term as that part of the total emitted energy that is stored in the dipole oscillators of the material rather than the dipole oscillators of the field. The negative sign indicates that the stored energy and field energy oscillate out of phase.

3.2 Plane wave reflection and refraction

In this section we follow, except for minor variations, the development of Born and Wolf, *Principles of Optics* (6th edition) because the present author could not see any way to improve on it.

In general, a plane wave incident at an interface between two media with different indices of refraction produces a reflected wave and a transmitted wave. These two new waves will propagate away from the interface at angles and with amplitudes to be determined. The three waves propagate in space and in time as $e^{i(\mathbf{k}\cdot\mathbf{r}-\omega t)}$. At their common point on the $z=0$ interface (Figure 3.2), the time phase factor for all three is the same and for all subsequent times this phase factor will evolve as $t-(\mathbf{r}\cdot\hat{\mathbf{k}}_{n}^{q}/v_{n})$, where $n=1,2,$

the index of the medium, $v_{1,2}$ the phase velocity in the medium, and $q = i, r, t$, the incident, reflected, or transmitted wave. Therefore, we can write

$$t - \frac{\mathbf{r} \cdot \hat{\mathbf{k}}_1^i}{v_1} = t - \frac{\mathbf{r} \cdot \hat{\mathbf{k}}_1^r}{v_1} = t - \frac{\mathbf{r} \cdot \hat{\mathbf{k}}_2^t}{v_2} \tag{3.68}$$

From Figure 3.2 we write any point along the interface $\mathbf{r} = x\hat{\mathbf{x}} + z\hat{\mathbf{z}}$ and write out the dot products explicitly

$$\frac{x\hat{k}_x^i + z\hat{k}_z^i}{v_1} = \frac{x\hat{k}_x^r + z\hat{k}_z^r}{v_1} = \frac{x\hat{k}_x^t + z\hat{k}_z^t}{v_2} \tag{3.69}$$

Since Equation 3.69 must hold for any coordinates x, z along the interface, we can write

$$\frac{\hat{k}_x^i}{v_1} = \frac{\hat{k}_x^r}{v_1} = \frac{\hat{k}_x^t}{v_2} \quad \text{and} \quad \frac{\hat{k}_z^i}{v_1} = \frac{\hat{k}_z^r}{v_1} = \frac{\hat{k}_z^t}{v_2} \tag{3.70}$$

The incident wave unit vector $\hat{\mathbf{k}}^i$ and a line normal to the interface (any line parallel to the z-axis) form the *plane of incidence*. Equation 3.70 shows that the incident, reflected, and transmitted waves lie in this plane. We take this plane to be the $X - Z$ plane as in Figure 3.2. Then all y components are null and the unit wave vector components can be written in terms of the angles $\theta_i, \theta_r, \theta_t$,

$$\begin{aligned} \hat{k}_x^i &= \sin \theta_i & \hat{k}_z^i &= \cos \theta_i \\ \hat{k}_x^r &= \sin \theta_r & \hat{k}_z^r &= \cos \theta_r \\ \hat{k}_x^t &= \sin \theta_t & \hat{k}_z^r &= \cos \theta_t \end{aligned} \tag{3.71}$$

We choose the incident wave to be travelling towards $z = 0$ from the negative z half-space. Since the reflected wave is travelling away from $z = 0$ in the same half-space, the z component of the reflected wave must be negative. From the foregoing (Equations 3.70 and 3.71) we have,

$$\sin \theta_r = \sin \theta_i \quad \text{but} \quad \cos \theta_r = -\cos \theta_i \tag{3.72}$$

So $\theta_r = \pi - \theta_i$ as indicated in Figure 3.2. Equations 3.70 and 3.72 constitute the *Law of Reflection*: the sin of the angle of incidence and the sin of the angle of reflection are equal and in the same plane. We can also see from Equations 3.70 and 3.72, and remembering that the phase velocity $v = c/n$,

$$\frac{\sin \theta_i}{\sin \theta_t} = \frac{v_1}{v_2} = \frac{n_2}{n_1} \tag{3.73}$$

Equation 3.73, together with the statement that incident and transmitted propagation angles are in a plane, constitute the *Law of Refraction* or Snell's law.

Fresnel theory

The laws of reflection and refraction determine propagation angles, but the amplitude, polarisation, and phase of these fields are also of interest. Expressions first derived by Augustin-Jean Fresnel (1788–1827) characterise these properties. We first focus on the amplitude of the incident plane-wave force field, which we write in the general phasor form of Equation 3.1:

$$\mathbf{A}(\mathbf{r}, t) = \mathbf{A}(\mathbf{r})e^{i\varphi}e^{-i\omega t} \tag{3.74}$$

where $\mathbf{A}(\mathbf{r}, t)$ is either the E-field or H-field of the three plane waves, incident, reflected, or transmitted. Next we group the amplitude and its spatially varying phase into a complex amplitude, $\tilde{\mathbf{A}} = \mathbf{A}(\mathbf{r})e^{i\varphi}$ and write the time-varying phase $i\omega t$ as in Equation 3.68.

$$\tau_i = t - \frac{\mathbf{r} \cdot \hat{\mathbf{k}}_i}{v_1} = t - \frac{x \sin \theta_i + z \cos \theta_i}{v_1} \tag{3.75}$$

The complex amplitudes of all three waves are resolved into components parallel and perpendicular to the plane of incidence as shown in Figure 3.2. We then write the cartesian components of the incident E-field as

$$E_x^{(i)} = -\tilde{E}_\| \cos \theta_i e^{-i\omega \tau_i}, \qquad E_y^{(i)} = \tilde{E}_\perp e^{-i\omega \tau_i}, \qquad E_z^{(i)} = \tilde{E}_\| \sin \theta_i e^{-i\omega \tau_i} \tag{3.76}$$

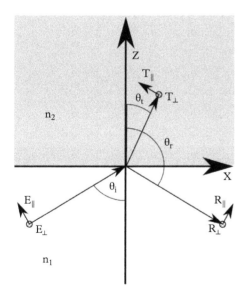

Figure 3.2 *Plane wave amplitudes, components parallel and perpendicular to the plane of incidence, for the incident E-field ($E_\|, E_\perp$), reflected E-field ($R_\|, R_\perp$), and transmitted E-field ($T_\|, T_\perp$) and angles $\theta_{i,r,t}$ of incidence, reflection, and transmission. The three waves scatter in the $X - Z$ plane at the interface between the indices of refraction n_1, n_2. The materials are supposed lossless, non-dispersive, with $n_2 > n_1$.*

We find the cartesian components of the H-field by using the right-hand orthogonality relations between $\tilde{\mathbf{E}}$ and $\tilde{\mathbf{H}}$:

$$\tilde{\mathbf{H}} = \sqrt{\frac{\varepsilon}{\mu}}\,\hat{\mathbf{k}} \times \tilde{\mathbf{E}} \tag{3.77}$$

We are usually not concerned with magnetic materials and set $\mu = 1$. Then the incident H-field cartesian components are

$$H_x^{(i)} = -\sqrt{\varepsilon_1}\,\tilde{E}_\perp \cos\theta_i e^{-i\omega\tau_i}, \qquad H_y^{(i)} = -\sqrt{\varepsilon_1}\,\tilde{E}_\parallel e^{-i\omega\tau_i}, \qquad H_z^{(i)} = \sqrt{\varepsilon_1}\,\tilde{E}_\perp \sin\theta_i e^{-i\omega\tau_i} \tag{3.78}$$

Now we write the complex amplitudes of the reflected and transmitted fields as $\tilde{R}$ and $\tilde{T}$ and their cartesian components as

$$E_x^{(r)} = -\tilde{R}_\parallel \cos\theta_r e^{-i\omega\tau_r}, \qquad E_y^{(r)} = \tilde{R}_\perp e^{-i\omega\tau_r}, \qquad E_z^{(r)} = \tilde{R}_\parallel \sin\theta_r e^{-i\omega\tau_r} \tag{3.79}$$

and

$$H_x^{(r)} = -\sqrt{\varepsilon_1}\,\tilde{R}_\perp \cos\theta_r e^{-i\omega\tau_r}, \qquad H_y^{(r)} = -\sqrt{\varepsilon_1}\,\tilde{R}_\parallel e^{-i\omega\tau_r}, \qquad H_z^{(r)} = \sqrt{\varepsilon_1}\,\tilde{R}_\perp \sin\theta_r e^{-i\omega\tau_r} \tag{3.80}$$

where

$$\tau_r = t - \frac{\mathbf{r}\cdot\hat{\mathbf{k}}_r}{v_1} = t - \frac{x\sin\theta_r + z\cos\theta_r}{v_1} \tag{3.81}$$

For the transmitted E-field

$$E_x^{(t)} = -\tilde{T}_\parallel \cos\theta_t e^{-i\omega\tau_t}, \qquad E_y^{(t)} = \tilde{T}_\perp e^{-i\omega\tau_t}, \qquad E_z^{(t)} = \tilde{T}_\parallel \sin\theta_t e^{-i\omega\tau_t} \tag{3.82}$$

and the transmitted H-field

$$H_x^{(t)} = -\sqrt{\varepsilon_2}\,\tilde{T}_\perp \cos\theta_t e^{-i\omega\tau_t}, \qquad H_y^{(t)} = -\sqrt{\varepsilon_2}\,\tilde{T}_\parallel e^{-i\omega\tau_t}, \qquad H_z^{(t)} = \sqrt{\varepsilon_2}\,\tilde{T}_\perp \sin\theta_t e^{-i\omega\tau_t} \tag{3.83}$$

with

$$\tau_t = t - \frac{\mathbf{r}\cdot\hat{\mathbf{k}}_t}{v_2} = t - \frac{x\sin\theta_t + z\cos\theta_t}{v_2} \tag{3.84}$$

Field continuity conditions at the interface dictate that the tangential (x and y) components of the three fields must be continuous,

$$E_x^{(i)} + E_x^{(r)} = E_x^{(t)} \qquad E_y^{(i)} + E_y^{(r)} = E_y^{(t)} \tag{3.85}$$

$$H_x^{(i)} + H_x^{(r)} = H_x^{(t)} \qquad H_y^{(i)} + H_y^{(r)} = E_y^{(t)} \tag{3.86}$$

and substituting Equations 3.76, 3.78, 3.79, 3.80, 3.82 and 3.83 into 3.85 and 3.86, while remembering that $\cos \theta_r = \cos(\pi - \theta_i) = -\cos \theta_i$, we have

$$\cos \theta_i (E_\| - R_\|) = \cos \theta_t T_\| \tag{3.87}$$

$$E_\perp + R_\perp = T_\perp \tag{3.88}$$

$$\sqrt{\varepsilon_1} \cos \theta_i (E_\perp - R_\perp) = \sqrt{\varepsilon_2} \cos \theta_t T_\perp \tag{3.89}$$

$$\sqrt{\varepsilon_1} (E_\| + R_\|) = \sqrt{\varepsilon_2} T_\| \tag{3.90}$$

We note that Equations 3.87 and 3.90 relate only components parallel to, and Equations 3.88 and 3.89 only components perpendicular to, the plane of incidence. The two equations involving parallel components are coupled, and the two equations involving perpendicular components are coupled, but the two sets are independent. Components parallel and perpendicular to the plane of incidence are said to have 'p-polarisation' and 's-polarisation', respectively. The use of the terms p- and s-polarisation come from the German *parallel* and *senkrecht*. The two sets of relations are also referred to as TM (transverse magnetic) and TE (transverse electric) polarisations. We can take linear combinations of Equations 3.88 and 3.89 to obtain expressions for $R_\perp$ and $T_\perp$ in terms of the incident E-field:

$$R_\perp = \frac{n_1 \cos \theta_i - n_2 \cos \theta_i}{n_1 \cos \theta_t + n_2 \cos \theta_t} E_\perp \tag{3.91}$$

$$T_\perp = \frac{2n_1 \cos \theta_i}{n_1 \cos \theta_i + n_2 \cos \theta_t} E_\perp \tag{3.92}$$

Similarly, linear combinations of Equations 3.87 and 3.90 result in

$$R_\| = \frac{n_2 \cos \theta_i - n_1 \cos \theta_t}{n_2 \cos \theta_i + n_1 \cos \theta_t} E_\| \tag{3.93}$$

$$T_\| = \frac{2n_1 \cos \theta_i}{n_2 \cos \theta_i + n_1 \cos \theta_t} E_\| \tag{3.94}$$

These expressions for the reflected and transmitted amplitudes in terms of the incident E-field amplitude are called the *Fresnel* relations. They can be recast[2] , using the law of

---

[2] The unitless expressions in brackets in Equations 3.95–3.98 and in parentheses in Equations 3.99–3.102 are often referred to as the *Fresnel coefficients*, r, t. We will make extensive use of them in the subsequent discussion of reflection and refraction at a material interface. They are not to be confused with R, T which have field amplitude units.

refraction, Equation 3.73, and standard trigonometry identities, to eliminate n_1, n_2,

$$R_\perp = -\left\{\frac{\sin(\theta_i - \theta_t)}{\sin(\theta_i + \theta_t)}\right\} E_\perp \tag{3.95}$$

$$T_\perp = \left\{\frac{2\sin\theta_t\cos\theta_i}{\sin(\theta_i + \theta_t)}\right\} E_\perp \tag{3.96}$$

$$R_\| = \left\{\frac{\tan(\theta_i - \theta_t)}{\tan(\theta_i + \theta_t)}\right\} E_\| \tag{3.97}$$

$$T_\| = \left\{\frac{2\sin\theta_t\cos\theta_i}{\sin(\theta_i + \theta_t)\cos(\theta_i - \theta_t)}\right\} E_\| \tag{3.98}$$

For the common case of normal incidence the expressions Equations 3.91–3.94 simplify to,

$$R_\perp = -\left(\frac{n-1}{n+1}\right) E_\perp \tag{3.99}$$

$$T_\perp = \left(\frac{2}{n+1}\right) E_\perp \tag{3.100}$$

$$R_\| = \left(\frac{n-1}{n+1}\right) E_\| \tag{3.101}$$

$$T_\| = \left(\frac{2}{n+1}\right) E_\| \tag{3.102}$$

where $n = n_2/n_1$. Note that for s-polarisation the Fresnel coefficients have the property that $1 = t_\perp - r_\perp$. This relation does *not* hold for p-polarisation, but $1 = r_\| + t_\|$ does hold at normal incidence.

3.2.1   Plane wave power reflection and refraction

In Section 3.2 we were concerned with plane wave *amplitude* reflection and transmission at a boundary between two materials with different indices of refraction. Continuity of the tangential field components at the boundary lead to the laws of reflection and refraction, and the Fresnel relations. In this section we consider reflection and transmission of *power* at the boundary. In addition to tangential field continuity we make use of energy conservation between incident flux normal to the boundary and the sum of reflected and transmitted fluxes. From Equations 3.27, 3.46, and 3.66 we write the incident energy flux or power density (J/m²s) as

$$\mathscr{S} = \sqrt{\varepsilon_1\varepsilon_0}c\,|E|^2 \tag{3.103}$$

and the incident flux normal to the boundary is

$$\mathscr{F}^{(i)} = \mathscr{S}\cos\theta_i = \sqrt{\varepsilon_1\varepsilon_0}c\,|E|^2\cos\theta_i \tag{3.104}$$

The energy fluxes reflected and transmitted normal to the boundary are given by

$$\mathscr{F}^{(r)} = \sqrt{\varepsilon_1}\varepsilon_0 c \, |R|^2 \cos\theta_i = n_1 \varepsilon_0 c \, |R|^2 \cos\theta_i \qquad (3.105)$$

$$\mathscr{F}^{(t)} = \sqrt{\varepsilon_2}\varepsilon_0 c \, |T|^2 \cos\theta_t = n_2 \varepsilon_0 c \, |T|^2 \cos\theta_2 \qquad (3.106)$$

The *reflectivity* and *transmissivity* are defined[3] as the fractional reflected and transmitted fluxes,

$$\mathscr{R} = \frac{\mathscr{F}^{(r)}}{\mathscr{F}^{(i)}} = \frac{|R|^2}{|E|^2} \qquad \mathscr{T} = \frac{\mathscr{F}^{(t)}}{\mathscr{F}^{(i)}} = \frac{n_2 \cos\theta_t}{n_1 \cos\theta_i} \frac{|T|^2}{|E|^2} \qquad (3.107)$$

Invoking energy conservation across the boundary,

$$1 = \mathscr{R} + \mathscr{T} \qquad (3.108)$$

Now we take a closer look at the reflectivity and transmissivity as a function of incident polarisation. Let φ be the angle of the incident plane wave E-field with respect to the plane of incidence as shown in Figure 3.3. Then we have $\mathscr{F}_\parallel$ and $\mathscr{F}_\perp$, the energy fluxes parallel and perpendicular to the plane of incidence, expressed as

$$\mathscr{F}_\parallel^{(i)} = \mathscr{F}^{(i)} \cos^2\varphi_i \qquad (3.109)$$

$$\mathscr{F}_\perp^{(i)} = \mathscr{F}^{(i)} \sin^2\varphi_i \qquad (3.110)$$

and similarly for the reflected and transmitted fluxes. The reflectivity and transmissivity can then be resolved into components parallel and perpendicular to the plane of incidence,

$$\mathscr{R} = \mathscr{R}_\parallel \cos^2\varphi_i + \mathscr{R}_\perp \sin^2\varphi_i \qquad (3.111)$$

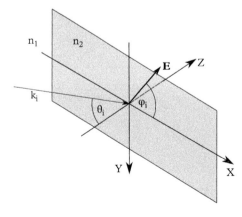

Figure 3.3 *Boundary in the $X-Y$ plane between two regions of indices of refraction n_1 and n_2. The plane of incidence is formed by the incoming wave vector k_i and the normal to the boundary along the z-axis. The angle θ_i is the angle of incidence between the incoming wave and the normal to the boundary. The angle φ_i lies between the incident linearly polarised E-field and the plane of incidence.*

[3] These quantities are also known as the *reflectance* and the *transmittance*.

where

$$\mathcal{R}_{\parallel} = \frac{\mathcal{F}_{\parallel}^{(r)}}{\mathcal{F}_{\parallel}^{(i)}} = \frac{|R_{\parallel}|^2}{|E_{\parallel}|^2} \tag{3.112}$$

$$\mathcal{R}_{\perp} = \frac{\mathcal{F}_{\perp}^{(r)}}{\mathcal{F}_{\perp}^{(i)}} = \frac{|R_{\perp}|^2}{|E_{\perp}|^2} \tag{3.113}$$

and

$$\mathcal{T}_{\parallel} = \frac{\mathcal{F}_{\parallel}^{(t)}}{\mathcal{F}_{\parallel}^{(i)}} = \frac{n_2 \cos\theta_t}{n_1 \cos\theta_i} \frac{|T_{\parallel}|^2}{|E_{\parallel}|^2} \tag{3.114}$$

$$\mathcal{T}_{\perp} = \frac{\mathcal{F}_{\perp}^{(t)}}{\mathcal{F}_{\perp}^{(i)}} = \frac{n_2 \cos\theta_t}{n_1 \cos\theta_i} \frac{|T_{\perp}|^2}{|E_{\perp}|^2} \tag{3.115}$$

From the expressions for the Fresnel amplitude relations, Equations 3.95–3.98 and Equations 3.114 and 3.115 the components of $\mathcal{R}$ and $\mathcal{T}$ can be written as

$$\mathcal{R}_{\parallel} = \left\{ \frac{\tan(\theta_i - \theta_t)}{\tan(\theta_i + \theta_t)} \right\}^2 \tag{3.116}$$

$$\mathcal{R}_{\perp} = \left\{ \frac{\sin(\theta_i - \theta_t)}{\sin\theta_i + \sin\theta_t} \right\}^2 \tag{3.117}$$

and

$$\mathcal{T}_{\parallel} = \frac{\sin 2\theta_i \sin 2\theta_t}{\sin^2(\theta_i + \theta_t) \cos^2(\theta_i - \theta_t)} \tag{3.118}$$

$$\mathcal{T}_{\perp} = \frac{\sin 2\theta_i \sin 2\theta_t}{\sin^2(\theta_i + \theta_t)} \tag{3.119}$$

Finally at normal incidence:

$$\mathcal{R} = \left\{ \frac{n-1}{n+1} \right\}^2 \tag{3.120}$$

$$\mathcal{T} = \frac{4n}{(n+1)^2} \tag{3.121}$$

It can be easily verified that $\mathcal{R}_{\perp} + \mathcal{T}_{\perp} = 1$ and that $\mathcal{R}_{\parallel} + \mathcal{T}_{\parallel} = 1$. In the special case where $\theta_i + \theta_t = \pi/2$, Equation 3.116 shows that the parallel component of the reflected power goes to zero. Furthermore, it can be seen from Figure 3.2, 3.3, and 3.4 that if the sum

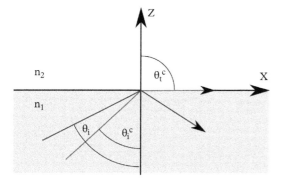

Figure 3.4 *Total internal reflection at the interface between two media with refractive indices $n_1 > n_2$. At the critical angle of incidence θ_i^c the transmitted wave travels as a surface wave along the interface boundary. The transmission angle θ_t^c is $\pi/2$. At incident angles greater than θ_i^c total reflection occurs and no power is transmitted to the n_2 half-space.*

of the angles of incidence and transmission angles is 90 degrees, then the angle between them must also be 90 degrees. Then the law of refraction reads

$$\frac{n_2}{n_1} = \frac{\sin\theta_i}{\sin\theta_t} = \frac{\sin\theta_i}{\sin(\pi/2-\theta_i)} = \frac{\sin\theta_i}{\cos\theta_i} = \tan\theta_i \qquad (3.122)$$

This special angle of incidence is called the *Brewster angle*. For an air/glass interface with the refractive index of glass equal to 1.5, the Brewster angle is $\theta_i = \tan^{-1}(1.5) \simeq 56$ degrees.

3.2.2 Total internal reflection

Consider again the law of refraction (Snell's law).

$$\frac{n_2}{n_1} = \frac{\sin\theta_i}{\sin\theta_t} \qquad \text{or} \qquad \sin\theta_t = \frac{n_1}{n_2}\sin\theta_i \qquad (3.123)$$

So far we assumed $n_1 < n_2$, the usual case when light in vacuum or in air impinges on a common isotropic dielectric material such as glass or fused quartz. Suppose, however, that the light is incident on the boundary from the high-index side so that $n_1 > n_2$. It is then possible that above some incident angle θ_i Snell's law yields a value for $\sin\theta_t$ greater than unity. The value of θ_i such that $\sin\theta_t$ is just equal to unity is called the *critical angle* θ_i^c and $\theta_t = \pi/2$. As we shall show below, at angles $\theta_i > \theta_i^c$, there is no power transmission and all the propagating flux is reflected at the boundary. When $\theta_i > \theta_i^c$, $\cos\theta_t$ becomes complex as can be seen from

$$\cos\theta_t = \sqrt{1-\sin^2\theta_t} = \sqrt{1-\left(\frac{n_1}{n_2}\right)^2\sin^2\theta_i} = \pm i\sqrt{\left(\frac{n_1}{n_2}\right)^2\sin^2\theta_i - 1} \qquad (3.124)$$

Although no power is transmitted to the low-index side of the interface, Equations 3.82, 3.83 and 3.84 show that evanescent E- and H-fields do exist there. For convenience we rewrite these equations here for the transmitted E-field components and substitute Equations 3.123 and 3.124 for $\sin\theta_t$ and $\cos\theta_t$:

$$E_x^{(t)} = -\tilde{T}_{\|}\cos\theta_t e^{-i\omega\tau_t}, \qquad E_y^{(t)} = \tilde{T}_{\perp}e^{-i\omega\tau_t}, \qquad E_z^{(t)} = \tilde{T}_{\|}\sin\theta_t e^{-i\omega\tau_t} \qquad (3.125)$$

where τ_t is given by

$$\tau_t = t - \frac{\mathbf{r}\cdot\hat{\mathbf{k}}_t}{v_2} = t - \frac{x\sin\theta_t + z\cos\theta_t}{v_2} \qquad (3.126)$$

$$\tau_t = t - \frac{\left(\frac{n_1}{n_2}\sin\theta_i\right)x + \mp i\left(\sqrt{\left(\frac{n_1}{n_2}\right)^2\sin^2\theta_i - 1}\right)z}{v_2} \qquad (3.127)$$

Clearly the argument of the phase of the transmitted E-field components in the $+z$ direction (above the interface in Figure 3.4) becomes a real negative number if the negative root of the imaginary term in Equation 3.127 is chosen; the field amplitudes fall off exponentially as $+z$ increases. Choosing the positive root results in exponential fall-off in the $-z$ direction. At the critical angle, the wave does propagate along x at the boundary because the x-component of the phase term in Equation 3.127 is unity.

We can show that beyond the critical angle, all the light energy is reflected by using the Fresnel relations, Equations 3.93 and 3.94, and substituting Equation 3.124 for $\cos\theta_t$. Since $\cos\theta_t$ is purely imaginary, taking the absolute value of $R_{\|}$ and $R_{\perp}$ results in

$$|R|_{\|}^2 = |E|_{\|}^2 \qquad \text{and} \qquad |R|_{\perp}^2 = |E|_{\perp}^2 \qquad (3.128)$$

So the sum of the incident energy parallel and perpendicular to the plane of incident is totally reflected at the boundary.

We can also calculate the time-averaged energy flux across the boundary in the z direction by evaluating the z-component of the Poynting vector. From Equations 3.82 and 3.83 we form S_z from the relevant components of the E- and H-fields:

$$|\mathbf{S}| = \mathbf{E}\times\mathbf{H}^* \qquad (3.129)$$

$$S_z^{(t)} = E_x^{(t)}H_y^{(t)} - E_y^{(t)}H_x^{(t)} \qquad (3.130)$$

We want to evaluate the time-average of the real part of S_z at the boundary $z = 0$:

$$\text{Re}[S_z(z=0)] = \frac{1}{2}[S_z + S_z^*] \qquad (3.131)$$

Substituting the relevant terms from Equations 3.82 and 3.83 into Equations 3.130 and 3.131 we find

$$\text{Re}[S_z] = \frac{1}{2}\left[\left(T_\parallel^2 \sqrt{\varepsilon_2} \cos\theta_t e^{-2i\omega\tau} + T_\perp^2 \sqrt{\varepsilon_2} \cos\theta_t e^{-2i\omega\tau}\right) + \left(T_\parallel^{*2} \sqrt{\varepsilon_2} \cos^*\theta_t e^{2i\omega\tau} + T_\perp^{*2} \sqrt{\varepsilon_2} \cos^*\theta_t e^{2i\omega\tau}\right)\right] \tag{3.132}$$

and with $\cos\theta_t$ pure imaginary when $\theta_i > \theta_i^c$,

$$\text{Re}[S_z] = \frac{1}{2}\left\{\sqrt{\varepsilon_2} \cos\theta_t \left[\left(T_\parallel^2 + T_\perp^2\right)e^{-2i\omega\tau} - \left(T_\parallel^{*2} + T_\perp^{*2}\right)e^{2i\omega\tau}\right]\right\} \tag{3.133}$$

where, at $z = 0$,

$$\tau = t - \frac{x\sin\theta_t}{v_2} = t - \frac{n_1}{n_2}\frac{x\sin\theta_i}{v_2} \tag{3.134}$$

The four terms in Equation 3.133 simply oscillate harmonically in t with period $T = 1/2\omega$. We take the time average over a time t' long compared to T:

$$\langle\text{Re}[S_z]\rangle = \frac{1}{2t'}\int_{-t'}^{t'} \text{Re}[S_z(t)]dt' \tag{3.135}$$

and find that

$$\lim_{t'\to\infty} \langle\text{Re}[S_z]\rangle \to 0 \tag{3.136}$$

Therefore at times long compared to an optical period, no energy flux crosses the boundary when $\theta_i \geq \theta_i^c$.

3.2.3   Reflection and transmission at a material interface

Reflection and transmission of light at surfaces is encountered frequently in the study of light–matter interaction. We discuss here a few simple cases in which a plane wave, incident from a half-space of vacuum ($\varepsilon_0, \mu_0, \sigma = 0$), impinges at the surface of a material characterised by ε, μ, σ. The cases are organised according to the nature (real or complex) and magnitude of the permittivity and conductivity. As before, we consider the material non-magnetic and keep the permeability as μ_0.

3.2.3.1   *Normal incidence on a general material*

We posit a plane wave propagating from left to right along z, with E-field linearly polarised along x (s-polarisation) and H-field along y as shown in Figure 3.1. The interface is in the $x-y$ plane and situated at $z = 0$. The half-space to the left ($z < 0$) is vacuum, and the half-space to the right ($z > 0$) is characterised by $\varepsilon, \mu_0, \sigma$. The incident E- and H-phasor fields, parallel to the $x-y$ plane, are given by

$$\mathbf{E_i} = E_0 e^{ik_0 z} \, \hat{\mathbf{x}} \tag{3.137}$$

$$\mathbf{H_i} = H_0 e^{ik_0 z} \, \hat{\mathbf{y}} = \frac{E_0}{Z_0} e^{ik_0 z} \, \hat{\mathbf{y}} \tag{3.138}$$

where Z_0 is the impedance of free space, $Z_0 = \sqrt{\mu_0/\varepsilon_0}$. At the interface, a fraction of the wave is reflected and a fraction is transmitted. Similarly to Section 3.2, we write the reflected wave[4] as

$$\mathbf{E_r} = -r E_0 e^{-ik_0 z} \, \hat{\mathbf{x}} \tag{3.139}$$

$$\mathbf{H_r} = r \frac{E_0}{Z_0} e^{-ik_0 z} \, \hat{\mathbf{y}} \tag{3.140}$$

where r is the *reflection coefficient*. The transmitted wave is given by

$$\mathbf{E_t} = t E_0 e^{ikz} \, \hat{\mathbf{x}} \tag{3.141}$$

$$\mathbf{H_t} = t \frac{E_0}{Z} e^{ikz} \, \hat{\mathbf{y}} \tag{3.142}$$

where t is the *transmission coefficient*. In the material to the right, the propagation parameter k is given by Equation 3.39 and rewritten here for convenience,

$$k = \frac{2\pi}{\lambda_0} \left[\varepsilon' + i \left(\varepsilon'' + \frac{\sigma}{\omega \varepsilon_0} \right) \right]^{1/2}$$

At $z = 0$ the total parallel field amplitudes on either side of the boundary must be continuous across it. Therefore:

$$E_i + E_r = E_t \tag{3.143}$$

$$E_i - r E_i = t E_i \tag{3.144}$$

$$H_i + r H_i = t H_i \tag{3.145}$$

$$\frac{E_i}{Z_0} + r \frac{E_i}{Z_0} = t \frac{E_i}{Z} \tag{3.146}$$

---

[4] The choice of phase for the reflected wave is a matter of convention. The reflected waves could also be written $\mathbf{E_r} = r E_0 e^{-ik_0 z} \hat{\mathbf{x}}$ and $\mathbf{H_r} = -r (E_0/Z_0) e^{-k_0 z} \hat{\mathbf{y}}$. The first convention, Equations 3.139 and 3.140, is often used in physical optics while the second is common in electrical engineering. The second choice results in a change of sign for the reflection coefficient with the consequence that $r = (Z - Z_0)/(Z_0 + Z)$ and $1 = t - r$.

From Equations 3.144 and 3.146 we have

$$1 - r = t \tag{3.147}$$

$$\frac{1}{Z_0} + \frac{r}{Z_0} = \frac{t}{Z} \tag{3.148}$$

From these two relations we find:

$$1 = r + t \tag{3.149}$$

$$r = \frac{Z_0 - Z}{Z_0 + Z} \tag{3.150}$$

$$t = \frac{2Z}{Z_0 + Z} \tag{3.151}$$

3.2.3.2 Lossless dielectric

For a lossless dielectric $\varepsilon'' = 0$ and $\sigma = 0$ so the propagation vector becomes $k = (2\pi/\lambda_0)\,\varepsilon'$. The Poynting vector on the left-hand side of the boundary is

$$\mathbf{S}_{z<0} = (\mathbf{E}_i + \mathbf{E}_r) \times (\mathbf{H}_i^* + \mathbf{H}_r^*) \tag{3.152}$$

$$= \left(E_0 e^{ik_0 x} - rE_0 e^{-ik_0 x}\right) \times \left(\frac{E_0}{Z_0} e^{-ik_0 z} + r\frac{E_0}{Z_0} e^{ik_0 z}\right) \hat{\mathbf{z}}$$

$$= \frac{E_0^2}{Z_0} \left[\left(1 - r^2\right) - i2r \sin\left(2k_0 z\right)\right] \hat{\mathbf{z}} \tag{3.153}$$

At $z = 0$ the energy flux from the left is

$$\mathbf{S}_- = \frac{E_0^2}{Z_0} \left(1 - r^2\right) \hat{\mathbf{z}} \tag{3.154}$$

and since $r < 1$, the net energy flux propagates from left to right. In the region $z < 0$ the oscillatory term in Equation 3.153 is due to the standing wave set up by reflection of the plane wave at the boundary. Notice that the phase of the standing wave is in quadrature with respect to the incident wave. The Poynting vector on the right-hand side is

$$\mathbf{S}_{z>0} = (\mathbf{E}_t) \times (\mathbf{H}_t^*) \hat{\mathbf{z}} \tag{3.155}$$

$$= t^2 \frac{E_0^2}{Z} \hat{\mathbf{z}} \tag{3.156}$$

The energy flux across the boundary should be continuous, and by using the expressions for r and t in terms of the impedances, Equations 3.147 and 3.148, the continuity can be verified.

3.2.3.3 Lossy dielectric

In the case of a lossy dielectric the propagation vector becomes

$$k = \frac{2\pi}{\lambda_0} \left(\varepsilon' + i\varepsilon''\right)^{1/2} \tag{3.157}$$

$$= \frac{2\pi}{\lambda_0}(\eta + i\kappa) \tag{3.158}$$

and the wave impedance becomes complex:

$$Z = \sqrt{\frac{\mu_0}{\varepsilon}} = \sqrt{\frac{\mu_0}{\varepsilon_0 \left(\varepsilon' + i\varepsilon''\right)}} = \frac{Z_0}{\eta + i\kappa} \tag{3.159}$$

Therefore, from Equations 3.150 and 3.151 the reflection and transmission coefficients become complex, and the parallel field component continuity conditions at the boundary now give

$$t = \frac{2}{(\eta + 1) + i\kappa} \tag{3.160}$$

$$r = \frac{(\eta - 1) + i\kappa}{(\eta + 1) + i\kappa} \tag{3.161}$$

Figure 3.5 shows a plane wave incident from the left on a slightly absorbing slab of SiO. The shortened wave length and slight loss of amplitude due to absorption can be discerned within the material. The Poynting vector in the $z < 0$ region is

$$\mathbf{S}_{z<0} = \frac{|E_0|^2}{Z_0} \left(1 + r^* e^{i2k_0 z} - r e^{-i2k_0 z} - |r|^2\right) \hat{\mathbf{z}} \tag{3.162}$$

and in the $z > 0$ region

$$\mathbf{S}_{z>0} = |t|^2 \frac{|E_0|^2}{Z^*} e^{-2\kappa z} \hat{\mathbf{z}} = \frac{|E_0|^2}{Z_0} \cdot \frac{4(\eta - i\kappa)}{(\eta + 1)^2 + \kappa^2} e^{-2\kappa z} \hat{\mathbf{z}} \tag{3.163}$$

At the boundary ($z = 0$) the flux from the $z < 0$ region is

$$\mathbf{S}_{z=0}^- = \frac{|E_0|^2}{Z_0} \left(1 + r^* - r - |r|^2\right) = \frac{|E_0|^2}{Z_0} \cdot \frac{4(\eta - i\kappa)}{(\eta + 1)^2 + \kappa^2} \tag{3.164}$$

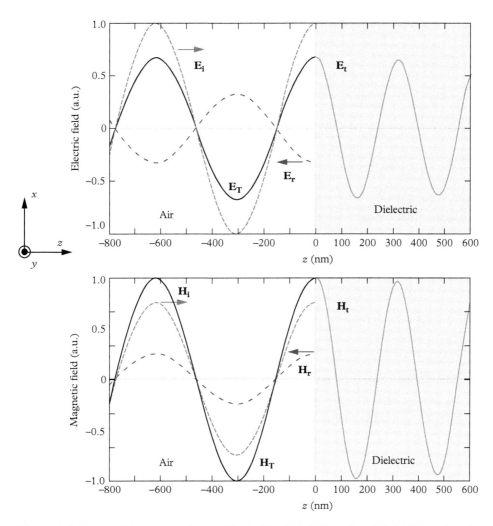

Figure 3.5 *Top panel shows the real part of the incident E-field E_i, reflected E-field E_r and total E-field E_T in air to the left of the air-dielectric interface. The wave length of the propagating wave in air is $\lambda_0 = 620$ nm. Within the dielectric the transmitted E-field E_t is shown. The dielectric is SiO with index of refraction $n = \eta + i\kappa = 1.969 + i0.01175$. Bottom panel shows the real part of the incident H-field H_i, reflected H-field H_r, and total H-field H_T in air to the left of the air-dielectric interface. Within the dielectric the transmitted H-field H_t is shown. Note the shortening of the wave length and the slight decrease in amplitude as the wave penetrates into the lossy dielectric.*

and from the $z > 0$ region

$$S^+_{z=0} = \frac{|E_0|^2}{Z_0} \cdot \frac{4(\eta - i\kappa)}{(\eta + 1)^2 + \kappa^2} \tag{3.165}$$

So we see that the complex Poynting vector is conserved across the boundary. When the Poynting vector is complex, we interpret the energy flux or the power flow as the cycle-averaged real part of the Poynting vector. The net energy flux traversing the boundary from the $z < 0$ region is therefore:

$$P^- = \frac{1}{2}\text{Re}[S^-] = \frac{|E_0|^2}{Z_0} \cdot \frac{2\eta}{(\eta + 1)^2 + \kappa^2} \tag{3.166}$$

and the energy flux penetrating the lossy dielectric on the $z > 0$ side of the boundary is

$$P^+ = \frac{1}{2}\text{Re}[S^+] = \frac{|E_0|^2}{Z_0} \cdot \frac{2\eta}{(\eta + 1)^2 + \kappa^2}e^{-2\kappa z} \tag{3.167}$$

Note that the incident energy flux on the boundary from the left is

$$P_i = \frac{1}{2}\text{Re}\left[E_0 e^{ik_0 z} \cdot \frac{E_0^*}{Z_0}e^{-ik_0 z}\right] = \frac{|E_0|^2}{2Z_0} \tag{3.168}$$

and the reflected flux is

$$P_r = \frac{1}{2}\text{Re}\left[-rE_0 e^{-ik_0 z} \cdot \frac{r^* E_0^*}{Z_0}e^{ik_0 z}\right] = -|r|^2\frac{|E_0|^2}{2Z_0} \tag{3.169}$$

The sum of the incident and reflected power

$$P_T = P_i + P_r = \frac{|E_0|^2}{2Z_0}\left(1 - |r|^2\right) = \frac{|E_0|^2}{Z_0} \cdot \frac{2\eta}{(\eta + 1)^2 + \kappa^2} \tag{3.170}$$

accords with the net power flow calculated from the superposed incident and reflected fields (Equations 3.162 and 3.166).

3.2.3.4   *Good conductor (low frequency regime)*

A plane wave incident on a perfect conductor and on a good conductor are shown in Figures 3.6 and 3.7. The difference between 'perfect' and 'good' is essentially the magnitude of the conductivity, σ. When $\sigma \to \infty$ a conductor approaches the perfect conductor idealisation. In the case of a good conductor

$$k = \frac{2\pi}{\lambda_0}\left(\varepsilon' + i\frac{\sigma}{\omega\varepsilon_0}\right)^{1/2} \tag{3.171}$$

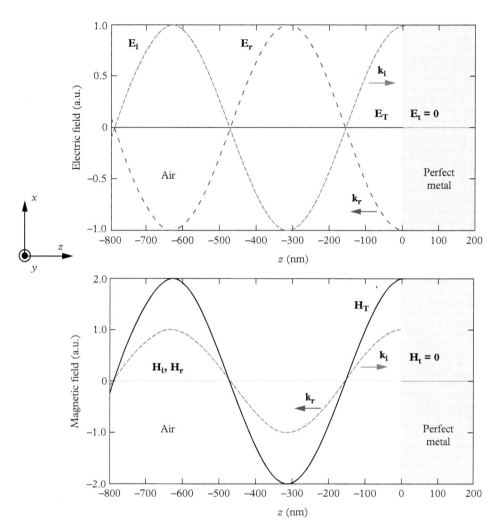

Figure 3.6 *Top panel shows the real part of the incident E-field E_i, reflected E-field E_r, and total E-field E_T in air to the left of the air-perfect metal interface. The wave length of the propagating wave in air is $\lambda_0 = 632$ nm. Within the perfect conductor ($\sigma \to \infty$) no electromagnetic field penetrates. Bottom panel shows the real part of the incident H-field H_i, reflected H-field H_r, and total H-field H_T in air to the left of the air-dielectric interface. Within the perfect conductor the transmitted H-field H_t is null.*

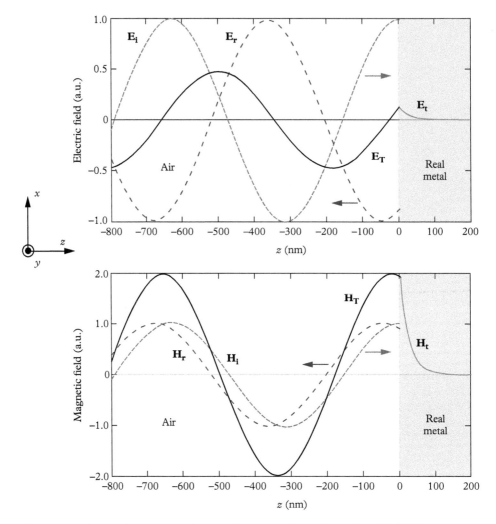

Figure 3.7 *Top panel shows the real part of the incident E-field E_i, reflected E-field E_r, and total E-field E_T in air to the left of the air-dielectric interface. The wave length of the propagating wave in air is $\lambda_0 = 632$ nm. Within the dielectric the transmitted E-field E_t is shown. Bottom panel shows the real part of the incident H-field H_i, reflected H-field H_r, and total H-field H_T in air to the left of the air-dielectric interface. Within the dielectric the transmitted H-field H_t is shown.*

As we saw in the case of silver metal, for good conductors $\sigma/(\omega\varepsilon_0)$ is on the order of thousands while ε' is on the order of unity. We can then, to a quite good approximation, write

$$k \simeq \frac{2\pi}{\lambda_0}\left(i\frac{\sigma}{\omega\varepsilon_0}\right)^{1/2} \qquad (3.172)$$

and recalling from Equation 3.52 that the skin depth is defined as

$$\delta = \frac{\lambda_0}{2\pi} \left(\frac{2\omega\varepsilon_0}{\sigma} \right)^{1/2}$$

we can express k as

$$k \simeq \frac{\sqrt{2i}}{\delta} = \pm\frac{1+i}{\delta} \tag{3.173}$$

The sign of k is chosen so that the wave amplitude decays exponentially as it propagates into the good conductor in the $z > 0$ half space. Wave impedance in the conductor is

$$Z = \sqrt{\frac{\mu_0}{\varepsilon}} = \sqrt{\frac{\mu_0}{\varepsilon_0}} \cdot i^{-1/2} \left(\frac{\omega\varepsilon_0}{\sigma} \right)^{1/2} = Z_0 \cdot \left[(1-i) \left(\frac{k_0\delta}{2} \right) \right] \tag{3.174}$$

The reflection and transmission coefficients are

$$r = \frac{2 - (k_0\delta)^2}{2 + 2k_0\delta + (k_0\delta)^2} + i\frac{2k_0\delta}{2 + 2k_0\delta + (k_0\delta)^2} \tag{3.175}$$

$$\simeq 1 - ik_0\delta \tag{3.176}$$

$$t = \frac{2\left[(k_0\delta)^2 + k_0\delta \right]}{2 + 2k_0\delta + (k_0\delta)^2} - i\frac{2k_0\delta}{2 + 2k_0\delta + (k_0\delta)^2} \tag{3.177}$$

$$\simeq (1-i)k_0\delta \tag{3.178}$$

At the boundary ($z = 0$), the flux from the $z < 0$ region is

$$S^-_{z=0} = \frac{|E_0|^2}{Z_0} \cdot \left(1 + r^* - r - |r|^2 \right)$$

$$= \frac{|E_0|^2}{Z_0} \cdot \frac{4k_0\delta}{2 + 2k_0\delta + (k_0\delta)^2}(1-i) \tag{3.179}$$

and the flux penetrating into the conductor ($z > 0$) is

$$S_{z>0} = \frac{|E_0|^2 \, |t|^2}{Z^*} e^{-(2/\delta)z}\hat{\mathbf{z}}$$

$$= \frac{|E_0|^2}{Z_0} \cdot \frac{4k_0\delta}{2 + 2k_0\delta + (k_0\delta)^2}(1-i)e^{-(2/\delta)z}\hat{\mathbf{z}} \tag{3.180}$$

At the $z = 0$ boundary, the flux from the $z > 0$ region is

$$S_{z=0}^+ = \frac{|E_0|^2 |t|^2}{Z^*}$$

$$= \frac{|E_0|^2}{Z_0} \cdot \frac{4k_0\delta}{2 + 2k_0\delta + (k_0\delta)^2}(1 - i) \tag{3.181}$$

and it is clear that the complex Poynting vectors $S_{z=0}^- = S_{z=0}^+$. Because the skin depth is only a few nanometres we see that the energy flux penetrates significantly only a few tens of nanometres. As in the case of the lossy dielectric, the power penetrating into the metal is given by

$$P^+ = \frac{1}{2}\mathrm{Re}\left[S_{z=0}^+\right] \tag{3.182}$$

$$= \frac{|E_0|^2}{Z_0} \cdot \frac{2k_0\delta}{2 + 2k_0\delta + (k_0\delta)^2} \tag{3.183}$$

And the power entering the metal from the $z < 0$ side is

$$P^- = \frac{1}{2}\mathrm{Re}\left[S_{z=0}^-\right] \tag{3.184}$$

$$= \frac{|E_0|^2}{Z_0} \cdot \frac{2k_0\delta}{2 + 2k_0\delta + (k_0\delta)^2} \tag{3.185}$$

So we see that $P^- = P^+$ and the power across the boundary is conserved. Furthermore, we can confirm that the sum of the incident power P_i and reflected power P_r is equal to the power transmitted across the boundary:

$$P_i + P_r = \frac{1}{2}\frac{|E_0|^2}{Z_0}\left(1 - |R|^2\right) \tag{3.186}$$

$$= \frac{|E_0|^2}{Z_0} \cdot \frac{2k_0\delta}{2 + k_0\delta + (k_0\delta)^2} \tag{3.187}$$

Now, from the earlier example in Section 3.1.4.2 for an air–silver metal interface, the value for $k_0 = 1.22 \times 10^7$ m^{-1}, the skin depth $\delta = 2.6 \times 10^{-9}$ m, and the product $k_0\delta = 3.2 \times 10^{-2}$. Therefore:

$$P^- = P^+ \simeq \frac{|E_0|^2}{Z_0} \cdot k_0\delta \tag{3.188}$$

The cycle-averaged incident power is

$$P_i = \frac{|E_0|^2}{2Z_0} \tag{3.189}$$

and therefore the fraction of the incident power entering the metal is

$$\frac{P^+}{Pi} = 2k_0\delta = 6.4 \times 10^{-2} \tag{3.190}$$

About 6% of the incident power is transmitted to the metal.
 The reflected power is given by

$$P_r = -|r|^2\frac{|E_0|^2}{2Z_0} \tag{3.191}$$

and the fraction of the incident power reflected is

$$\frac{P_r}{P_i} \simeq (1 - k_0\delta) = 96.8 \times 10^{-2} \tag{3.192}$$

We see that about 97% of the incident power is reflected.

3.2.3.5 *Current density and Joule heating*

We can easily calculate the magnitude and phase of the current density induced in the real metal by using Ohm's law and the field continuity conditions at the interface. From Ohm's law we have

$$\mathbf{J} = \sigma E_t \hat{\mathbf{x}} \tag{3.193}$$

and from the continuity condition

$$E_i(1 - r) = E_i t = E_t$$

For a high-conductivity material we have

$$Z = Z_0 \cdot \frac{1}{\sqrt{\varepsilon_m}} = Z_0\sqrt{\frac{\omega\varepsilon_0}{i\sigma}} = Z_0 e^{-i\pi/4}\sqrt{\frac{\omega\varepsilon_0}{\sigma}} \tag{3.194}$$

where the expression for the conductor permittivity ε_m is obtained from Equation 3.172. Since $|Z| \ll |Z_0|$, the transmission coefficient t can be written as

$$t = \frac{2Z}{Z_0 + Z} \simeq \frac{2Z}{Z_0} = 2\,e^{-i\pi/4}\sqrt{\frac{\omega\varepsilon_0}{\sigma}} \tag{3.195}$$

So the transmitted and incident E-field amplitudes are related by

$$E_t = 2\, e^{-i\pi/4} \sqrt{\frac{\omega\varepsilon_0}{\sigma}}\, E_0 \tag{3.196}$$

We see that the transmitted E-field exhibits a phase lag with respect to the incident E-field by $\pi/4$, and Equation 3.193 shows that the current density induced in the metal also lags the incident E-field by $\pi/4$.

Both the current density $\mathbf{J}$ and the transmitted E-field are rapidly attenuated exponentially as they penetrate the conductor. From Equation 3.173 we can write the propagation parameter in the conductor in terms of the skin depth:

$$\mathbf{J}(z) = 2\,\sqrt{\sigma\omega\varepsilon_0}\, E_0\, e^{i(z/\delta - \pi/4)}\, e^{-z/\delta}\,\hat{\mathbf{x}} \tag{3.197}$$

$$\mathbf{E_t}(z) = 2\,\sqrt{\frac{\omega\varepsilon_0}{\sigma}}\, E_0 e^{i(z/\delta - \pi/4)}\, e^{-z/\delta}\,\hat{\mathbf{x}} \tag{3.198}$$

The power density ($\mathrm{W\,m^{-3}}$) associated with 'Joule heating' in the conductor is given by

$$\frac{dP_{\mathrm{Joule}}}{dV} = \mathbf{E_t}\cdot\mathbf{J}^* \tag{3.199}$$

In order to compare this Joule heating power to P^+ or P^- we need first to integrate over the penetration depth in the z direction, then integrate over a cross-sectional area of one square metre in the $x-y$ plane in order to calculate the volume power dissipated in the conductor. This power dissipation corresponds to the third term on the right in the power-balance, Equation 3.65:

$$P_{\mathrm{Joule}} = \frac{1}{2}\mathrm{Re}\left[\int_0^1 \int_0^1 \int_0^\infty \mathbf{E_t}\cdot\mathbf{J}^*\, dx\,dy\,dz\right]$$

$$= 2\,\omega\varepsilon_0\, |E_0|^2\, \frac{\delta}{2}$$

$$= \frac{|E_0|^2}{Z_0}\cdot k_0\delta \tag{3.200}$$

and we see that this result agrees with our earlier power calculation, Equation 3.188, based on the Poynting vector energy flux entering the conductor normal to its surface.

3.2.4 Plane waves in a lossy, conductive medium (high frequency limit)

In Sections 3.1.4 and 3.2.3 we developed the physics of a plane electromagnetic wave interacting at the surface of, and within, a good conductor (most often encountered as a real metal). This development assumed that the conductivity σ is real. We show here that, although this assumption is adequate for radio and microwave frequencies, the conductivity must be considered complex in the optical range of the electromagnetic spectrum.

We start with the wave equation for the E-field component of an electromagnetic wave propagating in a medium, Equation 3.36, that we rewrite here for convenience:

$$\frac{\partial^2 E_x}{\partial z^2} + \mu\varepsilon\omega^2 \left(1 + i\frac{\sigma}{\omega\varepsilon}\right) E_x = 0 \tag{3.201}$$

and assume, as usual, a plane wave solution as in Equation 3.29:

$$E_x = E_0 e^{i(k_z z - \omega t)} \tag{3.202}$$

Substituting Equation 3.202 into Equation 3.201 results in

$$k_z^2 = \mu\varepsilon\omega^2 + i\mu\sigma\omega \tag{3.203}$$

As before, we assume non-magnetic materials ($\mu = \mu_0$) and consider the permittivity to be complex, $\varepsilon = \varepsilon_0 (\varepsilon' + i\varepsilon'')$. But now we consider the conductivity to be complex as well, $\sigma = \sigma' + i\sigma''$. Now we rewrite k_z as k_m to emphasise wave propagation inside the metallic conductor; and recognise that it is complex as well:

$$k_m = k'_m + ik''_m \tag{3.204}$$

Substituting the complex quantities into Equation 3.203 and gathering the real and imaginary parts,

$$k'^2_m - k''^2_m = \mu_0\varepsilon_0\varepsilon'\omega^2 - \mu_0\sigma''\omega$$
$$= \mu_0\varepsilon_0\omega^2 \left[\varepsilon' - \frac{\sigma''}{\varepsilon_0\omega}\right] = k_0^2 \left[\varepsilon' - \frac{\sigma''}{\omega\varepsilon_0}\right] = \alpha^2 \tag{3.205}$$

$$2k'_m k''_m = \mu_0\varepsilon_0\varepsilon''\omega^2 + \mu_0\sigma'\omega$$
$$= \mu_0\varepsilon_0\omega^2 \left[\varepsilon'' + \frac{\sigma'}{\varepsilon_0\omega}\right] = k_0^2 \left[\varepsilon'' + \frac{\sigma'}{\omega\varepsilon_0}\right] = \beta^2 \tag{3.206}$$

The two coupled equations, Equations 3.205 and 3.206, can be solved for the real and imaginary parts of k_m:

$$k'_m = \pm\frac{\alpha}{\sqrt{2}}\sqrt{1 \pm \sqrt{1 + (\beta/\alpha)^4}} \tag{3.207}$$

$$k''_m = \pm\frac{\alpha}{\sqrt{2}}\sqrt{-1 \pm \sqrt{1 + (\beta/\alpha)^4}} \tag{3.208}$$

We also know that the propagation parameter within the metal k_m is related to the free-space propagation parameter k_0 through the complex index of refraction n in the metal:

$$\left(\frac{k_m}{k_0}\right)^2 = n^2 = (\eta + i\kappa)^2 \tag{3.209}$$

But n is also related to the real and complex 'effective' dielectric constants in the metal by

$$n^2 = \varepsilon'_{\text{eff}} + i\varepsilon''_{\text{eff}} \tag{3.210}$$

The relations between η, κ and $\varepsilon'_{\text{eff}}, \varepsilon''_{\text{eff}}$ are

$$\eta = \sqrt{\frac{\varepsilon'_{\text{eff}}}{2}}\sqrt{1 \pm \sqrt{1 + \left(\varepsilon''_{\text{eff}}/\varepsilon'_{\text{eff}}\right)}} \tag{3.211}$$

$$\kappa = \sqrt{\frac{\varepsilon'_{\text{eff}}}{2}}\sqrt{1 \pm \sqrt{-1 + \left(\varepsilon''_{\text{eff}}/\varepsilon'_{\text{eff}}\right)}} \tag{3.212}$$

But:

$$k'_m = k_0\eta \tag{3.213}$$
$$k''_m = k_0\kappa \tag{3.214}$$

Then comparing Equations 3.213 and 3.214 to Equations 3.207 and 3.208 we see that

$$\varepsilon'_{\text{eff}} = \varepsilon' - \frac{\sigma''}{\omega\varepsilon_0} \tag{3.215}$$

$$\varepsilon''_{\text{eff}} = \varepsilon'' + \frac{\sigma'}{\omega\varepsilon_0} \tag{3.216}$$

The two terms $\varepsilon', \varepsilon''$ are the real and imaginary parts of the metal dielectric constant excluding the highly polarisable conduction electrons. They represent the response of the much more tightly bound valence electrons to the transmitted plane wave. The terms involving σ', σ'' represent the frequency-dependent response of the conduction electrons to the transmitted wave. In high-conductivity materials these latter two terms dominate the response. Now real metals are dispersive which means that $\varepsilon'_{\text{eff}}$ and $\varepsilon''_{\text{eff}}$ depend on the frequency. A simple dispersion relation can be found if the metal conductor is modelled as a harmonically oscillating free-electron gas. The harmonic motion is driven at the incident wave frequency ω and damped at a rate Γ, inserted into the model phenomenologically and corresponding to acoustic and radiative dissipation.

3.2.4.1   *Damped harmonic oscillator model for conduction current*

We posit that the equation of motion of the conduction electrons is governed by harmonic acceleration driven by a plane-wave electromagnetic field propagating in the metal and polarised along the x direction:

$$m_e \frac{d^2 x}{dt^2} + m_e \Gamma \frac{dx}{dt} = eE_t = eTE_0 e^{ikmz - \omega t} \tag{3.217}$$

where m_e, e are the electron mass and charge, Γ a phenomenological damping constant, and $eE_t = eTE_0 e^{i(kmz - \omega t)}$ the driving force on the conduction current. Dropping the common $e^{-i\omega t}$ factor, we have for the solutions of position, velocity, and acceleration:

$$x = -\frac{1}{m_e \left(\omega^2 + i\Gamma\omega\right)} eTE_0 \tag{3.218}$$

$$\frac{dx}{dt} = i\frac{\omega}{m_e \left(\omega^2 + i\Gamma\omega\right)} eTE_0 \tag{3.219}$$

$$\frac{d^2 x}{dt^2} = \frac{\omega^2}{m_e \left(\omega^2 + i\Gamma\omega\right)} eTE_0 \tag{3.220}$$

Now the amplitude of the conduction current density is given by

$$\mathcal{J}_c = eN_e \frac{dx}{dt} = eN_e \frac{\omega}{m_e} \left(\Gamma\omega - i\omega^2\right) eTE_0 \tag{3.221}$$

where the electron density in the metal conduction band N_e is related to the resonance frequency of the oscillating electrons ω_p by

$$\omega_p^2 = \frac{e^2 N_e}{m_e \varepsilon_0} \tag{3.222}$$

The resonant frequency is called the 'bulk plasmon frequency'. The conduction current $\mathcal{J}_c$ can now be written as

$$\mathcal{J}_c = \frac{\omega_p^2}{\Gamma - i\omega} \varepsilon_0 TE_0 \tag{3.223}$$

But from the standard constitutive relation we have

$$\mathcal{J}_c \hat{\mathbf{x}} = \sigma E_t \hat{\mathbf{x}} = \frac{1}{\varepsilon_0} \sigma \varepsilon_0 TE_0 \hat{\mathbf{x}} \tag{3.224}$$

Comparing Equations 3.224 and 3.223, and separating real and imaginary parts, we see that

$$\frac{\sigma}{\varepsilon_0} = \frac{\Gamma}{\left(\frac{\Gamma^2}{\omega_p^2} + \frac{\omega^2}{\omega_p^2}\right)} + i\frac{\omega}{\left(\frac{\Gamma^2}{\omega_p^2} + \frac{\omega^2}{\omega_p^2}\right)} \tag{3.225}$$

Now from Equation 3.222 we can estimate the bulk plasmon frequency and compare it to ω and Γ. A high-conductivity metal such as silver exhibits a conduction electron density $N_e \simeq 6 \times 10^{28}$ m^{-3} and therefore $\omega_p \simeq 1.3 \times 10^{16}$ s^{-1}. The damping rate Γ due to radiative loss and acoustic coupling is typically $\sim 10^{14}$ s^{-1}, and ω in the visible and near IR is $\sim 10^{15}$ s^{-1}. Therefore, we can write

$$\frac{\sigma}{\varepsilon_0} \simeq \Gamma \frac{\omega_p^2}{\omega^2} + i \frac{\omega_p^2}{\omega} \tag{3.226}$$

Thus, at relatively low frequencies (radio and microwave) the real term dominates, and the conductivity is said to be 'ohmic'. As the driving frequency into the optical range, however, the imaginary term dominates. At $\omega \sim 10^{15}$ s^{-1} the real term is only about 5% of the imaginary term. We can, from Equations 3.226, 3.215, and 3.216, write the frequency dependence of the 'effective' dielectric constants in the metal as

$$\varepsilon'_{\text{eff}} = \left(\varepsilon' - \frac{\omega_p^2}{\omega^2} \right) \tag{3.227}$$

$$\varepsilon''_{\text{eff}} = \left(\varepsilon'' + \frac{\Gamma \omega_p^2}{\omega^3} \right) \tag{3.228}$$

The valence electrons in a metal are very tightly bound with a large energy gap between the valence ground and excited states. Therefore, we can to good approximation write $\varepsilon' = 1$ and $\varepsilon'' = 0$. The effective, frequency-dependent dielectric constants then simplify to

$$\varepsilon'_{\text{eff}} \simeq \left(1 - \frac{\omega_p^2}{\omega^2} \right) \tag{3.229}$$

$$\varepsilon''_{\text{eff}} \simeq \left(\frac{\Gamma \omega_p^2}{\omega^3} \right) \tag{3.230}$$

Returning to Equation 3.226 we see that, at high frequency

$$\sigma \rightarrow i\sigma'' = i \frac{\omega_p^2}{\omega} \varepsilon_0 \tag{3.231}$$

and the E-field in the conductor becomes

$$E_t = TE_0 \rightarrow 2e^{-i\pi/2} \sqrt{\frac{\omega \varepsilon_0}{\sigma''}} E_0 \tag{3.232}$$

The E-field in the conductor now lags the incident driving field by $\pi/2$ and the conduction current, from Equation 3.224, becomes

$$\mathcal{J}_c \to i\frac{\omega_p^2}{\omega}\varepsilon_0 E_t = i\frac{\omega_p^2}{\omega}\varepsilon_0\,TE_0 = 2\sqrt{\sigma''\omega\varepsilon_0}E_0 = 2\omega_p\varepsilon_0 E_0 \qquad (3.233)$$

We see that, in the optical frequency regime, the conduction current induced in the metal is in phase with the driving field, proportional to the incident amplitude, and independent of the driving field frequency.

3.3 Summary

In this chapter we have reviewed the principal physical attributes of plane waves. We started with introducing the phasor form of the vector field and then showed how Maxwell's equations could be decoupled to form second-order differential equations of each vector field E and B. An important class of solutions to these equations take the phasor form. From there we introduced the scalar Helmholtz equations for the spatial dependence of the field amplitudes. Then we discussed plane waves in lossy dielectrics and good conductors in the low frequency limit. Energy densities and energy flux of plane waves were then presented, taking into account dissipative loss. The standard Fresnel relations reflection and transmission of amplitude and power were then discussed and applied to total internal reflection, reflection, and transmission at a material interface under lossy conditions. Expressions for plane waves travelling in a lossy conductive medium (real metals) at both a low and high frequency limit were then developed and applied to the damped harmonic oscillator model for the conduction current.

3.4 Further reading

1. M. Born and E Wolf, *Principles of Optics*, 6th edition, Pergamon Press (1980).
2. M. Mansuripur, *Field, Force, Energy and Momentum in Classical Electrodynamics*, Bentham e-Books (2011).
3. D. J. Griffiths, *Introduction to Electrodynamics*, 3rd edition, Pearson Addison Wesley, Prentice Hall (1999).
4. J.-P. Pérez, R. Carles, and R. Fleckinger, *Électromagnétism Fondements et applications*, 3ème édition, Masson (1997).
5. H. A. Haus, *Waves and Fields in Optoelectronics*, Prentice Hall (1984).
6. D. M. Pozar, *Microwave Engineering*, 3rd edition, John Wiley & Sons (2005).

4

Energy Flow in Polarisable Matter

4.1 Poynting's theorem in polarisable material

Recall from Chapter 2, Equation 2.60 that the energy flow across the surface of a spatial volume is

$$\frac{d\mathcal{E}_{\text{mech}}}{dt} + \frac{d\mathcal{E}_{\text{em}}}{dt} = -\oint \mathbf{E} \times \mathbf{H} \cdot d\mathbf{a} \tag{4.1}$$

and that the integrand on the right-hand side is identified with the energy flux crossing the surface,

$$\mathbf{S} = \mathbf{E} \times \mathbf{H} \tag{4.2}$$

with $\mathbf{S}$ called the Poynting vector. The two terms on the left are the rate of change of 'mechanical' energy stored in any material included in the bounded volume and the rate of change of 'field' energy associated with the E- and B-fields within the same volume. Let us deconstruct some of the characteristics of the Poynting vector to bring out the explicit time-dependent terms. Firstly,

$$\nabla \cdot \mathbf{S} = \nabla \cdot (\mathbf{E} \times \mathbf{H}) \tag{4.3}$$
$$= -\mathbf{E} \cdot (\nabla \times \mathbf{H}) + \mathbf{H} \cdot (\nabla \times \mathbf{E}) \tag{4.4}$$

From the Maxwell–Faraday law and the Maxwell–Ampère law, Equations 2.27 and 2.28, we have

$$\nabla \times \mathbf{E} = -\frac{\partial \mathbf{B}}{\partial t} = -\mu_0 \frac{\partial \mathbf{H}}{\partial t} \tag{4.5}$$

and

$$\nabla \times \mathbf{H} = \frac{\partial \mathbf{D}}{\partial t} + \mathbf{J}_{\text{free}} \tag{4.6}$$

Light-Matter Interaction. Second Edition. John Weiner and Frederico Nunes.
© John Weiner and Frederico Nunes 2017. Published 2017 by Oxford University Press.

so that

$$\nabla \cdot \mathbf{S} = -\mathbf{E} \cdot \left(\frac{\partial \mathbf{D}}{\partial t} + \mathbf{J}_{\text{free}} \right) + \mathbf{H} \cdot \left(-\mu_0 \frac{\partial \mathbf{H}}{\partial t} \right) \tag{4.7}$$

$$= -\left(\mathbf{E} \cdot \mathbf{J}_{\text{free}} + \mathbf{E} \cdot \frac{\partial \mathbf{D}}{\partial t} + \mu_0 \mathbf{H} \cdot \frac{\partial \mathbf{H}}{\partial t} \right) \tag{4.8}$$

Now we specify the case where there are no free currents ($\mathbf{J}_{\text{free}} = 0$), but the material within the enclosed volume is polarisable. Then the displacement field is

$$\mathbf{D} = \varepsilon_0 \mathbf{E} + \mathbf{P}$$

and

$$-\nabla \cdot \mathbf{S} = \left(\mathbf{E} \cdot \frac{\partial \mathbf{D}}{\partial t} + \mu_0 \mathbf{H} \cdot \frac{\partial \mathbf{H}}{\partial t} \right) \tag{4.9}$$

$$= \left[\frac{1}{2} \varepsilon_0 \cdot \frac{\partial}{\partial t} (\mathbf{E} \cdot \mathbf{E}) + \frac{1}{2} \mu_0 \frac{\partial}{\partial t} (\mathbf{H} \cdot \mathbf{H}) \right] + \mathbf{E} \cdot \frac{\partial \mathbf{P}}{\partial t}$$

$$= \frac{\partial}{\partial t} \left[\frac{1}{2} \left(\varepsilon_0 E^2 + \mu_0 H^2 \right) \right] + \mathbf{E} \cdot \frac{\partial \mathbf{P}}{\partial t} \tag{4.10}$$

The term in brackets on the right-hand side of Equation 4.10 is the field energy and the second term is the time rate of change of the mechanical energy of the material. We see that the flow of energy into or out of any material included within the volume is controlled by the time dependence of the polarisation field in that material.

4.2 Harmonically driven polarisation field

The polarisation can be considered a density of harmonic dipole oscillators characterised by a 'natural' frequency ω_0 and a damping frequency γ. If a harmonically oscillating external electric field $\mathbf{E}$ is applied to this dipole density, the equation of motion of the response will be

$$\frac{\partial^2 \mathbf{P}}{\partial t^2} + \gamma \frac{\partial \mathbf{P}}{\partial t} + \omega_0^2 \mathbf{P} = \omega_d^2 \varepsilon_0 \mathbf{E} \tag{4.11}$$

where ω_d^2 is a 'strength of external field coupling' parameter and $\mathbf{E}$ is the external driving field. The coupling parameter ω_d clearly is the frequency of the driving field $\mathbf{E}$. After a few lines of tedious but straight-forward algebra we find that

$$\mathbf{E} \cdot \frac{\partial \mathbf{P}}{\partial t} = \frac{1}{\varepsilon_0 \omega_d^2} \frac{1}{2} \frac{\partial}{\partial t} \overbrace{\left[\underbrace{\left(\frac{\partial \mathbf{P}}{\partial t}\right)^2}_{\text{kinetic}} + \underbrace{\omega_0^2 \mathbf{P}^2}_{\text{potential}} \right]}^{\text{stored}} + \overbrace{\frac{\gamma}{\varepsilon_0 \omega_d^2} \left(\frac{\partial \mathbf{P}}{\partial t}\right)^2}^{\text{dissipative}} \tag{4.12}$$

Remembering that for a polarisable but non-magnetic material the bound current density is given by,

$$\mathbf{J}_{\text{bound}} = \frac{\partial \mathbf{P}}{\partial t} \tag{4.13}$$

and

$$\frac{d\mathcal{E}_{\text{mech}}}{dt} = \mathbf{E} \cdot \mathbf{J}_{\text{bound}} \tag{4.14}$$

we see immediately that Equation 4.12 represents the flow of mechanical energy into and out of the ensemble of electric dipoles constituting the material of the system. Three terms constitute this mechanical energy: the kinetic and potential energy stored in the dipoles, and the dissipated mechanical energy associated with the damping rate γ as denoted in Equation 4.12.

4.3   Drude–Lorentz dispersion

Equation 4.12 describes the rate of change of mechanical energy in terms of the time dependence of the polarisation field, but the response of matter to an external driving field varies with the frequency. If the response to the driving field is linear, the polarisation field is proportional to the applied electric field.

$$\mathbf{P} = \varepsilon_0 \chi \mathbf{E} \tag{4.15}$$

where χ is the linear *susceptibility* of the material. Strictly speaking, the susceptibility associated with $\mathbf{P}$ should be written χ_e to distinguish it from the magnetic susceptibility χ_m, but here we drop the distinction since it is obvious from the context. The susceptibility is frequency dependent and can be complex, $\chi(\omega) = \chi'(\omega) + i\chi''(\omega)$. The relation characterising the material frequency response is called the material *dispersion* and is usually expressed as the frequency dependence of the complex dielectric constant $\varepsilon(\omega) = \varepsilon'(\omega) + i\varepsilon''(\omega)$. For linear materials the relation between susceptibility and dielectric constant is given by

$$\chi(\omega) = \varepsilon(\omega) - 1 \tag{4.16}$$

A simple dispersion relation for metals is the Drude–Lorentz model that expresses the frequency dependence of the dielectric constant:

$$\varepsilon(\omega) = 1 + \frac{\omega_p^2}{\omega_0^2 - \omega^2 - i\gamma\omega} \tag{4.17}$$

In Equation 4.17, ω is the driving frequency, ω_0 is the resonance frequency, γ is the irrecoverable rate of energy dissipation to 'heat', and ω_p is the bulk plasmon frequency of the electrons in the conduction band of the metal. Separating the real and imaginary parts of ε,

$$\varepsilon(\omega) = \left[1 + \frac{\omega_p^2(\omega_0^2 - \omega^2)}{(\omega_0^2 - \omega^2)^2 + (\gamma\omega)^2} \right] + i \left[\frac{\gamma\omega_p^2\omega}{(\omega_0^2 - \omega^2)^2 + (\gamma\omega)^2} \right] \tag{4.18}$$

$$= \varepsilon' + i\varepsilon''$$

Now, from the constitutive relation between the displacement field and the electric field, and the definition of the displacement field, we can write

$$\mathbf{D} = \varepsilon_0\varepsilon\mathbf{E} \tag{4.19}$$

$$= \varepsilon_0\mathbf{E} + \mathbf{P} \tag{4.20}$$

So we can write the polarization field in terms of the dielectric constant and the electric field as

$$\mathbf{P} = (\varepsilon - 1)\varepsilon_0\mathbf{E} \tag{4.21}$$

Therefore, we can express the factor in square brackets in the stored energy part of Equation 4.12:

$$\left(\frac{\partial \mathbf{P}}{\partial t} \right)^2 + \omega_0^2\mathbf{P}^2 = (\varepsilon - 1)^2\varepsilon_0^2 \left[\left(\frac{\partial \mathbf{E}}{\partial t} \right)^2 + \omega_0^2\mathbf{E}^2 \right] \tag{4.22}$$

The right-hand side of Equation 4.22 is complex due to the $(\varepsilon - 1)^2$ factor. In order to develop expressions for the stored and dissipated energies, it is therefore convenient to express the fields in complex representation and calculate the absolute value of the energy terms. In complex notation the E-field and its time derivative are written as

$$\mathbf{E} = \mathbf{E}_0 e^{-i\omega t} \quad \text{and} \quad \frac{\partial \mathbf{E}}{\partial t} = -i\omega\mathbf{E}$$

and the absolute value of the stored energy factor is

$$\left| \left(\frac{\partial \mathbf{P}}{\partial t} \right)^2 \right| + \omega_0^2|\mathbf{P}^2| = |(\varepsilon - 1)^2|\varepsilon_0^2 \left[\left| \left(\frac{\partial \mathbf{E}}{\partial t} \right)^2 \right| + \omega_0^2|\mathbf{E}^2| \right]$$

$$\left(\left| \frac{\partial \mathbf{P}}{\partial t} \right| \right)^2 + \omega_0^2|\mathbf{P}|^2 = \varepsilon_0^2 \left(\omega_0^2 + \omega^2 \right) \left[(\varepsilon' - 1)^2 + (\varepsilon'')^2 \right] E_0^2 \tag{4.23}$$

Now from the dispersion relation Equation 4.17 we can also write

$$(\varepsilon' - 1)^2 + (\varepsilon'')^2 = \frac{\omega_p^4 \left(\omega_0^2 - \omega^2\right)^2}{\left[\left(\omega_0^2 - \omega^2\right)^2 + (\gamma\omega)^2\right]^2} + \frac{\gamma^2 \omega_p^4 \omega^2}{\left[\left(\omega_0^2 - \omega^2\right)^2 + (\gamma\omega)^2\right]^2} \qquad (4.24)$$

$$= \frac{\omega_p^4}{\left(\omega_0^2 - \omega^2\right)^2 + (\gamma\omega)^2} \qquad (4.25)$$

Therefore, from Equations 4.22, 4.23, and 4.25 we have the absolute value of the stored energy from Equation 4.12

$$\frac{1}{2\varepsilon_0\omega_p^2}\left[\left(\left|\frac{\partial\mathbf{P}}{\partial t}\right|\right)^2 + \omega_0^2|\mathbf{P}|^2\right] = \frac{\varepsilon_0}{2}\left[\frac{\omega_p^2\left(\omega_0^2 + \omega^2\right)}{\left(\omega_0^2 - \omega^2\right)^2 + (\gamma\omega)^2}\right]E_0^2 \qquad (4.26)$$

By 'stored energy' we mean the part of the energy density that flows from the fields into the dipoles, is stored there, and can flow back to the fields.

Making analogous substitutions we have for the dissipative part of $d\mathscr{E}_{\text{mech}}/dt$,

$$\frac{\gamma}{\varepsilon_0\omega_p^2}\left(\frac{\partial|\mathbf{P}|}{\partial t}\right)^2 = \frac{\gamma\omega_p^2\omega^2}{\left(\omega_0^2 - \omega^2\right)^2 + (\gamma\omega)^2}\varepsilon_0 E_0^2$$

$$= \varepsilon''\varepsilon_0\omega E_0^2 \qquad (4.27)$$

This dissipative loss is irreversible and cannot be recovered by the fields.

Returning to the absolute value of the stored energy, the factor in square brackets on the right-hand side of Equation 4.26 can be written as the sum of two fractions,

$$\frac{\omega_p^2(\omega^2 + \omega_0^2)}{\left(\omega_0^2 - \omega^2\right)^2 + (\gamma\omega)^2} = \frac{\omega_p^2\left(\omega_0^2 - \omega^2\right)}{\left(\omega_0^2 - \omega^2\right)^2 + (\gamma\omega)^2} + \frac{2\omega^2\omega_p^2}{\left(\omega_0^2 - \omega^2\right)^2 + (\gamma\omega)^2}$$

and from Equation 4.18 we can identify the first term as

$$\varepsilon' - 1 \qquad \text{and the second as} \qquad \frac{2\omega}{\gamma}\varepsilon''$$

Therefore:

$$\frac{1}{2\varepsilon_0\omega_p^2}\left[\left(\frac{\partial|\mathbf{P}|}{\partial t}\right)^2 + \omega_0^2|\mathbf{P}|^2\right] = \frac{\varepsilon_0}{2}\left[(\varepsilon' - 1) + \frac{2\omega}{\gamma}\varepsilon''\right]E_0^2 \qquad (4.28)$$

4.3.1 Stored energy density

From Equations 4.10 and 4.12, and Stokes' theorem, we can write the energy conservation theorem as

$$-\int_A \mathbf{S}\cdot d\mathbf{a} = \int_V \frac{\partial}{\partial t}\left\{\frac{1}{2}\left[\varepsilon_0 E^2 + \mu_0 H^2 + \frac{1}{\varepsilon_0\omega_p^2}\left(\left(\frac{\partial \mathbf{P}}{\partial t}\right)^2 + \omega_0^2 \mathbf{P}^2\right)\right]\right\}dV +$$

$$\int_V \frac{\gamma}{\varepsilon_0\omega_p^2}\left(\frac{\partial \mathbf{P}}{\partial t}\right)^2 dV \qquad (4.29)$$

Then taking the absolute value of the right-hand side of Equation 4.29, and substituting
Equation 4.27 and Equation 4.28 into it, we have

$$-\int_A \mathbf{S}\cdot d\mathbf{a} = \int_V \frac{\partial}{\partial t}\left\{\frac{1}{2}\left[\varepsilon_0\left(\varepsilon' + \frac{2\omega}{\gamma}\varepsilon''\right)E^2 + \mu_0 H^2\right]\right\}dV +$$

$$\int_V \varepsilon_0\varepsilon''\omega E^2 dV \qquad (4.30)$$

and writing the magnetic field energy in terms of the electric field energy

$$H^2 = \frac{\varepsilon_0}{\mu_0}|\varepsilon|E^2 = \frac{\varepsilon_0}{\mu_0}\sqrt{\varepsilon'^2 + \varepsilon''^2}E^2 \qquad (4.31)$$

the stored energy density U_s within the curly brackets on the right-hand side of
Equation 4.30 can be written

$$U_s = \frac{\varepsilon_0}{2}\left[\left(\varepsilon' + \frac{2\omega}{\gamma}\varepsilon''\right) + \sqrt{\varepsilon'^2 + \varepsilon''^2}\right]E^2 \qquad (4.32)$$

and averaging U_s over an optical cycle

$$\langle U_s\rangle = \frac{\varepsilon_0}{4}\left[\left(\varepsilon' + \frac{2\omega}{\gamma}\varepsilon''\right) + \sqrt{\varepsilon'^2 + \varepsilon''^2}\right]E_0^2 \qquad (4.33)$$

In the limit of a lossless material, ε'', $\gamma \to 0$,

$$\langle U_s\rangle = \frac{1}{2}\varepsilon_0\varepsilon' E_0^2 \qquad (4.34)$$

as expected.

Finally, we can express this conservative energy density in terms of the real and imaginary parts of the index of refraction, η and κ, respectively:

$$\varepsilon' = \left(\eta^2 - \kappa^2\right) \tag{4.35}$$

$$\varepsilon'' = 2\eta\kappa \tag{4.36}$$

So that

$$U_s = \varepsilon_0 \left(\eta^2 + \frac{2\omega\eta\kappa}{\gamma}\right) E^2 \tag{4.37}$$

and averaging U_s over an optical cycle,

$$\langle U_s \rangle = \frac{\varepsilon_0}{2} \left(\eta^2 + \frac{2\omega\eta\kappa}{\gamma}\right) E_0^2 \tag{4.38}$$

4.3.2   Dissipated energy density

The *rate* of energy density dissipation $\partial U_d / \partial t$ is given by the integrand of the second term on the right-hand side of Equation 4.30:

$$\frac{\partial U_d}{\partial t} = \frac{\gamma}{\varepsilon_0 \omega_p^2} \left(\frac{\partial \mathbf{P}}{\partial t}\right)^2 \tag{4.39}$$

Remembering that $\mathbf{P} = \varepsilon_0 (\varepsilon - 1)\mathbf{E}$, we have

$$\mathbf{P}^2 = \varepsilon_0^2 \left[\left(\varepsilon' - 1\right)^2 + \left(\varepsilon''\right)^2\right] E^2 \tag{4.40}$$

As usual, we take E to be a harmonically oscillating field, $E = E_0 e^{-i\omega t}$. Then:

$$\mathbf{P} = \varepsilon_0 \sqrt{(\varepsilon' - 1)^2 + (\varepsilon'')^2} E \tag{4.41}$$

and

$$\frac{\partial \mathbf{P}}{\partial t} = \varepsilon_0 \sqrt{(\varepsilon' - 1)^2 + (\varepsilon'')^2} \frac{\partial E}{\partial t} \tag{4.42}$$

So, substituting Equation 4.42 into Equation 4.39, we have

$$\frac{\partial U_d}{\partial t} = \frac{\gamma \varepsilon_0 \omega^2}{\omega_p^2} \left[\left(\varepsilon' - 1\right)^2 + \left(\varepsilon''\right)^2\right] E^2 \tag{4.43}$$

where E_0 is the electric field amplitude. This expression can be further simplified by noting that

$$\left(\varepsilon' - 1\right)^2 + \left(\varepsilon''\right)^2 = \frac{\omega_p^4}{\left(\omega_0^2 - \omega^2\right)^2 + (\gamma\omega)^2}$$

and therefore, taking into account Equation 4.27,

$$\frac{\partial U_d}{\partial t} = \varepsilon_0 \varepsilon'' \omega E^2 \tag{4.44}$$

Averaging over an optical cycle,

$$\left\langle \frac{\partial U_d}{\partial t} \right\rangle = \frac{1}{2}\varepsilon_0 \varepsilon'' \omega E_0^2 \tag{4.45}$$

and the average energy dissipated in one optical cycle $(T = 2\pi/\omega)$ is

$$\langle U_d \rangle = \pi \varepsilon_0 \varepsilon'' E_0^2 \tag{4.46}$$

4.3.3   Time dependence of stored and dissipated energy density

Representing the driving E-field as a real quantity, $\mathbf{E} = \mathbf{E_0}\cos\omega t$, we see from Equation 4.37 that the stored energy is

$$U_s = \varepsilon_0 \left(\eta^2 + \frac{2\omega\eta\kappa}{\gamma} \right) E_0^2 \cos^2\omega t \tag{4.47}$$

and is therefore in phase with the driving field energy flux, $S_z = E_x H_y \cos^2\omega t$. Therefore, the *rate of change* of stored energy is in quadrature:

$$\frac{dU_s}{dt} = -2\omega\varepsilon_0 \left(\eta^2 + \frac{2\omega\eta\kappa}{\gamma} \right)^2 E_0^2 \cos\omega t \sin\omega t \tag{4.48}$$

In contrast, from Equation 4.44, the rate of change of the dissipated energy density is in phase with the driving field energy flux

$$\frac{dU_d}{dt} = \varepsilon_0 \varepsilon'' \omega E_0^2 \cos^2\omega t \tag{4.49}$$

while the dissipated energy itself accumulates over the optical cycles

$$U_d = \varepsilon_0 \varepsilon'' E_0^2 \left(\frac{\omega t}{2} + \frac{\sin 2\omega t}{4} \right) \tag{4.50}$$

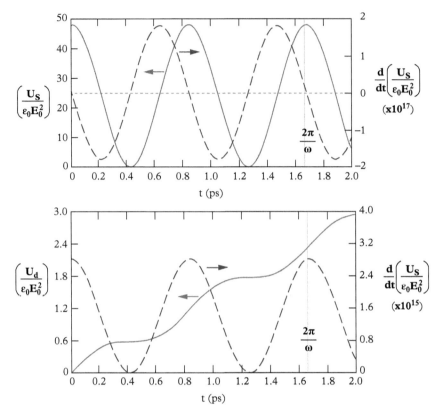

Figure 4.1 *Top panel: The full curve shows time dependence of stored energy (normalised to $\varepsilon_0 E_0^2$), Equation 4.47, and the dashed curve shows time dependence of the rate of change of stored energy, Equation 4.48. Bottom panel: The full curve shows accumulation of dissipated energy with time, Equation 4.50 and the dashed curve shows rate of change of dissipated energy (dissipated power), Equation 4.49. These graphs are plotted with the following parameters taken for silver metal: $\varepsilon' = -9.108$, $\varepsilon'' = 0.753$, $\omega = 3.768 \times 10^{15}$ s^{-1}, $(\lambda = 500$ nm$)$, $\gamma = 1.0 \times 10^{14}$ s^{-1}.*

Figure 4.1 shows the time dependence of the various constituents of the energy density in a polarisable medium.

4.3.4 Frequency dependence of stored and dissipated energy

Returning to the expression for the stored energy in terms of the real and imaginary parts of the dielectric constants, Equation 4.32, and taking the limit of negligible loss,

$$U_s \simeq \varepsilon_0 \varepsilon' E^2 \qquad (4.51)$$

The dispersion relation of the Drude-Lorentz model for ε' is

$$\varepsilon'(\omega) = \frac{\omega_p^2 \left(\omega_0^2 - \omega^2\right)}{\left(\omega_0^2 - \omega^2\right)^2 + (\gamma\omega)^2} + 1 \tag{4.52}$$

and therefore the frequency dependence of the stored energy in lossless media is

$$U_s(\omega) \simeq \varepsilon_0 \left[\frac{\omega_p^2 \left(\omega_0^2 - \omega^2\right)}{\left(\omega_0^2 - \omega^2\right)^2 + (\gamma\omega)^2} + 1 \right] E^2 \tag{4.53}$$

The stored energy exhibits the same 'dispersive' frequency dependence as a driven harmonic oscillator, in phase with the driving field below resonance, $\omega_0 - \omega > 0$ and π out of phase at frequencies above resonance, $\omega_0 - \omega < 0$. In the simplest case of a metal modelled with a single resonance at the bulk plasmon frequency, $\omega_0 = \omega_p$, the stored energy responds in phase at all frequencies 'to the red' of ω_p. For most common noble metals ω_p is in the blue or near ultraviolet regions of the spectrum. Therefore, for incident driving fields in the visible or near-infrared, the usual case in plasmonic studies, the stored energy is in phase with the incident energy flux.

The dispersion relation for the material loss term is

$$\varepsilon''(\omega) = \frac{\gamma\omega\omega_p^2}{\left(\omega_0^2 - \omega^2\right)^2 + (\gamma\omega)^2} \tag{4.54}$$

and, using Equation 4.50, the dissipative energy density is peaked at the resonance frequency ω_0, decreasing symmetrically as the driving frequency is tuned above or below resonance.

$$U_d = \varepsilon_0 \left[\frac{\gamma\omega\omega_p^2}{\left(\omega_0^2 - \omega^2\right)^2 + (\gamma\omega)^2} \right] E_0^2 \left(\frac{\omega t}{2} + \frac{\sin 2\omega t}{4} \right) \tag{4.55}$$

The dispersive and absorptive frequency profiles are simply those of a driven, damped harmonic oscillator. This behaviour is consistent with the polarisation of the material modelled as a density of dipoles.

4.4   Polarisation from polarisability

In Chapter 2 the polarisation vector field was introduced in Equations 2.15 and 2.17. The polarisation field **P** adds vectorially to the applied electric field (multiplied by the permittivity) $\varepsilon_0 \mathbf{E}$ to produce the total displacement field **D**, and in many cases, the polarisation field itself is proportional to the applied electric field. The proportionality constant, the susceptibility χ, is a property of the material; and the dielectric constant of the material is simply related to this susceptibility by $\varepsilon = 1 + \chi$. As the term suggests, the susceptibility is a measure of the response of the material to an external applied field. Since matter is actually composed of atoms and molecules, the macroscopic response must somehow be related to the distortion of local electric fields around the constituent microscopic particles. The goal of this section is to develop that relation.

4.4.1   Electric field inside a material

At the truly microscopic distance scale, comparable to the distance between atoms in condensed matter, the electric field due to electronic and nuclear charge densities is extremely complicated and constantly changing in time due to the orbital motion of electron charge densities around nuclei and the thermal motion of the nuclei themselves. The calculation of this field is hopeless, but in fact, it is not the field we really seek anyway. What we really want is the field at a fixed point inside the material due to two influences: the external E-field applied to the bulk material and the E-field internal to the material arising from its polarisation. Figure 4.2 is a sketch of the various elements of the problem. The figure shows a slab of material with susceptibility χ, dielectric constant ε, placed between two conducting plates across which an electric field $\mathbf{E}_{ext}$ is applied. The field at a point internal to the material can be considered the sum of two superposed fields. The source of the first field is the charge density induced at the surface of the material by the applied external field. This surface charge density produces a uniform polarisation field $\mathbf{P}$ opposite in direction to the applied E-field. The E-field due to this polarisation field is $\mathbf{E}_{int} = -\mathbf{P}/\varepsilon_0$. The source of the second field are the microscopic induced dipoles themselves. As we have already discussed above, at a spatial resolution

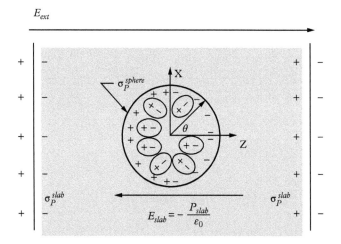

Figure 4.2 *External electric field E_{ext} is applied to two external conducting parallel plates. The material slab with dielectric constant ε and susceptibility χ exhibits a polarisation field $-P$ the source of which is the surface charge density σ_P induced at the slab-plate interface. The volume of the sphere inside the slab is large with respect to the size of individual dipoles comprising the material but small compared to the bulk volume of the slab itself.*

of an individual dipole, the E-field is neither practically calculable nor useful. We seek instead the field at a specific point due to the average influence of many induced dipoles. To this end we construct a sphere whose volume is large with respect to the volume of an individual dipole but negligible compared to the bulk volume. Using Gauss's law, we can compute this E-field $\mathbf{E}_P$ at the centre of the sphere due to the bound charge density around the surface. The net field at the centre of the sphere, experienced by a molecule at that site, is then $\mathbf{E}_{ext} - \mathbf{P}/\varepsilon_0 + \mathbf{E}_P$.

The displacement field $\mathbf{D} = \varepsilon_0 \mathbf{E}_{ext}$ is perpendicular to the interface and therefore must be continuous across the slab boundary, and therefore, the E-field at the point of interest due to the external applied field is

$$\frac{\mathbf{D}}{\varepsilon_0} = \mathbf{E}_{ext} + \frac{\mathbf{P}}{\varepsilon_0} \tag{4.56}$$

The surface charge density at the two slab interfaces σ_P^{slab} is the source of a polarisation field at all points within the slab equal to $-\mathbf{P}/\varepsilon_0$. Now we invoke Gauss's law to calculate the E-field $\mathbf{E}_P$ at centre of the sphere from the polarisation charge density σ_P^{sphere} at the sphere surface. This bound charge density is determined from the polarisation component normal to the sphere, and therefore the distribution of charge density around the surface goes as $\sigma_P^{sphere}(\theta) = |\mathbf{P}| \cos\theta$ where θ is the angle between the z-axis and the surface normal as shown in Figure 4.2. The E-field at the sphere centre is then

$$E_P = \frac{1}{4\pi\varepsilon_0} \int_S \frac{\sigma_P^{sphere} \cos\theta}{r_0^2} dS \tag{4.57}$$

where r_0 is the sphere radius. The surface differential is $dS = r_0^2 \sin\theta \, d\theta \, d\varphi$, and integrating over the azimuthal angle φ we have

$$dS = 2\pi r_0^2 \sin\theta \, d\theta \tag{4.58}$$

Putting $\sigma_P^{sphere}(\theta)$ and dS into Equation 4.57 we have

$$E_P = \frac{|\mathbf{P}| 2\pi r_0^2}{4\pi\varepsilon_0 r_0^2} \int_0^\pi \sin\theta \cos^2\theta \, d\theta \tag{4.59}$$

$$= \frac{|\mathbf{P}|}{3\varepsilon_0} \tag{4.60}$$

and

$$\mathbf{E}_P = \frac{\mathbf{P}}{3\varepsilon_0} \tag{4.61}$$

The effective E-field at the centre of the sphere, experienced by a material molecule placed at that position, is therefore

$$\mathbf{E}_{\text{eff}} = \frac{\mathbf{D}}{\varepsilon_0} - \frac{\mathbf{P}}{\varepsilon_0} + \mathbf{E}_P \tag{4.62}$$

$$= \mathbf{E}_{ext} + \frac{\mathbf{P}}{\varepsilon_0} - \frac{\mathbf{P}}{\varepsilon_0} + \frac{\mathbf{P}}{3\varepsilon_0} \tag{4.63}$$

$$\mathbf{E}_{\text{eff}} = \mathbf{E}_{ext} + \frac{\mathbf{P}}{3\varepsilon_0} \tag{4.64}$$

Note that the effective E-field does not depend on the radius of the sphere. Any sphere will yield the same result as long as the spherical volume is not comparable to an individual dipole element. The only question that remains is the possible contribution of the E-fields of individual dipole elements inside the sphere. Since the material was initially assumed isotropic with no net polarisation in the absence of an external field, we can safely assume that there is no E-field contribution to the dipole elements themselves.

4.4.2   Polarisation and polarisability

Now we want to use $\mathbf{E}_{\text{eff}}$ to obtain the macroscopic polarisation field from the microscopic dipole moment induced in each molecule. The connection from the micro- to the macro-world is through the *polarisability*, a property of matter at the molecular level. The polarisability α is defined through the relation,

$$\mathbf{p} = \alpha \varepsilon_0 \mathbf{E}_{\text{eff}} \tag{4.65}$$

where $\mathbf{p}$ is the induced dipole moment of an individual molecule. Since the macroscopic polarisation $\mathbf{P}$ can be interpreted as a dipole density, we can write

$$\mathbf{P} = N\mathbf{p} = N\alpha\varepsilon_0\mathbf{E}_{\text{eff}} = N\alpha\varepsilon_0\left(\mathbf{E} + \frac{\mathbf{P}}{3\varepsilon_0}\right) \tag{4.66}$$

where N is the number of molecules per unit volume. Remembering that for linear response

$$\mathbf{P} = \varepsilon_0(\varepsilon - 1)\mathbf{E} \tag{4.67}$$

We can set Equation 4.66 equal to Equation 4.67 and find

$$\frac{\varepsilon - 1}{\varepsilon + 2} = \frac{N\alpha}{3} = \frac{N_0\,\rho_m\,\alpha}{3M} \tag{4.68}$$

where in the last term on the right, N_0 is Avogadro's constant, ρ_m is the density, and M is the molecular mass. This expression is called the Clausius–Mossotti relation. It tells us how we can calculate the dielectric constant ε, a macroscopic property, from the microscopic polarisability, α. As such, it furnishes a key bridge between microscopic and macroscopic physics of matter.

4.5 Summary

In this chapter we have considered energy flow in electrically polarisable matter. In Section 4.1 we used the Poynting vector and identified the energy flow associated with the 'mechanical' energy of the material. In Section 4.2 we identified the stored and dissipative parts of this energy flow, and in Section 4.3 we introduced the Drude–Lorentz model, commonly used to characterise the polarisation dispersion of many common materials. In the rest of the section we considered the time and frequency dependence, and phase relation of the stored and dissipated energy through an optical cycle. Finally, in Section 4.4 we developed the relation between macroscopic polarisation and microscopic polarisability to arrive at the very useful Clausius-Mossotti relation.

4.6 Further reading

1. M. Born and E Wolf, *Principles of Optics*, 6th edition, Pergamon Press (1980).
2. M. Mansuripur, *Field, Force, Energy and Momentum in Classical Electrodynamics*, Bentham e-Books (2011).
3. J. S. Stratton, *Electromagnetic Theory*, McGraw-Hill Book Company (1941).
4. D. J. Griffiths, *Introduction to Electrodynamics*, 3rd edition, Pearson Addison Wesley, Prentice Hall (1999).
5. J.-P. Pérez, R. Carles, R. Fleckinger, *Électromagnétism Fondements et applications*, 3ème édition, Masson (1997).

5

The Classical Charged Oscillator and the Dipole Antenna

5.1 Introduction

Active devices involving oscillating charge confined along linear or pyramidal structures at the nanoscale constitute a major research and development effort. Here we present the basic physics and engineering of the oscillating dipole and some real antennas and antenna arrays whose principles are also pertinent to the nanoscale.

5.2 The proto-antenna

The classical charge dipole, $p_0 = qa$, oscillating at frequency ω (the electromagnetic fields of which we considered in Section 2.8), emits radiation of wavelength $\lambda = 2\pi c/\omega$ and can therefore be considered a prototypical or microscopic antenna radiative source. We recall Equations 2.151 and 2.153, the far-field solutions to Maxwell's equations for a small dipole, $a \ll \lambda$:

$$\mathbf{E}_\theta = -\frac{p_0 \sin\theta}{4\pi\varepsilon_0} \frac{\omega^2}{rc^2} \cos\left[\omega\left(t - \frac{r}{c}\right)\right]\hat{\boldsymbol{\theta}} \tag{5.1}$$

$$\mathbf{B}_\varphi = -\frac{\mu_0 p_0 \sin\theta}{4\pi} \frac{\omega^2}{rc} \cos\left[\omega\left(t - \frac{r}{c}\right)\right]\hat{\boldsymbol{\varphi}} \tag{5.2}$$

and taking account of $\mu_0 = 1/\varepsilon_0 c^2$, we note that the electric and magnetic field amplitudes are related by $B = E/c$. We can associate a characteristic current i_c of the dipole with length a oscillating at frequency ω,

$$i_c = \frac{\omega p_0}{a} \tag{5.3}$$

Light-Matter Interaction. Second Edition. John Weiner and Frederico Nunes.

and a characteristic electric field E_c

$$E_c = \left(\frac{\mu_0}{\varepsilon_0}\right)^{1/2} \frac{i_c a \sin\theta}{2\lambda r} \tag{5.4}$$

where λ is the wavelength of light at frequency ω. We recall from Equation 2.156 that the cycle-averaged energy flux emitted by the oscillator is given by

$$\langle S \rangle = \frac{\mu_0 p_0^2 \omega^4}{32\pi^2 c}\left(\frac{\sin^2\theta}{r^2}\right) \tag{5.5}$$

or in terms of i_c and E_c,

$$\langle S \rangle = \frac{1}{8}\left(\frac{\mu_0}{\varepsilon_0}\right)^{1/2}\left(\frac{i_c a \sin\theta}{\lambda r}\right)^2 \tag{5.6}$$

Figure 5.1 shows the distribution of radiated power density from a point dipole. The cycle-averaged total power radiated over all space is

$$\langle \mathscr{P} \rangle = \frac{1}{4\pi\varepsilon_0}\frac{\langle \omega^4 p_0^2 \rangle}{3c^3} = \frac{2\pi}{3}\left(\frac{\mu_0}{\varepsilon_0}\right)^{1/2}\left(\frac{a}{\lambda}\right)^2\frac{i_c^2}{2} \tag{5.7}$$

where we have used the fact that $\langle \omega^4 p_0^2 \rangle = \omega^4 p_0^2/2$. Now from a unit analysis, it is easy to show that

$$\left(\frac{\mu_0}{\varepsilon_0}\right)^{1/2} \simeq 377 \quad \text{ohms} \tag{5.8}$$

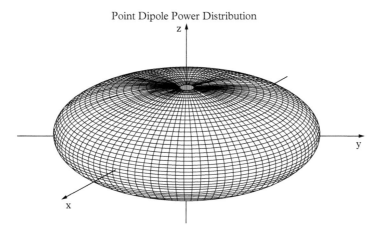

Point Dipole Power Distribution

Figure 5.1 *Angular distribution of power density (power radiated through a unit area) from a radiating point dipole oriented along z.*

and the cycle-averaged power emitted by the oscillator can be set equal to the average power required to drive the oscillator

$$\langle \mathscr{P} \rangle = \frac{1}{2} R_c i_c^2 \tag{5.9}$$

where R_c is the characteristic oscillator impedance to radiation,

$$R_c = \frac{2\pi}{3} \left(\frac{\mu_0}{\varepsilon_0} \right)^{1/2}$$

$$= 790 \quad \text{ohms}$$

and for a given dipole length and wavelength $(a \ll \lambda)$, the radiative impedance R_r is

$$R_r = \frac{2\pi}{3} \left(\frac{\mu_0}{\varepsilon_0} \right)^{1/2} \left(\frac{a}{\lambda} \right)^2 \tag{5.10}$$

Finally, the directional gain of any antenna, $G(\theta)$, is defined as the ratio of power emitted in a given direction to the total radiated power. In the case of the dipole proto-antenna, from Equations 5.5 and 5.7,

$$G(\theta) = 4\pi r^2 \frac{\langle S \rangle}{\langle \mathscr{P} \rangle} = \frac{3}{2} \sin^2 \theta \tag{5.11}$$

5.3   Real antennas

We consider a real, practical antenna consisting of charge oscillating along a straight conductor of finite length, comparable to the wavelength of the emission. We suppose that the electromagnetic field resulting from such an antenna can be analysed as a linear combination of proto-antennas arranged along the length of the conductor. Figure 5.2 shows the antenna aligned along the z-axis. At point M the field is due to the linear superposition of contributions from the proto-antennas arranged along z. One of these point dipoles is shown at position P along z. The E-field at M for one proto-antenna is

$$\mathbf{E} = \left(\frac{\mu_0}{\varepsilon_0} \right)^{1/2} \frac{\sin\theta}{2\lambda PM} i(z) dz\, e^{\left[-i\omega \left(t - \frac{PM}{c} \right) \right]} \hat{\boldsymbol{\theta}} \tag{5.12}$$

We seek to write the distance PM in terms of r and then integrate the resulting expression for $\mathbf{E}$ over the length of the antenna. Invoking the law of cosines we can write

$$PM = \left(r^2 + OP^2 - 2r\, OP \cos\theta \right)^{1/2} = r \left(1 + \left(\frac{OP}{r} \right)^2 - \frac{2OP \cos\theta}{r} \right)^{1/2} \tag{5.13}$$

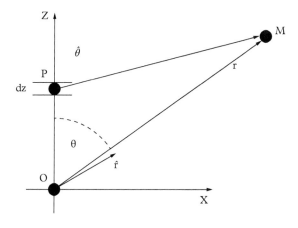

Figure 5.2 *Geometric disposition for determining the E-field of a line antenna of length R at point M by integrating contributions from point dipoles aligned along z.*

At distances far from the antenna where $r \gg OP$ we have $PM \to r$. Then substituting this result into Equation 5.12, and using $k = \omega/c$, we have

$$\mathbf{E}(M) \simeq \left(\frac{\mu_0}{\varepsilon_0}\right)^{1/2} \frac{\sin\theta}{2\lambda r}\left[\int_{-R/2}^{+R/2} i_c(z)e^{ik_z z}\,dz\right]e^{-[\omega(t-\frac{r}{c})]}\hat{\boldsymbol{\theta}} \qquad (5.14)$$

where

$$k_z = k\cos\theta = \left(\frac{2\pi}{\lambda}\right)$$

5.3.1   The half-wave antenna

We posit that the antenna length is $\lambda/2$ and that a proto-dipole oscillator source, located at the midpoint, is driving current to the antenna extremities. A standing wave is set up along the antenna such that

$$i_c(z) = i_c \cos\left(\frac{2\pi z}{\lambda}\right) = i_c \cos(kz) \qquad (5.15)$$

Then the integral in Equation 5.14 becomes

$$\mathscr{I}_c(k) = \int_{-\lambda/4}^{\lambda/4} i_c \cos\left(\frac{2\pi z}{\lambda}\right) e^{ik\cos\theta z}\,dz \qquad (5.16)$$

and the resulting E-field at P is

$$\mathbf{E}_{\mathrm{hw}}(P) = -\left(\frac{\mu_0}{\varepsilon_0}\right)^{1/2}\left(\frac{i_c}{2\pi r}\right)\frac{\cos\left[\pi\cos\left(\frac{\theta}{2}\right)\right]}{\sin\theta}\cos\left[\omega\left(t-\frac{r}{c}\right)\right]\hat{\boldsymbol{\theta}} \qquad (5.17)$$

The cycle-averaged energy flux for the half-wave dipole antenna, $\langle S \rangle_{\text{hw}}$, is

$$\langle S \rangle_{\text{hw}} = \frac{1}{8\pi^2 \varepsilon_0 c} \left(\frac{i_c^2}{r^2} \right) \frac{\cos^2 \left[\pi \cos \left(\frac{\theta}{2} \right) \right]}{\sin^2 \theta} \tag{5.18}$$

and the total cycle-averaged power of the half-wave antenna is

$$\langle \mathscr{P} \rangle_{\text{hw}} = \frac{1}{4\pi \varepsilon_0 c} i_c^2 \int_0^\pi \frac{\cos^2 \left[\pi \cos \left(\frac{\theta}{2} \right) \right]}{\sin \theta} \, d\theta \tag{5.19}$$

$$\langle \mathscr{P} \rangle_{\text{hw}} = 36.5 i_c^2$$

$$\langle \mathscr{P} \rangle_{\text{hw}} = \frac{1}{2} i_c^2 R_r$$

and therefore the radiative impedance of the half-wave antenna is $R_r = 73$ ohms. A typical FM radio emission frequency is 100 MHz, and from the above calculation, we see why a typical FM radio antenna consists of a 75 ohm cable about 1.5 metres in length.

5.3.2   Array of half-wave antennas

The directionality of the emission for a single half-wave antenna is given by Equation 5.18, and Figure 5.3 shows that it is only slightly more peaked along the $\theta = \pi/2$ direction than the proto-dipole oscillator. The emission can be made more directional by implementing an array of half-wave dipole antennas along the x-axis as indicated in

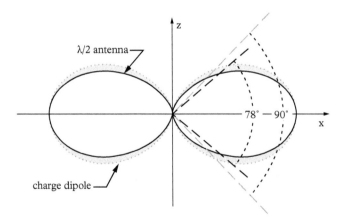

Figure 5.3 *Plot of the radiation distribution, normalised to the maximum at $\theta = \pi/2$, from a linear half-wave antenna and single point dipole. Angles 78° and 90° correspond to half-power points for half-wave antenna and point dipole, respectively.*

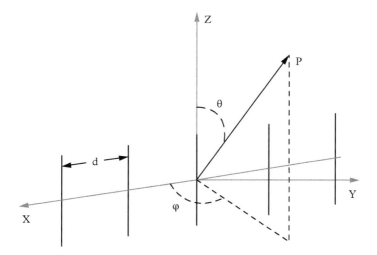

Figure 5.4 *Five λ/2 antennas arranged in a linear array along the x-axis, each element of the array separated by a distance d.*

Figure 5.4. The E-field of the array at any point is essentially the linear superposition of the E-fields of the individual elements, each element offset by a phase shift related to the array pitch d along x. Suppose we have $N = 2n + 1$ array elements symmetrically arranged along $\pm x$ with n elements on each side of the origin. Then the field at point P is simply,

$$\mathbf{E}_{\text{array}}(P) = \mathbf{E}_{\text{hw}}(P) \sum_{m=-n}^{m=n} e^{-im\varphi}$$

$$= \mathbf{E}_{\text{hw}}(P) e^{in\varphi} \sum_{m=0}^{m=(N-1)} e^{-im\varphi}$$

$$= \mathbf{E}_{\text{hw}}(P) e^{in\varphi} \frac{1 - e^{-iN\varphi}}{1 - e^{-i\varphi}}$$

$$= \mathbf{E}_{\text{hw}}(P) \frac{\sin(N\varphi/2)}{\sin(\varphi/2)} \qquad (5.20)$$

where φ is the azimuthal phase shift associated with the displacement a along x. Figure 5.5 shows the angular distribution of the E-field of a five-element linear array as shown in Figure 5.4:

$$\varphi = k_x a = \frac{2\pi}{\lambda} \sin\theta \cos\varphi \, a$$

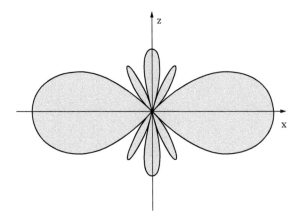

Figure 5.5 *Angular E-field distribution of a half-wave antenna array with five elements.*

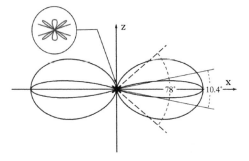

Figure 5.6 *Angular distribution of radiated power from the five-element array shown in Figure 5.4. Note that the power directed along x with half-power points of 10.4° compared to a single half-wave antenna with half-power points at 78°.*

Figure 5.5 was determined from an oscillation frequency of $\nu = 1 \times 10^9$ Hz and a wavelength $\lambda = 0.3$ m. The separation between the elements of the array is $d = \lambda/2 = 0.15$ m. We see that the maximum of the principal lobes of intensity occur for $\theta = \pi/2$ but we see small adjacent lobes, reminiscent of a diffraction pattern. This result is hardly surprising since Equation 5.20 is the same as for linear diffraction grating of N elements.

The corresponding radiated power distribution of the five-element array is shown in Figure 5.6. The concentration of the radiated power along x is clearly evident. The angular aperture of the half-power points in the five-element array is only 10.4° compared to 78° for the single-element antenna.

5.4 Summary

This relatively short chapter reintroduces point dipole radiation and expressions for the energy flux and power radiated. The idea of radiative impedance and the impedance of free space is introduced. The discussion is then extended to real antennas in which the oscillating charge travels along lengths comparable to the wavelength. The important half-wave antenna, and arrays of half-wave antennas, then show how enhanced directionality can be achieved.

5.5 Further reading

1. J. D. Kraus and R. J. Marhefka, *Antennas for All Applications*, 3rd edition, McGraw-Hill (2002).

2. J.-P. Pérez, R. Carles, and R. Fleckinger, *Électromagnétisme Fondements et applications*, Chapter 20, Masson (1997).

3. A. Sommerfeld, *Partial Differential Equations in Physics*, Chapter VI, Academic Press (1964).

6

Classical Black-body Radiation

6.1 Field modes in a cavity

Now that we have established the essential vocabulary of radiation fields and the equations of motion governing them, we can begin our discussion of light–matter interaction by applying our new language to a straight forward problem from the classical theory of radiation. What we seek to do is calculate the energy density inside a bounded conducting volume. We will then use this result to describe the interaction of light with a collection of two-level atoms inside the cavity.

The basic physical idea is to consider that the electrons inside the conducting volume boundary oscillate as a result of thermal motion, and through dipole radiation, set up electromagnetic standing waves inside the cavity. Because the cavity walls are conducting, the electric field **E** must be zero there. Our task is twofold: first to count the number of standing waves that satisfy this boundary condition as a function of frequency; second, to assign an energy to each wave, and thereby determine the spectral distribution of energy density in the cavity.

After decoupling the electric and magnetic fields in Equations 2.27 and 2.28 we arrive at an expression that describes the equation of motion of an E-field wave,

$$\nabla^2 \mathbf{E} = \frac{1}{c^2} \frac{\partial^2 \mathbf{E}}{\partial t^2}$$

propagating in a charge-free space containing no E-field sources,

$$\nabla \cdot \mathbf{E} = 0$$

Harmonic standing-wave solutions factor into oscillatory temporal and spatial terms. Now, respecting the boundary conditions for a three-dimensional box with sides of length L, we have for the components of **E**:

$$
\begin{aligned}
E_x(x, t) &= E_{0x} e^{-i\omega t} \cos(k_x x) \sin(k_y y) \sin(k_z z) \\
E_y(y, t) &= E_{0y} e^{-i\omega t} \sin(k_x x) \cos(k_y y) \sin(k_z z) \\
E_z(z, t) &= E_{0z} e^{-i\omega t} \sin(k_x x) \sin(k_y y) \cos(k_z z)
\end{aligned}
\tag{6.1}
$$

Light-Matter Interaction. Second Edition. John Weiner and Frederico Nunes.
© John Weiner and Frederico Nunes 2017. Published 2017 by Oxford University Press.

where, again, **k** is the wave vector of the light, with amplitude

$$|\mathbf{k}| = \frac{2\pi}{\lambda} \tag{6.2}$$

and components

$$k_x = \frac{\pi n}{L} \quad n = 0, 1, 2, \ldots$$

and similarly for k_y, k_z. Notice that the cosine and sine factors for the E_x field component show that the transverse field amplitudes E_y, E_z have nodes at 0 and L, as they should, and similarly for E_y and E_z. In order to calculate the mode density, we begin by constructing a three-dimensional orthogonal lattice of points in **k** space as shown in Figure 6.1. The separation between points along the k_x, k_y, k_z axes is $\frac{\pi}{L}$, and the volume associated with each point is therefore

$$V = \left(\frac{\pi}{L}\right)^3$$

Now the volume of a spherical shell of radius $|\mathbf{k}|$ and thickness dk in this space is $4\pi k^2 dk$. However, the periodic boundary conditions on **k** restrict k_x, k_y, k_z to positive values, so the effective shell volume lies only in the positive octant of the sphere. The number of points is therefore just this volume divided by the volume per point:

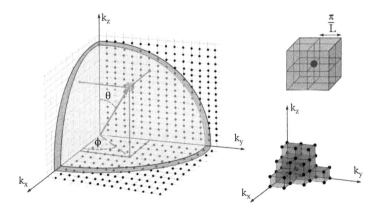

Figure 6.1 *Mode points in **k** space. Right panel shows one-half of the volume surrounding each point. Left panel shows one-eighth of the volume of spherical shell in this **k** space.*

$$\text{Number of } k \text{ points in spherical shell} = \frac{\frac{1}{8}\left(4\pi k^2 dk\right)}{\left(\frac{\pi}{L}\right)^3} = \frac{1}{2}L^3\frac{k^2 dk}{\pi^2}$$

Remembering that there are two independent polarisation directions per k point, we find that the number of radiation modes between k and dk is,

$$\text{Number of modes in spherical shell} = L^3\frac{k^2 dk}{\pi^2} \qquad (6.3)$$

and the spatial density of modes in the spherical shell is

$$\frac{\text{Number of modes in shell}}{L^3} = d\rho(k) = \frac{k^2 dk}{\pi^2}$$

We can express the spectral mode density, mode density per unit k, as

$$\frac{d\rho(k)}{dk} = \rho_k = \frac{k^2}{\pi^2}$$

and therefore the mode number as

$$\rho_k dk = \frac{k^2}{\pi^2} dk$$

with ρ_k as the *mode density* in k space. The expression for the mode density can be converted to frequency space, using the relations

$$k = \frac{2\pi}{\lambda} = \frac{2\pi \nu}{c} = \frac{\omega}{c}$$

and

$$\frac{d\nu}{dk} = \frac{c}{2\pi}$$

Clearly,

$$\rho_\nu d\nu = \rho_k dk$$

and therefore

$$\rho_\nu d\nu = \frac{8\pi \nu^2 d\nu}{c^3}$$

The density of oscillator modes in the cavity increases as the square of the frequency. Now the average energy per mode of a collection of oscillators in thermal equilibrium,

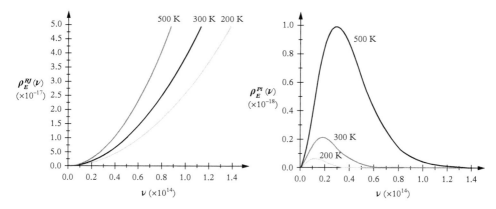

Figure 6.2 *Left panel: Rayleigh–Jeans black-body energy density distribution as a function of frequency, showing the rapid divergence as frequencies tend towards the ultraviolet (the ultraviolet catastrophe). Right panel: Planck black-body energy density distribution showing correct high-frequency behaviour.*

according to the principal of equipartition of energy, is equal to $k_B T$, where k_B is the Boltzmann constant. We conclude, therefore, that the *energy density* in the cavity is

$$\rho_E^{RJ}(\nu)d\nu = \frac{8\pi \nu^2 k_B T d\nu}{c^3} \tag{6.4}$$

which is known as the *Rayleigh–Jeans law of black-body radiation*; and, as Figure 6.2 shows, leads to the unphysical conclusion that energy storage in the cavity increases as the square of the frequency without limit. This result is sometimes called the 'ultra-violet catastrophe' since the energy density increases without limit as oscillator frequency increases towards the ultraviolet region of the spectrum. We achieved this result by multiplying the number of modes in the cavity by the average energy per mode. Since there is nothing wrong with our mode counting, the problem must be in the use of the equipartition principle to assign energy to the oscillators.

6.2   Planck mode distribution

We can get around this problem by first considering the mode excitation probability distribution of a collection of oscillators in thermal equilibrium at temperature T. This probability distribution P_i comes from statistical mechanics and can be written in terms of the Boltzmann factor $e^{-\epsilon_i/k_B T}$ and the partition function $q = \sum_{i=0}^{\infty} e^{-\epsilon_i/k_B T}$:

$$P_i = \frac{e^{-\epsilon_i/k_B T}}{q}$$

Now, Planck suggested that instead of assigning the average energy $k_B T$ to every oscillator, this energy could be assigned in discrete amounts, proportional to the frequency, such that

$$\epsilon_i = \left(n_i + \frac{1}{2} \right) h\nu$$

where $n_i = 0, 1, 2, 3 \ldots$ and the constant of proportionality $h = 6.626 \times 10^{-34}$ J·s. We then have

$$P_i = \frac{e^{-h\nu/2k_B T} e^{-n_i h\nu/k_B T}}{e^{-h\nu/2k_B T} \sum\limits_{n_i=0}^{\infty} e^{-n_i h\nu/k_B T}} = \frac{\left(e^{-h\nu/k_B T} \right)^{n_i}}{\sum\limits_{n_i=0}^{\infty} \left(e^{-h\nu/k_B T} \right)^{n_i}} \tag{6.5}$$

$$= \left(e^{-h\nu/k_B T} \right)^{n_i} \left(1 - e^{-h\nu/k_B T} \right) \tag{6.6}$$

where we have recognised that $\sum\limits_{n_i=0}^{\infty} \left(e^{-h\nu/k_B T} \right)^{n_i} = 1 / \left(1 - e^{-h\nu/k_B T} \right)$. The average energy per mode then becomes

$$\bar{\epsilon} = \sum_{i=0}^{\infty} P_i \epsilon_i =$$

$$\sum_{n_i=0}^{\infty} \left(e^{-h\nu/k_B T} \right)^{n_i} \left(1 - e^{-h\nu/k_B T} \right) (n_i) h\nu = \frac{h\nu}{e^{h\nu/k_B T} - 1} \tag{6.7}$$

and we obtain the Planck energy density in the cavity by substituting $\bar{\epsilon}$ from Equation 6.7 for $k_B T$ into Equation 6.4:

$$\rho_E^{Pl}(\nu) d\nu = \frac{8\pi h}{c^3} \nu^3 \frac{1}{e^{h\nu/k_B T} - 1} d\nu \tag{6.8}$$

This result, plotted in Figure 6.2, is much more satisfactory than the Rayleigh–Jeans result since the energy density has a bounded upper limit and the distribution agrees with experiment.

6.3   The Einstein *A* and *B* coefficients

Let us consider a two-level atom or collection of atoms inside the conducting cavity. We have N_1 atoms in the lower level E_1, and N_2 atoms in the upper level E_2. Light interacts with these atoms through resonant stimulated absorption and emission, $E_2 - E_1 = \hbar\omega_0$, the rates of which, $B_{12}\rho_\omega$, $B_{21}\rho_\omega$, are proportional to the spectral energy density ρ_ω of the cavity modes. Atoms populated in the upper level can also emit light 'spontaneously' at a rate A_{21} that depends only on the density of cavity modes (i.e. the volume of the cavity). This phenomenological description of light absorption and emission can be described by rate equations first written down by Einstein. These rate equations were meant to interpret measurements in which the spectral width of the radiation sources was broad compared to a typical atomic absorption line width, and the source spectral flux $\bar{I}_\omega$ (W/m²·Hz) was weak compared to the saturation intensity of a resonant atomic transition. Although modern laser sources are, according to these criteria, both narrow and intense, the spontaneous rate coefficient A_{21} and the stimulated absorption coefficient B_{12} are still often used in the spectroscopic literature to characterise light–matter interaction in atoms and molecules. These Einstein rate equations describe the energy flow between the atoms in the cavity and the field modes of the cavity, assuming of course, that total energy is conserved:

$$\frac{dN_1}{dt} = -\frac{dN_2}{dt} = -N_1 B_{12}\rho_\omega + N_2 B_{21}\rho_\omega + N_2 A_{21} \tag{6.9}$$

At thermal equilibrium we have a steady-state condition $\frac{dN_1}{dt} = -\frac{dN_2}{dt} = 0$ with $\rho_\omega = \rho_\omega^{th}$ so that

$$\rho_\omega^{th} = \frac{A_{21}}{\left(\frac{N_1}{N_2}\right)B_{12} - B_{21}}$$

and the Boltzmann distribution controlling the distribution of the number of atoms in the lower and upper levels,

$$\frac{N_1}{N_2} = \frac{g_1}{g_2}e^{-(E_1-E_2)/kT}$$

where g_1 and g_2 are the degeneracies of the lower and upper states, respectively. So:

$$\rho_\omega^{th} = \frac{A_{21}}{\left(\frac{g_1}{g_2}e^{\hbar\omega_0/kT}\right)B_{12} - B_{21}} = \frac{\frac{A_{21}}{B_{21}}}{\left(\frac{g_1}{g_2}e^{\hbar\omega_0/kT}\right)\frac{B_{12}}{B_{21}} - 1} \tag{6.10}$$

But this result has to be consistent with the Planck distribution, Equation 6.8:

$$\rho_E^{Pl}(\nu)\,d\nu = \frac{8\pi h}{c^3}\nu_0^3\frac{1}{e^{h\nu_0/k_BT}-1}\,d\nu \tag{6.11}$$

$$\rho_E^{Pl}(\omega)\,d\omega = \frac{\hbar}{\pi^2 c^3}\omega_0^3\frac{1}{e^{\hbar\omega_0/k_BT}-1}\,d\omega \tag{6.12}$$

Therefore, comparing these last two expressions with Equation 6.10, we must have

$$\frac{g_1}{g_2}\frac{B_{12}}{B_{21}} = 1 \tag{6.13}$$

and

$$\frac{A_{21}}{B_{21}} = \frac{8\pi h}{c^3} v_0^3 \tag{6.14}$$

or

$$\frac{A_{21}}{B_{21}} = \frac{\hbar \omega_0^3}{\pi^2 c^3} \tag{6.15}$$

These last two equations show that if we know one of the three rate coefficients, we can always determine the other two.

It is worthwhile to compare the spontaneous emission rate A_{21} to the stimulated emission rate B_{21}:

$$\frac{A_{21}}{B_{21}\rho_\omega^{th}} = e^{\hbar\omega_0/kT} - 1 \tag{6.16}$$

which shows that for $\hbar\omega_0$, much greater than kT (visible, UV, and X-ray), the spontaneous emission rate dominates; but for regions of the spectrum much less than kT (far IR, microwaves, and radio waves), the stimulated emission process is much more important. It is also worth mentioning that even when stimulated emission dominates, spontaneous emission is always present.

6.4 Summary

This chapter discusses black-body radiation by showing how correct mode counting in a conducting cavity leads to a disturbing conclusion if energy is partitioned equally among all modes—the conclusion being that the energy density goes to infinity with increasing frequency. We then show how Planck saved the day (and energy conservation) by suggesting the quantisation of energy packets; the size of which was proportional to the frequency, and the population probability of which falls off exponentially. The chapter closes with an identification of the phenomenological Einstein A and B coefficients with the Planck distribution.

6.5 Further reading

1. T. S. Kuhn, *Black-Body Theory and the Quantum Discontinuity, 1894–1912*, The University of Chicago Press (1978).
2. P. W. Milonni, *The Quantum Vacuum*, Academic Press (1994).

7

Surface Waves

7.1 Introduction

We treat here an important wave phenomenon: the surface wave. It occurs in many physical systems, both mechanical and electromagnetic. Mechanical surface waves can exist at the interface between two media with different densities, such as the oceans and earth. Ordinary sea waves and tsunamis are surface waves. Seismic events can produce both longitudinal and shear waves at the earth's surface, resulting in earthquakes. Electromagnetic surface waves can occur at the interface between dielectric and conductive media or between two dielectrics satisfying appropriate boundary conditions. Air-salt water or glass-metal interfaces support electromagnetic surface waves. Here we focus on electromagnetic waves at the interface between common dielectrics and noble metals, such as air or glass and silver or gold. We will characterise these waves by their distinctive properties of dispersion, spatial field distribution, and polarisation. These properties will be used to analyse useful phenomena such as wave guiding within subwavelength structures and spatial light localisation below the diffraction limit.

7.2 History of electromagnetic surface waves

The impetus for the study of electromagnetic surface waves began with the development of radio communication near the end of the nineteenth century. Ordinary Maxwellian 'space' waves cannot directly propagate beyond the horizon, although Marconi had been able to demonstrate long-range radio communication, including transatlantic messaging between 1899 and 1901. A possible explanation for this successful over-the-horizon transmission, was the excitation and propagation of electromagnetic waves at the air-earth or air-sea interface that could be 'guided' over the curvature of the planet. It was therefore necessary to find solutions to Maxwell's equations that could explain the observations. Stimulated by Arnold Sommerfeld's analysis of electromagnetic wave propagation along a single conducting wire in 1899, his student Johann Zenneck presented a paper on planar surface electromagnetic waves in 1907. Zenneck found that Maxwell's equations had a solution corresponding to a wave coupled to a flat surface interface at the boundary between a dielectric (air) and a medium with finite conductivity

Light-Matter Interaction. Second Edition. John Weiner and Frederico Nunes.
© John Weiner and Frederico Nunes 2017. Published 2017 by Oxford University Press.

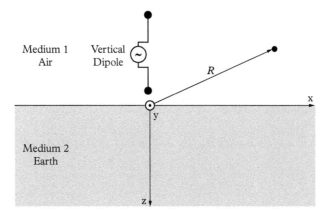

Figure 7.1 *Schematic arrangement for Zenneck wave generation.*

(salt water). The configuration used by Zenneck is illustrated in Figure 7.1. Soon after Zenneck's article appeared, Sommerfeld again published an influential paper in 1909 in which he analysed the propagation of radio waves along the surface of the earth. He considered a vertical oscillating dipole (radio antenna) perpendicularly positioned to the interface between two media: one dielectric, the other conducting. Sommerfeld studied secondary waves produced by the discontinuity at the interface between air and land, considered a weakly conductive material. His analysis showed that the Hertz vector $\mathbf{\Pi}$ that describes the electromagnetic field in each medium is given by the sum of two fields, P and Q, corresponding to two waves. The wave Q represents a space propagating wave and P a surface-guided wave. For long distances R from the source, Q varies as $(1/R)$, where $R = \sqrt{(r^2 + z^2)}$ (see Figure 7.1 for a definition of the coordinates), while P varies asymptotically as $1/\sqrt{r}$ and is confined to the earth's surface. Over the course of the next two decades a controversy arose over the existence of the so-called Zenneck waves because Sommerfeld's original analysis contained a sign error, and it was found that it was difficult to excite the surface-guided wave with practical antenna sources, or to detect and distinguish the true surface wave from the spatially propagating wave very close to the surface. In any event, as far as long-distance wireless communication was concerned, the whole question became irrelevant when it was discovered that reflection from the ionosphere was the principal agent responsible for over-the-horizon radio transmission.

7.3   Plasmon surface waves at optical frequencies

Recent interest in electromagnetic surface waves stems from their potential applications in nanoscale opto-electronic devices. In contrast to the Sommerfeld–Zenneck analyses emphasising radio frequencies, modern studies at optical frequencies usually take Raether's influential treatise, *Surface Plasmons on Smooth and rough Surfaces and on Gratings*, as their point of departure. The modern terminology for the Zenneck wave at the interface between a metal and a dielectric is the surface plasmon or the surface plasmon-polariton (SPP). At optical frequencies the surface plasmon exhibits strong spatial localisation at the dielectric-conductor interface and can be simply, and efficiently, excited by appropriately polarised light incident on surfaces decorated with features (slits, holes, and grooves, etc.) at the subwavelength scale (tens to hundreds of nanometres). Here, we consider the interface between two media: one dielectric and the other a metallic conductor. Each material is characterised by a relative permittivity, ε_r: ε_r^d for the dielectric, and ε_r^m for the metal. In order to lighten notation we drop the subscript r and write the relative permittivities (dielectric constants) as ε_d and ε_m for the dielectric and metal, respectively. The metal also exhibits a high conductivity, σ, and obeys 'Ohm's law' $\mathbf{E} = \sigma \mathbf{J}$, where $\mathbf{E}$ is the electric field applied to the metal and $\mathbf{J}$ is the current density induced in it. We consider the dielectric medium as essentially transparent (lossless) so that $\varepsilon_d = \varepsilon_d'$ where ε_d' is real. In the case of metals, absorptive loss in the optical region of the electromagnetic spectrum is significant, the dielectric constant is complex, and $\varepsilon_m = \varepsilon_m' + i\varepsilon_m''$, where the imaginary term represents absorptive loss. Figure 7.2 shows two orthogonal linear polarisation orientations of E-M fields incident on the metal surface.[1] The plane of incidence is defined by the incident and reflected plane-wave propagation vectors $\mathbf{k}$, and in Figure 7.2, this plane is aligned with the $x-z$ plane. With $\mathbf{H}$ perpendicular to the plane of incidence (right-hand panel, Figure 7.2), the polarisation is called 'transverse-magnetic' (TM). TM polarisation restricts the incident wave to field components E_x, E_z, H_y. With $\mathbf{E}$ perpendicular to the same plane (left-hand panel, Figure 7.2), the polarisation is 'transverse-electric' (TE). The field components of a TE polarised wave are H_x, H_z, E_y. We will consider continuity conditions at the surface for each case.

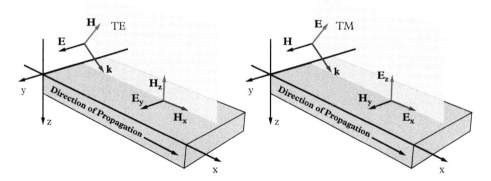

Figure 7.2 *Configuration of field components for* TE *and* TM *polarised plane waves incident on a dielectric-metal interface.*

[1] Although in Chapter 2 the direction of propagation is always along z, in this chapter we must consider both the incident propagating wave and the surface wave. The incident space-propagating wave is taken to be incident along z, while the surface wave propagates along x.

We start with the two Maxwell curl equations, the Faraday law, and the Maxwell–Ampère law, Equations 2.27 and 2.28, and posit two solutions: E- and H-fields characterised by the propagation vector $\mathbf{k} = (2\pi/\lambda)\hat{\mathbf{k}}$, where λ is the wavelength, and oscillating harmonically in time with frequency ω:

$$\mathbf{E}(\mathbf{r}, t) = \mathbf{E}_0(\mathbf{r})e^{i(\mathbf{k}\cdot\mathbf{r}-\omega t)} \tag{7.1}$$

$$\mathbf{H}(\mathbf{r}, t) = \mathbf{H}_0 e^{i(\mathbf{k}\cdot\mathbf{r}-\omega t)} \tag{7.2}$$

Application of the curl operation leads to the following relations:

$$\frac{\partial E_y}{\partial z} = -(i\omega\mu_0)H_x \tag{7.3}$$

$$\frac{\partial H_x}{\partial z} - \frac{\partial H_z}{\partial x} = -i(\omega\varepsilon_0\varepsilon)E_y \tag{7.4} \qquad \text{TE polarisation}$$

$$\frac{\partial E_y}{\partial x} = (i\omega\mu_0)H_z \tag{7.5}$$

and

$$\frac{\partial H_y}{\partial z} = (i\omega\varepsilon_0\varepsilon)E_x \tag{7.6}$$

$$\frac{\partial E_x}{\partial z} - \frac{\partial E_z}{\partial x} = (i\omega\mu_0)H_y \tag{7.7} \qquad \text{TM polarisation}$$

$$\frac{\partial H_y}{\partial x} = -(i\omega\varepsilon_0\varepsilon)E_z \tag{7.8}$$

The incident and reflected waves propagate in the dielectric medium characterised by the dielectric constant ε_d. The propagation vector in free space is denoted k_0, and within the dielectric, $k_d = k_0 n$, where n is the index of refraction. The dielectric constant and the index of refraction of any medium are related by $\varepsilon = n^2$. Since the incident $k = k_i$ is in the $x - z$ plane we can write

$$k_i^2 = k_x^2 + k_z^2 \tag{7.9}$$

$$= \varepsilon_d k_0^2 \tag{7.10}$$

Similarly, on the metal side of the interface for the transmitted propagation vector k^m, we have

$$k_t^2 = k_x^2 + k_z^2 \tag{7.11}$$

$$= \varepsilon_m k_0^2 \tag{7.12}$$

7.3.1   Surface waves with TE polarisation

In TE polarisation we find the wave solution for E_y by applying the posited plane-wave expression, Equation 7.1, to the relations Equations 7.3, 7.4 and 7.5. The result (in the dielectric medium) is

$$E_y^d(x, z) = E_{0y}^d e^{i[(k_x^d x + k_z^d z) - \omega t]} \tag{7.13}$$

From this E-field component, the H-field components on the dielectric side of the interface are readily obtained:

$$H_x^d = -\frac{i}{\omega\mu_0}\frac{\partial E_y^d}{\partial z} = \frac{k_z^d}{\omega\mu_0}E_y^d \tag{7.14}$$

$$H_z^d = \frac{i}{\omega\mu_0}\frac{\partial E_y^d}{\partial x} = -\frac{k_x^d}{\omega\mu_0}E_y^d \tag{7.15}$$

The same analysis applies to the metal side of the interface with the result

$$E_y^m(x, z) = E_{0y}^m e^{i[(k_x^m x + k_z^m z) - \omega t]} \tag{7.16}$$

and

$$H_x^m = -\frac{i}{\omega\mu_0}\frac{\partial E_y^m}{\partial z} = \frac{k_z^m}{\omega\mu_0}E_y^m \tag{7.17}$$

$$H_z^m = \frac{i}{\omega\mu_0}\frac{\partial E_y^m}{\partial x} = -\frac{k_x^m}{\omega\mu_0}E_y^m \tag{7.18}$$

At the interface, $z = 0$, continuity conditions for H_x and E_y require that

$$H_x^d(z = 0) = H_x^m(z = 0) \tag{7.19}$$
$$E_y^d(z = 0) = E_y^m(z = 0) \tag{7.20}$$

and therefore, at the boundary,

$$k_z^d = k_z^m \tag{7.21}$$
$$\sqrt{\varepsilon_d}k_0 = \sqrt{\varepsilon_m}k_0 \tag{7.22}$$

But this last equation can only be valid for $k_0 = 0$ since the dielectric constants on the two sides of the boundary can never be equal. *Therefore there can be no wave solution at the boundary for TE polarisation.*

7.3.2   Surface waves with TM polarisation

With TM polarisation we have three field components, H_y, E_x, E_z. We find the wave solution for H_y on the dielectric side of the interface:

$$H_y^d(x, z) = H_{y0}^d e^{i[(k_x x + k_z z) - \omega t]} \tag{7.23}$$

Then using Equations 7.6 and 7.8, we find expressions for E_x^d and E_z^d:

$$E_x^d = \frac{k_z^d}{\omega \varepsilon_0 \varepsilon_d} H_y^d(x, z) \tag{7.24}$$

$$E_z^d = -\frac{k_x^d}{\omega \varepsilon_0 \varepsilon_d} H_y^d(x, z) \tag{7.25}$$

On the metal side of the interface we have

$$E_x^m = \frac{k_z^m}{\omega \varepsilon_0 \varepsilon_m} H_y^m(x, z) \tag{7.26}$$

$$E_z^m = -\frac{k_x^m}{\omega \varepsilon_0 \varepsilon_m} H_y^m(x, z) \tag{7.27}$$

Continuity conditions at the interface ($z = 0$) are

$$E_x^d(x, 0) = E_x^m(x, 0) \tag{7.28}$$
$$H_y^d(x, 0) = H_y^m(x, 0) \tag{7.29}$$
$$\varepsilon_d E_z^d(x, 0) = \varepsilon_m E_z^m(x, 0) \tag{7.30}$$

Then substituting Equations 7.24 and 7.25 and 7.26 and 7.27 into the continuity conditions we find that at the boundary

$$k_x^d = k_x^m = k_x \tag{7.31}$$

$$\frac{k_z^d}{\varepsilon_d} = \frac{k_z^m}{\varepsilon_m} \tag{7.32}$$

The wave vector relations on the dielectric and metal sides of the interface, Equations 7.10 and 7.12 allow us to write

$$\frac{(k_z^m)^2}{(k_z^d)^2} = \frac{\varepsilon_m k_0^2 - (k_x^m)^2}{\varepsilon_d k_0^2 - (k_x^d)^2} \tag{7.33}$$

Then, using Equations 7.31 and 7.32, we can eliminate the k_z components and obtain an expression for k_x, the magnitude of the surface wave vector propagating along the boundary, in terms of the free-space wave vector k_0 and the dielectric constants on each side of the interface:

$$k_x^s = \sqrt{\frac{\varepsilon_d \varepsilon_m}{\varepsilon_d + \varepsilon_m}} k_0 \tag{7.34}$$

The z components of the surface wave, on the dielectric and metallic sides of the boundary, can be easily obtained from

$$\left(k_z^{s,d}\right)^2 = \varepsilon_d k_0^2 - \left(k_x^s\right)^2 k_0^2 = \left(1 - \frac{\varepsilon_m}{\varepsilon_m + \varepsilon_d}\right) \varepsilon_d k_0^2 \tag{7.35}$$

$$\left(k_z^{s,m}\right)^2 = \varepsilon_m k_0^2 - \left(k_x^s\right)^2 k_0^2 = \left(1 - \frac{\varepsilon_d}{\varepsilon_m + \varepsilon_d}\right) \varepsilon_m k_0^2 \tag{7.36}$$

$$k_z^{s,d} = \pm \frac{\varepsilon_d}{\sqrt{\varepsilon_d + \varepsilon_m}} k_0 \tag{7.37}$$

$$k_z^{s,m} = \pm \frac{\varepsilon_m}{\sqrt{\varepsilon_d + \varepsilon_m}} k_0 \tag{7.38}$$

Note that if the real part of the dielectric constant in the metal is negative and if $\varepsilon_d + \varepsilon_m < 0$, then k_z^m, k_z^d will be imaginary. The E_z component of the surface wave on both sides of the interface will then be evanescent:

$$k_z^{s,d} \to \pm i \kappa_s^d = \pm i \left|\frac{\varepsilon_d}{\sqrt{\varepsilon_d + \varepsilon_m}}\right| k_0 \tag{7.39}$$

$$k_z^{s,m} \to \pm i \kappa_s^m = \pm i \left|\frac{\varepsilon_m}{\sqrt{\varepsilon_d + \varepsilon_m}}\right| k_0 \tag{7.40}$$

The choice of sign ($\pm$) ensures that the wave amplitude decreases exponentially with increasing distance ($\pm z$) from the interface. With the choice of coordinate axes as shown in Figure 7.2,

$$k_z^{s,d} \to i \kappa_s^d \qquad z > 0 \tag{7.41}$$

$$k_z^{s,m} \to -i \kappa_s^m \qquad z < 0 \tag{7.42}$$

and the components of the surface wave projecting onto the dielectric side take on the following forms,

$$H_y^{s,d}(x, z, t) = H_{0y} e^{\kappa_s^d z} e^{i(k_x^s x - \omega t)} \tag{7.43}$$

$$E_x^{s,d}(x, z, t) = -\frac{i\kappa_s^d}{\omega \varepsilon_0 \varepsilon_d} H_{0y} e^{\kappa_s^d z} e^{i(k_x^s x - \omega t)} \tag{7.44}$$

$$E_z^{s,d}(x, z, t) = -\frac{k_x^s}{\omega \varepsilon_0 \varepsilon_d} H_{0y} e^{\kappa_s^d z} e^{i(k_x^s x - \omega t)} \tag{7.45}$$

Similar expressions obtain on the metallic side with the appropriate change of sign for $i\kappa_s^m$ (Equation 9.104). Figure 7.3 shows how the electric and displacement fields of surface waves decrease exponentially on both dielectric and metal interfaces. In addition to the exponential decrease in the $\pm z$ direction, the plots in Figure 7.3 also show that the absolute magnitude of the dielectric constant on the metal side is typically much greater than on the dielectric side. Therefore, field penetration into the metal is limited to the *skin depth*. For gold or silver in the optical interval ($350 \leq \lambda_0 \leq 850$ nm) of the electromagnetic frequency spectrum, significant field penetration is of the order $20 - 30$ nm.

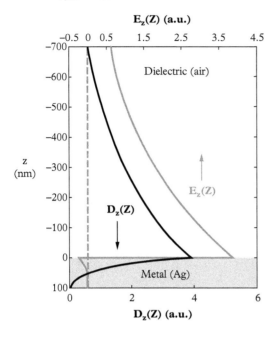

Figure 7.3 *Vertical (z) components of the surface wave electric field (E-field) and displacement field (D-field). Note that E_z is discontinuous at the surface while D_z is continuous. The units of the E-field and D-field are arbitrary.*

Furthermore, the E-field is *discontinuous* at the boundary, due to the presence of surface charge density, while the displacement field ($\mathbf{D} = \varepsilon_0\varepsilon\mathbf{E}$) takes into account the sign change in the dielectric constant across the boundary and is therefore continuous (although the *derivative* of the D-field with respect to the boundary normal is not continuous). Table 7.1 summarises the various components of the electromagnetic wave in the dielectric, the metal, and at the interface for TM polarisation.

Table 7.1 *Field components for TM Polarisation*

Region		$E_x(x, z, t)$	$E_z(x, z, t)$	$H_y(x, z, t)$
dielectric		$\dfrac{k_z^d}{\omega\varepsilon_0\varepsilon_d}H_{(x, z, t)}$	$-\dfrac{k_x^d}{\omega\varepsilon_0\varepsilon_d}H_y(x, z, t)$	$H_{y0}e^{i(k_x^d x + k_z^d z - \omega t)}$
metal		$\dfrac{k_z^m}{\omega\varepsilon_0\varepsilon_m}H_y(x, z, t)$	$-\dfrac{k_x^m}{\omega\varepsilon_0\varepsilon_m}H_y(x, z, t)$	$H_{y0}e^{i(k_x^m x + k_z^m z - \omega t)}$
interface	d side	$-\dfrac{i\kappa_s^d}{\omega\varepsilon_0\varepsilon_d}H_y(x, z, t)$	$-\dfrac{k_x^s}{\omega\varepsilon_0\varepsilon_d}H_y(x, z, t)$	$H_{0y}e^{\kappa_s^d z}e^{i(k_x^s x - \omega t)}$
	m side	$\dfrac{i\kappa_s^m}{\omega\varepsilon_0\varepsilon_m}H_y(x, z, t)$	$-\dfrac{k_x^s}{\omega\varepsilon_0\varepsilon_m}H_y(x, z, t)$	$H_{0y}e^{-\kappa_s^m z}e^{i(k_x^s x - \omega t)}$

Equation 7.34 shows that k_x^s is complex since ε_m is complex. The three components of the surface wave E_x^s, E_z^s, H_y^s propagate in the x direction at the interface. Taking E_x^s as an example,

$$E_x^s(x) = E_{0x}^s e^{ik_x^s x} = E_{0x}^s e^{ik_0 n_s x} \tag{7.46}$$

and the argument of the exponential will have a real and imaginary term since $k_x^s = k_0 n_s$ is complex. The factor $n_s = \eta_s + i\kappa_s$ is the complex surface index of the refraction at the interface. The imaginary term represents the exponential loss in amplitude due to dissipation in the metal as the wave propagates along the surface. From the definition of the surface index of refraction, $n_s = k_x^s/k_0$, the complex dielectric constant of the metal, $\varepsilon_m = \varepsilon_m' + i\varepsilon_m''$ and Equation 7.34, we can determine the corresponding complex index of refraction:

$$n_s = \sqrt{\frac{\varepsilon_m \varepsilon_d}{\varepsilon_m + \varepsilon_d}} \tag{7.47}$$

$$n_s^2 = \frac{\varepsilon_m' \varepsilon_d}{\varepsilon_m' + \varepsilon_d} \cdot \frac{1 + i\frac{\varepsilon_m''}{\varepsilon_m'}}{1 + i\frac{\varepsilon_m''}{\varepsilon_m' + \varepsilon_d}} \tag{7.48}$$

For the two metals commonly used in plasmonic devices, silver and gold, $|\varepsilon_m''/\varepsilon_m'| \ll 1$ over the optical range of interest, and this characteristic can be used to simplify Equation 7.48. Figure 7.4 and Figure 7.5 show a plot of the real and imaginary terms of the dielectric constant of silver metal as a function of wavelength over the optical range, and Figure 7.6 plots the ratio of $|\varepsilon_m''/\varepsilon_m'| \ll 1$ for the same silver data. Separating Equation 7.48 in real and imaginary terms while dropping fractional terms quadratic in ε_m'', we find that

$$n_s^2 = \frac{\varepsilon_m' \varepsilon_d}{\varepsilon_m' + \varepsilon_d} \cdot \left[1 + i\frac{\varepsilon_m'' \varepsilon_d}{\varepsilon_m' \left(\varepsilon_m' + \varepsilon_d \right)} \right] \tag{7.49}$$

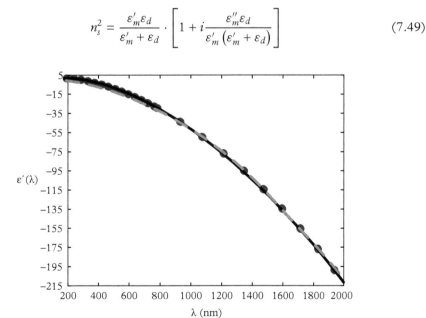

Figure 7.4 *Real part of the dielectric constant ε′ vs. wavelength for silver from the Johnson and Christy measurements (see Reference 8 in the reading list in Section 7.10).*

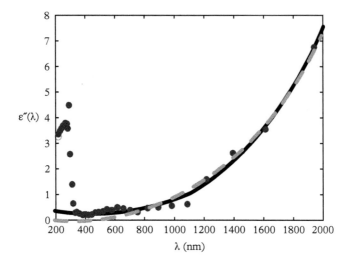

Figure 7.5 *Imaginary part of the dielectric constant ε'' vs. wavelength for silver from the Johnson and Christy measurements (Reference 8. in Section 7.10).*

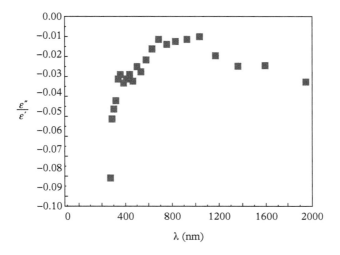

Figure 7.6 *Ratio $\varepsilon''/\varepsilon'$ vs. wavelength for silver from the Johnson and Christy measurements (Reference 8 in Section 7.10).*

and from the definition of the complex index of refraction,

$$n_s^2 = \eta^2 - \kappa_s^2 + i2\eta\kappa_s \tag{7.50}$$

Setting the real and imaginary terms in Equations 7.49 and 7.50 equal,

$$\eta_s^2 - \kappa_s^2 = \frac{\varepsilon_m' \varepsilon_d}{\varepsilon_m' + \varepsilon_d} \tag{7.51}$$

$$2\eta_s\kappa_s = \frac{\varepsilon_m' \varepsilon_m'' \varepsilon_d}{\varepsilon_m' \left(\varepsilon_m' + \varepsilon_d\right)^2} \tag{7.52}$$

Equations 7.51 and 7.52 can be decoupled and solved for η_s and κ_s, the two components of the surface complex index of refraction. The result is:

$$\eta_s = \sqrt{\frac{\varepsilon_m' \varepsilon_d}{\varepsilon_m' + \varepsilon_d}} \tag{7.53}$$

$$\kappa_s = \frac{1}{2} \frac{\varepsilon_m''}{\left(\varepsilon_m'\right)^2} \left(\frac{\varepsilon_m' \varepsilon_d}{\varepsilon_m' + \varepsilon_d}\right)^{3/2} \tag{7.54}$$

These expressions can be simplified further if the dielectric is vacuum or air. In that case, ε_d can be set equal to unity and

$$\eta_s = \sqrt{\frac{\varepsilon_m'}{\varepsilon_m' + 1}} \tag{7.55}$$

$$\kappa_s = \frac{1}{2\sqrt{\varepsilon_m'}} \cdot \frac{\varepsilon_m''}{\left(\varepsilon_m' + 1\right)^{3/2}} \tag{7.56}$$

As a typical example of a surface wave at a metal-dielectric interface, we take $\lambda_0 = 632$ nm, $\varepsilon_d = 1$ (air), $\varepsilon_m = -15.625 + i(1.04059)$ (silver), and calculate η_s, κ_s. From Equations 7.55 and 7.56 and the definition of k_0 we find that

$$\eta_s = 1.0336$$

$$\kappa_s = 2.3534 \times 10^{-3}$$

$$k_0 = 2\pi/\lambda_0 = 9.942 \times 10^6 \text{ m}^{-1}$$

Taking the 'effective length' x_{eff} to be the distance over which the amplitude of the surface waves falls to $1/e$ of the initial value, we find $x_{\text{eff}} = 42.74 \,\mu$m. Figure 7.7 shows the behaviour of the surface wave E_x-field amplitude at the interface ($z = 0$), propagating in the x direction. Figure 7.8 shows the variation of E_x, E_z near the metal-dielectric interface.

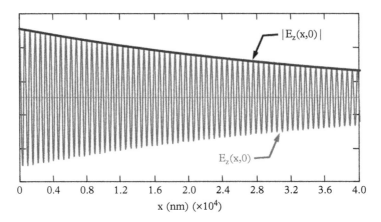

Figure 7.7 *The surface wave E_x-field amplitude at the interface ($z = 0$), propagating in the x direction. The distance along which the amplitude decreases to $1/e$ of the initial value at $x = 0$ is called the 'effective' length.*

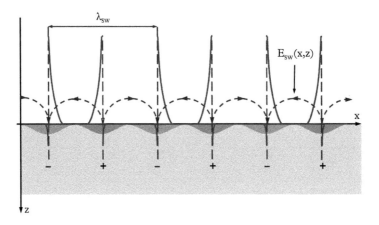

Figure 7.8 *Instantaneous distribution of the electric field E_x and E_z in the x and z directions, respectively. The signal variation of the field E_z along x is accompanied by the variation of electric charge (+,-) corresponding to the condition of E_z-field discontinuity. The representation of surface charge density along z is exaggerated for clarity. The charge density can only exist at the surface.*

7.4   Plasmon surface wave dispersion

7.4.1   The free-electron plasma

The dielectric constant of materials in general depends on the frequency of the electromagnetic fields with which they interact. The equation describing this dependence is called the dispersion relation, and the free-electron plasma model leads to a very simple dispersion relation for high-conductivity metals. The force on a free electron subject to an applied E-field in one dimension is

$$m_e \frac{d^2 x}{dt^2} = -eE_x \tag{7.57}$$

and if the E-field is time-harmonic, $\mathbf{E} = \mathbf{E}_0 e^{-i\omega t}$, we have:

$$-\omega^2 m_e x = -eE_x \tag{7.58}$$

The polarisation field $\mathbf{P}$ of the free-electron plasma, considered as a volume density of one-dimensional dipoles $p_x = -ex$, is

$$\mathbf{P} = -N_e \frac{e^2}{m_e \omega^2} \mathbf{E} \tag{7.59}$$

with N_e the electron concentration of the plasma. From the definition of the displacement field $\mathbf{D}$, and the constitutive relation between $\mathbf{D}$ and $\mathbf{E}$, we have

$$\mathbf{D} = \varepsilon_0 \mathbf{E} + \mathbf{P} \tag{7.60}$$

$$\mathbf{D} = \varepsilon_0 \varepsilon \mathbf{E} \tag{7.61}$$

and therefore the dielectric constant of the free-electron plasma is

$$\varepsilon(\omega) = 1 + \frac{P}{\varepsilon_0 E} = 1 - \frac{N_e e^2}{\varepsilon_0 m_e \omega^2} \tag{7.62}$$

The characteristic *plasma frequency* is defined by

$$\omega_p^2 = \frac{N_e e^2}{\varepsilon_0 m_e} \tag{7.63}$$

So finally, we have the lossless dielectric constant:

$$\varepsilon(\omega) = 1 - \frac{\omega_p^2}{\omega^2} \tag{7.64}$$

7.4.2 The Drude model of metals

The free-electron plasma model describes harmonic electron motion in the limit of a negligible restoring force acting on the electron, and no 'collisional' or dissipative losses. The Drude model does not introduce a finite restoring force, but does add dissipation to the equation of motion for the electrons in the plasma gas (when a linear restoring force *is* introduced into the conduction electron equation of motion, the theory is called the Drude–Lorentz model of metals). The equation of motion for the simple Drude model of the conduction electrons can still be described by a one-dimensional driven harmonic oscillator expression, where the driver issues from an electromagnetic field incident at the dielectric-metal boundary. If we assume the incident radiation is polarised along x, then the Drude model equation of motion can be written as

$$m_e \frac{d^2 x}{dt^2} + m_e \Gamma \frac{dx}{dt} = -eE(x,t) = -eE_{0m} e^{-i\omega t} \tag{7.65}$$

where m_e is the electron mass, Γ is a phenomenological damping constant, and $eE_{0m} e^{-i\omega t}$ is the driving force of an E-M field in the metal with amplitude E_{0m} and frequency ω. Again, dropping the common oscillatory phase term, the solutions for the position, velocity, and acceleration are:

$$x = \frac{1}{m_e \left(\omega^2 + i\Gamma\omega \right)} eE_{0m} \tag{7.66}$$

$$\frac{dx}{dt} = -\frac{i\omega}{m_e \left(\omega^2 + i\Gamma\omega \right)} eE_{0m} \tag{7.67}$$

$$\frac{d^2 x}{dt^2} = -\frac{\omega^2}{m_e \left(\omega^2 + i\Gamma\omega \right)} eE_{0m} \tag{7.68}$$

Again, we write the polarisation as a density of dipoles, $P_x = -N_e ex$, and now the dielectric constant is complex:

$$\varepsilon(\omega) = 1 - \frac{N_e e^2}{\varepsilon_0 m_e \left(\omega^2 + i\Gamma\omega \right)} = 1 - \frac{\omega_p^2}{\omega^2 + i\Gamma\omega} \tag{7.69}$$

Separating the real and imaginary terms,

$$\varepsilon(\omega) = \frac{\omega^2 - \omega_p^2 + \Gamma^2}{\omega^2 + \Gamma^2} + i \frac{\omega_p^2 \Gamma}{\omega(\omega^2 + \Gamma^2)} \tag{7.70}$$

If $\Gamma \ll \omega$, as is the usual case at optical frequencies,

$$\varepsilon(\omega) = \varepsilon'(\omega) + i\varepsilon''(\omega)$$

$$\varepsilon(\omega) \simeq 1 - \frac{\omega_p^2}{\omega^2} + i \frac{\omega_p^2 \Gamma}{\omega^3} \tag{7.71}$$

A summary of the Drude-model parameters of many common metals is given in Table 7.2.

Table 7.2 *Drude-model parameters for some common metals**

Metal	ω_p (rad/s)	Γ (s^{-1})	τ (s)	N_e (m^{-3})
Al	2.24×10^{16}	2.73×10^{13}	3.66×10^{-14}	1.56×10^{29}
Ag	1.37×10^{16}	5.56×10^{13}	1.80×10^{-14}	5.83×10^{28}
Au	1.37×10^{16}	4.05×10^{13}	2.47×10^{-14}	5.83×10^{28}
Co	6.03×10^{15}	5.56×10^{13}	1.80×10^{-14}	1.13×10^{28}
Cu	1.12×10^{16}	1.38×10^{13}	7.25×10^{-14}	3.90×10^{28}
Fe	6.22×10^{15}	2.77×10^{13}	3.61×10^{-14}	1.20×10^{28}
Mo	1.13×10^{16}	7.77×10^{13}	1.29×10^{-14}	3.97×10^{28}
Ni	7.43×10^{15}	6.63×10^{13}	1.51×10^{-14}	1.72×10^{28}
Pb	1.12×10^{16}	3.07×10^{13}	3.25×10^{-14}	3.90×10^{28}
Pd	8.29×10^{15}	2.34×10^{13}	4.28×10^{-14}	2.14×10^{28}
Pt	7.82×10^{15}	1.05×10^{13}	9.51×10^{-14}	1.90×10^{28}
Ti	3.83×10^{15}	7.20×10^{13}	1.39×10^{-14}	4.56×10^{27}
V	7.84×10^{15}	9.22×10^{13}	1.08×10^{-14}	1.91×10^{28}
W	9.75×10^{15}	9.18×10^{13}	1.09×10^{-14}	2.96×10^{28}

* Compiled from data in Reference 8. in Section 7.10

7.4.2.1 Drude model dispersion curves

The dispersive response to electromagnetic excitation is characterised by the frequency dependence of the dielectric constant of a material, $\varepsilon(\omega)$. Many common optical dielectrics (air, glass, and fused quartz) exhibit dielectric constants with near-zero imaginary terms (negligible absorption) and real terms that vary little with frequency over the visible range. We have just seen how metallic response can be described by simple physical models such as the free-plasma model or the Drude model. The behaviour of a material is usually summarised by the 'dispersion curve' that plots the energy or frequency of the exciting radiation against the propagation constant k of the light in the material. There is a simple chain of relations between the propagation constant $k(\omega)$ and the dielectric constant $\varepsilon(\omega)$. First,

$$k = \frac{2\pi}{\lambda} \quad \text{and} \quad \lambda = \frac{\lambda_0}{n} \tag{7.72}$$

where λ_0 is the free-space wavelength and n the index of refraction. Then, the material index of refraction can be complex, $n = \eta + i\kappa$, and is related to the dielectric constant ε by

$$n(\omega) = \sqrt{\varepsilon(\omega)} \tag{7.73}$$

$$\eta(\omega) + i\kappa(\omega) = \sqrt{\varepsilon'(\omega) + i\varepsilon''(\omega)} \tag{7.74}$$

and therefore,

$$\eta^2 - \kappa^2 = \varepsilon'(\omega) \tag{7.75}$$

$$2\eta\kappa = \varepsilon''(\omega) \tag{7.76}$$

Decoupling η and κ results in a quadratic equation in η:

$$\eta^4 - \eta^2 \varepsilon'(\omega) - \frac{1}{4}\left[\varepsilon''(\omega)\right]^2 = 0 \tag{7.77}$$

The solution to Equation 7.77 can be obtained directly from the elementary quadratic formula. The real part of the index of diffraction η can then be substituted into Equation 7.76 to obtain the imaginary part κ. Thus η, κ, and $\varepsilon', \varepsilon''$ can be readily interconverted. Figure 7.9 shows the dielectric constant terms and the refractive index terms, calculated from Eq. 7.71 as a function of ω/ω_p, where ω_p is the plasma frequency for silver metal, $\omega_p = 1.35 \times 10^{16}\,\mathrm{s}^{-1}$, and the relaxation rate Γ is taken to be $\Gamma = 1 \times 10^{14}\,\mathrm{s}^{-1}$.

The dispersion curve for a Drude metal can be obtained by starting with the expressions for the real and imaginary parts of the dielectric constant, Equation 7.71, and rescaling the frequency variable to $y = \omega/\omega_p$:

$$\varepsilon'_m = 1 - \frac{1}{y^2} \tag{7.78}$$

$$\varepsilon''_m = \frac{1}{y^3}\left(\frac{\Gamma}{\omega_p}\right) \tag{7.79}$$

and taking the propagation constant of the surface wave along the dielectric-metal interface as $k_s = k_0 n_s$. The factor n_s is the surface of index of refraction, the real and imaginary parts of which we write from Equations 7.53 and 7.54:

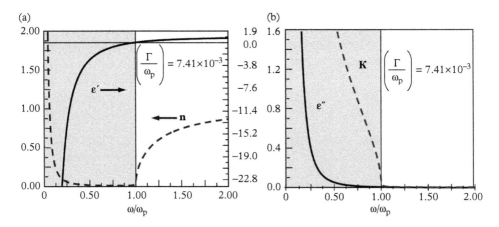

Figure 7.9 *Real (panel a) and imaginary (panel b) terms of the index of refraction and the dielectric constant calculated from the Drude model (Equation 7.71) vs. frequency ω normalised to ω_p, the plasma frequency for silver. The relaxation rate Γ is taken to be $\Gamma = 10^{14}\,s^{-1}$ and $\omega_p = 1.35 \times 10^{16}\,s^{-1}$.*

$$k_s = k_0 n_s = k_0 \left[\left(\frac{\varepsilon_d \varepsilon'_m}{\varepsilon_d + \varepsilon'_m} \right)^{1/2} + i \frac{\varepsilon''_m}{2(\varepsilon'_m)^2} \left(\frac{\varepsilon_d \varepsilon''_m}{\varepsilon_d + \varepsilon'_m} \right)^{3/2} \right] \tag{7.80}$$

$$k_s = k'_s + i k''_s \tag{7.81}$$

Thus:

$$k'_s = k_0 \left(\frac{\varepsilon_d \varepsilon'_m}{\varepsilon_d + \varepsilon'_m} \right)^{1/2} \tag{7.82}$$

$$k''_s = \frac{\varepsilon''_m}{2(\varepsilon'_m)^2} \left(\frac{\varepsilon_d \varepsilon''_m}{\varepsilon_d + \varepsilon'_m} \right)^{3/2} \tag{7.83}$$

At first neglecting the imaginary term, we define a new 'reduced' variable:

$$x = \frac{k'_s c}{\omega_p} = \frac{k'_s}{k_p} \tag{7.84}$$

and write Equation 7.82 as:

$$x = y \left(\frac{\varepsilon_d (y^2 - 1)}{y^2 (\varepsilon_d + 1) - 1} \right)^{1/2} \tag{7.85}$$

Now Equation 7.85 results in a quadratic equation in y, the solution of which is

$$y = \sqrt{ \frac{\varepsilon_d + (\varepsilon_d + 1)x^2 \pm \sqrt{\left(\varepsilon_d + (\varepsilon_d + 1)x^2 \right)^2 - 4\varepsilon_d x^2}}{2\varepsilon_d} } \tag{7.86}$$

The solution constitutes the dispersion curve (in 'reduced' units) of the frequency vs. the propagation parameter for the surface wave. The reduced variable x is defined in terms of the real part of the propagation parameter k'_s, and therefore, this expression describes only the real part of the dispersion curve. The solution consists of two branches, plotted in Figure 7.10. The lower branch describes the surface-wave dispersion from the long wavelength limit ($y \to 0$; $x \to 0$) to an asymptote ($y \to y_{sp} = \omega_{sp}/\omega_p$; $x \to \infty$). This frequency asymptote at the short wavelength limit corresponds to the *surface* plasma resonance where the surface charge density oscillates collectively over the entire surface and the phase velocity of the surface wave $d\omega/dk_s$ slows to near zero. An expression for this surface plasma asymptotic frequency can be obtained from Equation 7.86:

$$\omega \to \omega_{sp} = \frac{\omega_p}{\sqrt{1 + \varepsilon_d}} \tag{7.87}$$

This condition is often called the *surface plasmon resonance* and bears a simple relation to ω_p, the *bulk plasmon resonance* derived from the free-electron plasma model. Between

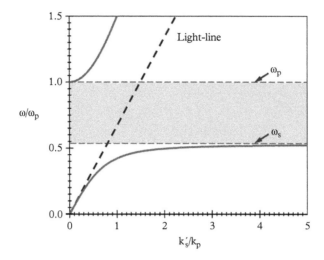

Figure 7.10 *Dispersion curve for a silver-glass interface. The real dielectric constant for glass is taken to be $\varepsilon' = 2.25$. The loss term in the metal is not included.*

these two frequencies there is no charge density surface wave or bulk wave propagation. This 'forbidden zone' is sometimes termed a *stopband*. At frequencies above ω_p, propagation once again occurs on the surface and in the bulk. This frequency limit corresponds to the point where $\varepsilon'_m \rightarrow 0$. Below ω_p, $\varepsilon'_m < 0$ and above ω_p, $\varepsilon'_m > 0$. Also shown in Figure 7.10 is the linear 'light line', the dispersion curve of light freely propagating in the dielectric medium, $\omega_{ll} = k_0 c/\eta$.

Inclusion of the imaginary part of the surface parameter k_s in the definition of the reduced variable x in Equation 7.84, and subsequent solution of Equations 7.85 and 7.86, modifies the dispersion curve and opens propagation in the stopband. The real and imaginary parts of the full dispersion curve are shown in Figure 7.11. Note that in panel (a) of this figure, the phase velocity

$$v_{\text{phase}} = \frac{\omega}{k'_s} \tag{7.88}$$

reverses sign in the stopband region. This property is a necessary characteristic of 'negative index' metamaterials. Panel (b) shows that, unfortunately for many potential applications, this region is also one of high absorption.

7.4.2.2   Conduction current in a Drude metal

The conduction current in the metal is given by

$$J_c = eN_e \frac{dx}{dt} = eN_e \frac{\omega}{m_e \left(\Gamma\omega - i\omega^2\right)} eE_{0m} \tag{7.89}$$

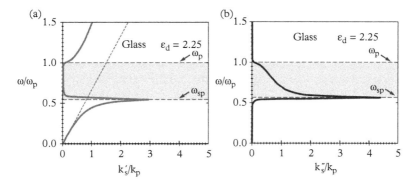

Figure 7.11 *Panel (a): dispersion curve for the real part of the surface propagation parameter k'_s. Panel (b): Dispersion curve for the imaginary part of the surface propagation parameter k''_s. Both propagation parameters have been normalised to $k_p = \omega_p/c$. Inclusion of the imaginary term results in finite propagation in the stopband (compare with Figure 7.10).*

and with the electron density related to the bulk plasma frequency, Equation 7.63, we have

$$J_c = \frac{\omega_p^2}{(\Gamma - i\omega)}\varepsilon_0 E_{0m} \tag{7.90}$$

From the standard constitutive relations

$$\mathbf{J_c} = \sigma\mathbf{E} = \frac{\sigma}{\varepsilon_0}\varepsilon_0\mathbf{E} \tag{7.91}$$

and so, comparing Equations 7.90 and 7.91, the expression for the complex conductivity can be expressed as

$$\frac{\sigma}{\varepsilon_0} = \frac{\Gamma}{\left(\frac{\Gamma^2}{\omega_p^2} + \frac{\omega^2}{\omega_p^2}\right)} + i\frac{\omega}{\left(\frac{\Gamma^2}{\omega_p^2} + \frac{\omega^2}{\omega_p^2}\right)} \tag{7.92}$$

We can compare the damping rate Γ to the plasma frequency ω_p if we can estimate the latter, and from Equation 7.63 we see that this requires knowledge of the conduction electron density. A typical 'good' metal, such as silver, exhibits an electron density $N_e \simeq 6 \times 10^{28}$ m^{-3}, and therefore, $\omega_p \simeq 1.5 \times 10^{16}$ s^{-1}. The damping due to radiative or collisional loss is typically $\sim 10^{14}$ s^{-1}, and therefore, $\Gamma \ll \omega_p$. Thus, in the optical regime we can write

$$\frac{\sigma}{\varepsilon_0} \simeq \frac{\Gamma\omega_p^2}{\omega^2} + i\frac{\omega_p^2}{\omega} \tag{7.93}$$

This last expression for the conductivity shows that at relatively low frequencies (RF, microwave, and radio wave) the real term dominates, and the conductivity is in phase with the driving field and 'ohmic'. As the frequency increases into the optical range, however, the electrons can no longer follow the driving field in phase, and the imaginary term dominates. At $\omega \simeq 10^{15}\,\mathrm{s^{-1}}$, typical of the visible optical regime, the real term is only about 5 per cent of the imaginary term. Figure 7.10 plots the dispersion relation of the Drude model, Equation 7.71.

7.5 Energy flux and density at the boundary

7.5.1 Energy flux normal to the boundary

As indicated in Equation 2.61, the Poynting vector is defined as

$$\mathbf{S} = \mathbf{E} \times \mathbf{H} \tag{7.94}$$

and the energy flux of an electromagnetic wave averaged over an optical cycle is given by

$$\mathbf{S} = \frac{1}{2}\mathrm{Re}[\mathbf{E} \times \mathbf{H}^*] \tag{7.95}$$

Figure 7.12 shows the x and z components of the energy flow on the two sides of the surface between a dielectric and a metal. Writing out these components explicitly from Table 7.1, we have, at normal incidence, the transmitted energy flux impinging at the surface from the dielectric into the metal along the z direction:

$$S_z^d = \frac{1}{2}\mathrm{Re}\left[E_x^d H_y^*\right] = \frac{1}{2}\mathrm{Re}\left[\frac{k_z^d}{\varepsilon_d \varepsilon_0 \omega}\right]H_{0y}^2 \tag{7.96}$$

$$S_z^m = \frac{1}{2}\mathrm{Re}\left[E_x^m H_y^*\right] = \frac{1}{2}\mathrm{Re}\left[\frac{k_z^m}{\varepsilon_m \varepsilon_0 \omega}\right]H_{0y}^2 \tag{7.97}$$

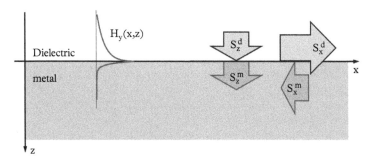

Figure 7.12 *Indication of the components of the Poynting vector of the dielectric-metal interface in which a wave propagates surface plasmon polariton (SPP).*

From the continuity relation at the surface, Equation 7.33, we see immediately that the energy flux is continuous across the boundary in the z direction. On the dielectric side, the permittivity $\varepsilon_0 \varepsilon_d$ and the propagation constant k_z^d are real (for a lossless material), and therefore, the expression for the energy flux is also real. On the metal side the dielectric constant is complex; $\varepsilon_m = \varepsilon'_m + i\varepsilon''_m$, as is $k_z^m = k_0 n_m = k_0(\eta_m + i\kappa_m)$. The metal index of refraction n_m is related to the metal dielectric constant through $n_m = \sqrt{\varepsilon_m}$, and therefore, the expression for the product of E_x^m and H_y^* in Equation 7.97 exhibits real and imaginary terms. The real part of Equation 7.97 represents the transmitted, exponentially decreasing energy flux into the metal. In order to obtain explicit expressions for n_m and k_m^z we need to examine the propagation of electromagnetic waves inside real metals.

7.5.2 Electromagnetics in metals

We start by writing the curl equations of Maxwell in terms of $\mathbf{E}$ and $\mathbf{B}$, the electric field and magnetic induction field:

$$\nabla \times \mathbf{E} = -\frac{\partial \mathbf{B}}{\partial t} \qquad \text{Faraday law} \qquad (7.98)$$

$$\nabla \times \mathbf{B} = \mu_m \varepsilon_m \frac{\partial \mathbf{E}}{\partial t} + \mu_m \mathbf{J}_c \qquad \text{Maxwell–Ampère law} \qquad (7.99)$$

where μ_m and ε_m are the permeability and permittivity of the metal, and $\mathbf{J}_c$ is the free charge current. For a high-conductivity metal such as silver or gold, $\mathbf{J}_c$ is directly proportional to $\mathbf{E}$ with the conductivity σ, and the proportionality constant, $\mathbf{J}_c = \sigma \mathbf{E}$. Equations 7.98 and 7.99 can be decoupled by first applying the curl operation to the Faraday equation,

$$\nabla \times (\nabla \times \mathbf{E}) = \nabla(\nabla \cdot \mathbf{E}) - \nabla^2 \mathbf{E} = \nabla \times \left(\frac{\partial \mathbf{B}}{\partial t}\right) = -\frac{\partial(\nabla \times \mathbf{B})}{\partial t} \qquad (7.100)$$

$$= -\mu_m \varepsilon_m \frac{\partial^2 \mathbf{E}}{\partial t^2} - \mu_m \sigma \frac{\partial \mathbf{E}}{\partial t} \qquad (7.101)$$

which together with $\nabla \cdot \mathbf{E} = 0$, results in a wave equation for $\mathbf{E}$ in the metal:

$$\nabla^2 \mathbf{E} = \mu_m \varepsilon_m \frac{\partial^2 \mathbf{E}}{\partial t^2} + \mu_m \sigma \frac{\partial \mathbf{E}}{\partial t} \qquad (7.102)$$

Similarly, applying the curl operation to Equation 7.99 produces the wave equation for $\mathbf{B}$:

$$\nabla^2 \mathbf{B} = \mu_m \varepsilon_m \frac{\partial^2 \mathbf{B}}{\partial t^2} + \mu_m \sigma \frac{\partial \mathbf{B}}{\partial t} \qquad (7.103)$$

Plane-wave solutions to these equations for waves propagating in the z direction can be written as

$$\mathbf{E}_m = \mathbf{E}_{0m} e^{i(k_z^m z - \omega t)} \tag{7.104}$$

$$\mathbf{B}_m = \mathbf{B}_{0m} e^{i(k_z^m z - \omega t)} \tag{7.105}$$

The amplitudes $\mathbf{E}_{0m}, \mathbf{B}_{0m}$ and the propagation vector k_z^m are generally complex quantities. Substituting these solutions back into the uncoupled wave equations, Equations 7.102 and 7.103, results in an expression for the propagation parameter in terms of the permeability, complex permittivity, and complex conductivity of the metal,

$$\left(k_z^m\right)^2 = \mu_m \varepsilon_m \omega^2 + i\mu_m \sigma \omega \tag{7.106}$$

with

$$k_z^m = k_r + ik_i \qquad \varepsilon_m = \varepsilon_r + i\varepsilon_i \qquad \sigma = \sigma_r + i\sigma_i \tag{7.107}$$

Equating real and imaginary terms after substitution into Equation 7.106 yields

$$k_r^2 - k_i^2 = \mu_0 \varepsilon_r \omega^2 - \mu_0 \sigma_i \omega \tag{7.108}$$

$$2k_r k_i = \mu_0 \varepsilon_i \omega^2 + \mu_0 \sigma_r \omega \tag{7.109}$$

where we have set $\mu_m = \mu_0$. Now writing the complex permittivity in terms of the dielectric constants, $\varepsilon_r = \varepsilon_0 \varepsilon_1$ and $\varepsilon_i = \varepsilon_0 \varepsilon_2$, we have

$$k_r^2 - k_i^2 = \mu_0 \varepsilon_0 \omega^2 \left(\varepsilon_1 - \frac{\sigma_i}{\varepsilon_0 \omega} \right) = k_0^2 \left(\varepsilon_1 - \frac{\sigma_i}{\varepsilon_0 \omega} \right) \tag{7.110}$$

$$2k_r k_i = \mu_0 \varepsilon_0 \omega^2 \left(\varepsilon_2 + \frac{\sigma_r}{\varepsilon_0 \omega} \right) = k_0^2 \left(\varepsilon_2 + \frac{\sigma_r}{\varepsilon_0 \omega} \right) \tag{7.111}$$

where $\mu_0 \varepsilon_0 = 1/c^2$ and $k_0 = \omega/c$ have been used. Setting

$$k_0^2 \left(\varepsilon_1 - \frac{\sigma_i}{\varepsilon_0 \omega} \right) = \beta^2 \tag{7.112}$$

$$k_0^2 \left(\varepsilon_2 + \frac{\sigma_r}{\varepsilon_0 \omega} \right) = \gamma^2 \tag{7.113}$$

we separate Equations 7.110 and 7.111:

$$k_r = \pm \frac{\beta}{\sqrt{2}} \sqrt{1 \pm \sqrt{1 + \left(\frac{\gamma}{\beta} \right)^4}} \tag{7.114}$$

$$k_i = \pm \frac{\beta}{\sqrt{2}} \sqrt{-1 \pm \sqrt{1 + \left(\frac{\gamma}{\beta} \right)^4}} \tag{7.115}$$

But the complex propagation parameter k_m can also be expressed in terms of the complex index of refraction, n_m:

$$k_m = k_0 n_m = k_0(\eta_m + i\kappa_m) \tag{7.116}$$

and n_m is related to the metal dielectric constant ε_m through

$$n_m = \sqrt{\varepsilon_m} = \sqrt{\varepsilon'_m + i\varepsilon''_m} \tag{7.117}$$

Again, equating real and imaginary terms,

$$\eta_m = \sqrt{\frac{\varepsilon'_m}{2}} \sqrt{1 \pm \sqrt{1 + \left(\frac{\varepsilon''_m}{\varepsilon'_m}\right)^4}} \tag{7.118}$$

$$\kappa_m = \sqrt{\frac{\varepsilon'_m}{2}} \sqrt{-1 \pm \sqrt{1 + \left(\frac{\varepsilon''_m}{\varepsilon'_m}\right)^4}} \tag{7.119}$$

and consequently

$$k_r = k_0 \eta_m = k_0 \sqrt{\frac{\varepsilon'_m}{2}} \sqrt{1 \pm \sqrt{1 + \left(\frac{\varepsilon''_m}{\varepsilon'_m}\right)^4}} \tag{7.120}$$

$$k_i = k_0 \kappa_m = k_0 \sqrt{\frac{\varepsilon'_m}{2}} \sqrt{-1 \pm \sqrt{1 + \left(\frac{\varepsilon''_m}{\varepsilon'_m}\right)^4}} \tag{7.121}$$

Comparing Equations 7.120 and 7.121 with Equations 7.114 and 7.115 shows that

$$\varepsilon'_m = \left(\varepsilon_1 - \frac{\sigma_i}{\varepsilon_0 \omega}\right) \tag{7.122}$$

$$\varepsilon''_m = \left(\varepsilon_2 + \frac{\sigma_r}{\varepsilon_0 \omega}\right) \tag{7.123}$$

Equations 7.122 and 7.123 separate the metal dielectric constant into two parts: first ε_1 and ε_2 represent the dispersive and absorptive response of the metal *excluding* the conduction electrons, and second $\sigma_{i,r}/\varepsilon_0\omega$ represents the conducting electrons contribution to the dielectric constant. Substituting for $\sigma_{i,r}/\varepsilon_0\omega$ from Equation 7.93, we have

$$\varepsilon'_m = \left(\varepsilon_1 - \frac{\omega_p^2}{\varepsilon_0 \omega^2}\right) \qquad \varepsilon''_m = \left(\varepsilon_2 + \frac{\Gamma \omega_p^2}{\varepsilon_0 \omega^3}\right) \tag{7.124}$$

If we consider the metal response, *excluding* the conduction electrons, to behave essentially as a lossless dielectric material, we can posit ε_1 and ε_2 as 1 and 0, respectively. This assumption is equivalent to the Drude model in the high frequency limit, Equation 7.71.

7.5.3 Energy flux along the boundary

The energy flux of the surface wave propagating along the x direction:

$$S_x^{s,d} = \frac{1}{2}\mathrm{Re}\left[\mathbf{E}^{s,d} \times \mathbf{H}^{s,d}\right] = \frac{1}{2}E_z^{s,d}\left(H_y^d\right)^* = \frac{1}{2}\frac{k_x^s}{\varepsilon_d\varepsilon_0\omega}H_{0y}^2 \tag{7.125}$$

$$S_x^{s,m} = \frac{1}{2}\mathrm{Re}\left[\mathbf{E}^{s,m} \times \mathbf{H}^{s,m}\right] = \frac{1}{2}E_z^{s,m}\left(H_y^m\right)^* = \frac{1}{2}\mathrm{Re}\left[\frac{k_x^s}{\varepsilon_m\varepsilon_0\omega}H_{0y}^2\right] \tag{7.126}$$

Note that the real part of the metal permittivity is usually a relatively large negative number (see Figure 7.4, for example), and therefore, as shown in Figure 7.12, $S_z^{s,m}$ has the *opposite* sign from $S_z^{x,d}$. The *net* energy flux is the sum of the $S_x^{s,d}$ and $S_x^{s,m}$ components.

7.6   Plasmon surface waves and waveguides

7.6.1 Introduction

The plasmon surface wave at a metal-dielectric interface can be considered a guided wave since it follows the surface. In this sense the interface itself acts as a one-dimensional (1-D) waveguide. We will take up a systematic discussion of transmission lines and waveguides in Chapter 8. Here we show that plasmon-like electromagnetic waves can propagate in the presence of two parallel interfaces separated by a sub-wavelength gap and obtain the properties of this 'gap' plasmon propagating mode. Figure 7.13 shows a slab waveguide with two exterior metallic cladding layers and one interior dielectric layer. The lowest order propagating mode of this guide resembles a surface-wave, SPP, bounded by the two outer metal claddings instead of the usual single surface. Sometimes this propagating mode is called a 'gap plasmon'. Other modes, symmetric (cosine-like) or antisymmetric (sine-like) with respect to the core centreline, are possible as the width of the core increases. If the two metal cladding layers are different (e.g. gold and silver), the guided wave will not be perfectly symmetric or anti-symmetric since the dispersion curves of different metals vary. The allowed modes can be determined by matching the fields in the transverse z direction at the cladding-core boundaries. The matching condition results in a transcendental equation that can be solved numerically for the propagation parameter q as a function of the core width a. Here we carry out in some detail the matching operation for TM modes as an illustrative example.

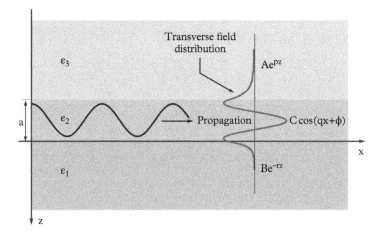

Figure 7.13 *Design of a planar waveguide formed by three zones of different permittivity: $\varepsilon_1, \varepsilon_3$, the 'cladding' or shell can be considered metallic, and ε_2, the 'core', a dielectric. This type of structure is sometimes termed a MIM waveguide (metal-insulator-metal). The complementary structure, IMI waveguides, are also possible. Propagating modes can be symmetric or antisymmetric with respect to the core centreline.*

7.6.2 Field matching at core-cladding boundaries

7.6.2.1 TM polarisation

For TM (transverse magnetic) polarisation we have three field components, H_y, E_x, E_z, in each of the three zones. From the coordinate setup in Figure 7.13 we see that H_y and E_x are parallel to the boundaries, and therefore, must be continuous across them. We assume a harmonic force-field wave $\mathbf{F}$ propagating in the x direction:

$$\mathbf{F}(\mathbf{r}, t) = \mathbf{F}(y, z)e^{i(k_x x - \omega t)} \tag{7.127}$$

and use the two curl equations of Maxwell to find relations among the relevant field components. We further assume that the materials are non-magnetic with relative permeability $\mu = 1$:

$$\frac{\partial H_y}{\partial z} = i\omega\varepsilon_0\varepsilon E_x \tag{7.128}$$

$$\frac{\partial E_x}{\partial z} - ikE_z = i\omega\mu_0 H_y \tag{7.129}$$

$$ikH_y = -i\omega\varepsilon_0\varepsilon E_z \tag{7.130}$$

Now we write down the real field components in the transverse z direction in each of the three zones indicated in Figure 7.13, while expressing the propagation in the x direction as e^{ikx}:

For $z < 0$

$$H_y = Be^{ikx}e^{pz}$$ (7.131)

$$E_x = \frac{-iB}{\omega\varepsilon_3\varepsilon_0}pe^{ikx}e^{pz} = \left(\frac{-ip}{\omega\varepsilon_3\varepsilon_0}\right)H_y$$ (7.132)

$$E_z = \frac{-Bk}{\omega\varepsilon_3\varepsilon_0}e^{ikx}e^{pz} = \left(\frac{-k}{\omega\varepsilon_3\varepsilon_0}\right)H_y$$ (7.133)

For $0 < z < a$

$$H_y = Ae^{ikx}\cos(qz + \varphi)$$ (7.134)

$$E_x = \frac{i}{\omega\varepsilon_2\varepsilon_0}qAe^{ikx}\sin(qz + \varphi)$$ (7.135)

$$E_z = \frac{-k}{\omega\varepsilon_2\varepsilon_0}Ae^{ikx}\cos(qz + \varphi) = \left(\frac{-k}{\omega\varepsilon_2\varepsilon_0}\right)H_y$$ (7.136)

For $z > a$

$$H_y = Ce^{ikx}e^{-rz}$$ (7.137)

$$E_x = \frac{iC}{\omega\varepsilon_1\varepsilon_0}re^{ikx}e^{-rz} = \left(\frac{ir}{\omega\varepsilon_1\varepsilon_0}\right)H_y$$ (7.138)

$$E_z = \frac{-Ck}{\omega\varepsilon_1\varepsilon_0}e^{ikx}e^{-rz} = \left(\frac{-k}{\omega\varepsilon_1\varepsilon_0}\right)H_y$$ (7.139)

Next we write the continuity conditions at each boundary for H_y and E_x.

For $z = 0$

$$B = A\cos\varphi$$ (7.140)

$$\left(\frac{-ip}{\omega\varepsilon_3\varepsilon_0}\right)B = \left(\frac{iq}{\omega\varepsilon_2\varepsilon_0}\right)A\sin\varphi$$ (7.141)

For $z = a$

$$A\cos(qa + \varphi) = Ce^{-ra}$$ (7.142)

$$\frac{iq}{\omega\varepsilon_2\varepsilon_0}A\sin(qa + \varphi) = C\frac{ir}{\omega\varepsilon_1\varepsilon_0}e^{-ra}$$ (7.143)

Rearranging and dividing Equation 7.141 by Equation 7.140 we have:

$$\tan\varphi = -\frac{\varepsilon_2 p}{\varepsilon_3 q}$$ (7.144)

and proceeding similarly with Equations 7.143 and 7.142 we have:

$$\tan(qa + \varphi) = \frac{\tan qa + \tan\varphi}{1 - \tan qa \tan\varphi} = \frac{\varepsilon_2 r}{\varepsilon_1 q}$$ (7.145)

Substituting the right-hand side of Equation 7.144 into Equation 7.145 results in the transcendental equation, the solutions to which define the propagating modes in the waveguide:

$$\tan qa = \frac{\frac{\varepsilon_2 r}{\varepsilon_1 q} + \frac{\varepsilon_2 p}{\varepsilon_3 q}}{1 - \frac{\varepsilon_2^2 pr}{\varepsilon_1 \varepsilon_3 q^2}} \tag{7.146}$$

The propagation parameters, p, q, and r are also related through

$$\frac{\partial E_x}{\partial z} - ikE_z = i\omega\mu_0 H_y \tag{7.147}$$

Substituting the relevant field components into this expression shows that

$$k^2 - p^2 = \varepsilon_3 k_0^2 \tag{7.148}$$
$$q^2 + k^2 = \varepsilon_2 k_0^2 \tag{7.149}$$
$$-r^2 + k^2 = \varepsilon_1 k_0^2 \tag{7.150}$$

where $k_0 = \omega/c$ is the propagation parameter of the free-space propagating wave with frequency ω. Rearranging these expressions, we can define an 'effective' index of refraction that characterises the propagating mode in the core:

$$p = \sqrt{k^2 - \varepsilon_3 k_0^2} = k_0\sqrt{n_{\text{eff}}^2 - n_3^2} \tag{7.151}$$
$$q = \sqrt{\varepsilon_2 k_0 - k^2} = k_0\sqrt{n_2^2 - n_{\text{eff}}^2} \tag{7.152}$$
$$r = \sqrt{k^2 - \varepsilon_1 k_0^2} = k_0\sqrt{n_{\text{eff}}^2 - n_1^2} \tag{7.153}$$

where $n_1 = \sqrt{\varepsilon_1}$, $n_2 = \sqrt{\varepsilon_2}$, $n_3 = \sqrt{\varepsilon_3}$, and $n_{\text{eff}} = k/k_0$.

7.6.2.2 TE polarisation

For TE (transverse electric) polarisation the relevant field components are E_y, H_x, H_z, and carrying through the matching procedure at $z = 0, z = a$, we find the transcendental equation,

$$\tan qa = \frac{\frac{r}{q} + \frac{p}{q}}{1 - \frac{pr}{q^2}} \tag{7.154}$$

As $qa \to 0$ we find a limiting expression in TM polarisation, Equation 7.146, for the channel width a,

$$a = -\frac{\varepsilon_1 p + \varepsilon_3 r}{\varepsilon_2 pr} \tag{7.155}$$

which is physically allowed since $\varepsilon_1, \varepsilon_3$ are negative. Therefore, even as $a \to 0$, a propagating mode always exists. In the case of TE polarisation the limiting expression is

$$a = -\frac{p+r}{pr} \tag{7.156}$$

and since the channel width cannot be negative, this result indicates that TE modes are not supported in the deep subwavelength limit.

7.7   Surface waves at a dielectric interface

7.7.1   Introduction

In the previous sections of this chapter we have focused almost entirely on electromagnetic surface waves at dielectric-metal interfaces, and we have developed the physics of charge density waves as the source of these surface plasmon-polariton fields. We have also observed that the interface between two media can also be considered a 1-D waveguide. In the present section we will again use boundary matching, as we did in Section 7.6.2, to find wave solutions confined to the line defined at the interface between two *dielectric* slabs, one of which is also in contact with a ground plane. We will also find the 1-D waveguide modes of these surface waves for both TM and TE polarisation.

The geometry is shown in Figure 7.14. We will consider the top dielectric, unbounded in the positive x direction, as air or vacuum; and the dielectric slab with thickness t might be glass or silicon. This arrangement can be realised practically on ridge waveguides and is similar to stripline or microstrip geometries popular in integrated circuit design. We study TM and TE modes separately and assume surface wave propagation in the $+z$ direction with propagation parameter β.

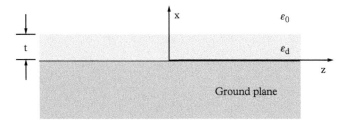

Figure 7.14 *Surface waves propagate along z at the interface between vacuum ε_0 and material dielectric ε_d. The ground plane truncates wave propagation in $-x$ direction and sets field amplitudes to zero at the ε_d-ground plane interface.*

7.7.2 TM modes

TM polarisation specifies the existence of E_x, E_z, H_y polarised amplitudes. We seek solutions to the scalar wave equation in the region above the $\varepsilon_0/\varepsilon_d$ interface and in the region within the ε_d dielectric slab. We match these solutions at the interface.

Above the interface the free-space propagation parameter k_0 in the x–z plane is related to k_x and β by $k_0^2 = k_x^2 + \beta^2$, and the 1-D wave equation in the space above the interface is given by

$$\left(\frac{\partial^2}{\partial x^2} + k_0^2 - \beta^2 \right) \mathscr{E}_z(x) = 0 \tag{7.157}$$

and within the dielectric slab,

$$\left(\frac{\partial^2}{\partial x^2} + k_d^2 - \beta^2 \right) \mathscr{E}_z(x) = 0 \tag{7.158}$$

where $k_d^2 = \varepsilon_d k_0^2$ and $E_z = \mathscr{E}_z(x) e^{i\beta z}$. We assume that there is no variation in the fields along the y direction. Within the dielectric slab, the general solution, in terms of linear combinations of complex travelling waves, is

$$\mathscr{E}_z(x) = A e^{ik_c x} \pm B e^{-ik_c x} \tag{7.159}$$

where

$$k_c^2 = k_d^2 - \beta^2 = \varepsilon_d k_0^2 - \beta^2 \tag{7.160}$$

Above the interface the E-field of the surface wave must decrease exponentially, and the form of the solution must be

$$\mathscr{E}_z^a(x) = C e^{-\kappa x} \tag{7.161}$$

where

$$\kappa^2 = \beta^2 - k_0^2 \tag{7.162}$$

and $\mathscr{E}_z^a(x)$ is the phasor amplitude in the air above the dielectric slab. We anticipate that the propagation velocity of the surface wave will be somewhat slower than a wave of the same frequency propagating in air since a significant fraction of the surface wave amplitude will be immersed in the dielectric slab. Therefore, we expect $\beta > k_0$ and Equation 7.162 is written so that κ is real and can be chosen with a positive sign. This choice assures that Equation 7.161 represents an exponentially decaying amplitude in the $+x$ direction.

At $x = 0$ the boundary condition requires $\mathscr{E}_z^d(0) = 0$. Therefore, $B = A$ and

$$\mathscr{E}_z^d(x) = \pm A \left(e^{ik_c x} - e^{-ik_c x} \right) = \pm A2i \sin k_c x \tag{7.163}$$

where $\mathscr{E}_z^d(x)$ is the amplitude for the surface-wave phasor $E_z(x) = \mathscr{E}_z(x)e^{i\beta z}$. At $x = t$ the amplitudes for the E-field amplitudes on both sides of the interface have to match:

$$\pm A2i \sin k_c t = C e^{-\kappa t} \tag{7.164}$$

The H-field amplitudes have to match as well, and we obtain the H-fields from the E-fields by using the relations obtained in Chapter 8, Section 8.5.1. In the dielectric slab we have

$$\mathscr{H}_y^d(x) = \frac{i}{k_c^2} \omega \varepsilon_0 \varepsilon_d \frac{\partial E_z}{\partial x}$$

$$= \pm \frac{i}{k_c} \omega \varepsilon_0 \varepsilon_d A2i \cos k_c x \tag{7.165}$$

and on the air side

$$\mathscr{H}_y^a(x) = -\frac{i}{\kappa} \omega \varepsilon_0 C e^{-\kappa x} \tag{7.166}$$

At the interface, H-field amplitudes must match. Since $\mathscr{H}_y^a(x)$ has a negative amplitude, we must choose the negative amplitude solution for $\mathscr{H}_y^d(x)$:

$$-\frac{i}{k_c} \omega \varepsilon_0 \varepsilon_d A2i \cos k_c t = -\frac{i}{\kappa} \omega \varepsilon_0 C e^{-\kappa t} \tag{7.167}$$

Now divide Equation 7.164 by Equation 7.167, the two matching equations at the interface. The result is

$$k_c \tan k_c t = \varepsilon_d \kappa \tag{7.168}$$

which provides a relation between the wave vector k_c, the thickness of the dielectric slab t, and the damping constant κ, for a given slab material with dielectric constant ε_d. Equations 7.159 and 7.160 specify that k_c is the wave vector component along x of that part of the surface wave within the dielectric slab. We can write κ in terms of k_c and k_0 by using the fact that the propagation parameter β along z must be the same for the two parts of the surface wave: one part in the dielectric slab and the other in the air. We can therefore eliminate β from Equations 7.160 and 7.162 resulting in the expression

$$k_c^2 + \kappa^2 = k_0^2(\varepsilon_d - 1) \tag{7.169}$$

In principle we could eliminate κ between Equations 7.168 and 7.169, and obtain an expression for k_c as a function of three parameters: the free-space wave vector k_0, the material dielectric constant ε_d, and the slab thickness t. However, Equation 7.168 is a transcendental equation with no simple analytic solution for k_c. The best we can do is find solutions for k_c numerically and plot them to gain some insight into the properties of the TM modes.

One way to plot the solutions is to divide both sides of Equation 7.168 by k_c and rearrange it as

$$\tan(k_c t) = \frac{\varepsilon_d \kappa}{k_c} \tag{7.170}$$

and consider k_c as an independent variable y. We can then plot both sides of the equation, and identify solutions to the transcendental equation as the points of intersection:

$$y_1 = \tan(k_c t) \tag{7.171}$$

$$y_2 = \frac{\varepsilon_d \kappa t}{(k_c t)} = \frac{\varepsilon_d \left[k_0^2 (\varepsilon_d - 1) - k_c^2 \right]^{1/2} t}{(k_c t)} \tag{7.172}$$

Figure 7.15 shows, schematically, the appearance of the two families of curves y_1 and y_2 when plotted against $k_c t$. Clearly, y_1 is just a plot of the tangent function with branches starting at $y_1 = 0$ for $k_c t = 0, \pi, 2\pi \dots$. It is also clear that only positive values of y_1 can provide legitimate solutions to Equation 7.168 since the numerator of y_2 consists of three factors, all of which are intrinsically positive. For fixed ε_d and k_0, y_2, as a function of $(k_c t)$, is a family of curves parametrically increasing with the slab thickness t. From Equation 7.172 we see that $y_2 \to \infty$ as $k_c t \to 0$ and that y_2 crosses 0, becoming pure imaginary, when $k_c = k_0 \sqrt{\varepsilon_d - 1}$. Intersections of y_2 with y_1 within the positive half of the first branch ($0 \le k_c t \le \pi/2$) constitute solutions to the transcendental equation, Equation 7.168, belonging to the lowest TM mode, TM_0. We see immediately that the TM_0 mode has no cutoff for k_c as $k_c t \to 0$. For fixed k_c, t and as k_0 increases, y_2 will begin to intersect y_1 both in the first and in the second positive branch ($\pi \le k_c t \le 3\pi/2$). These intersections in the second branch constitute TM_1 solutions to Equation 7.168. Higher order TM modes begin to propagate as the $k_c = k_0 \sqrt{\varepsilon_d - 1}$ upper limit increases with k_0. From Figure 7.15 it is clear that the cutoff points for $TM_1, TM_2, \dots$ occur at $k_c t = n\pi$ with $n = 1, 2, 3, \dots$. The general expression for the cutoff frequency (in Hz units) is given by,

$$v_c = \frac{k_c}{2\pi \sqrt{\mu \varepsilon}} \tag{7.173}$$

where $1/\sqrt{\mu \varepsilon}$ is the local, 'effective' speed of light. For the TM modes of this surface wave, therefore, we have:

$$v_c = \frac{nc}{2t\sqrt{\varepsilon_d - 1}} \qquad n = 0, 1, 2, \dots \tag{7.174}$$

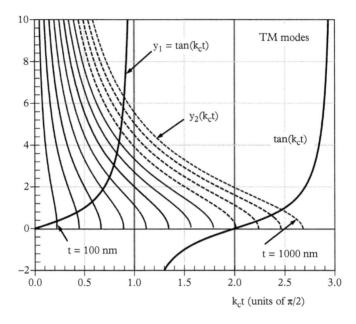

Figure 7.15 *The expressions y_1, y_2 plotted as a function of $k_c t$. The function y_2 is plotted for a range of thicknesses t from 100 nm to 1000 nm. Dashed curves show four representative solutions where the thickness t permits modes TM_0 and TM_1. Note that TM_0 has no cutoff thickness but TM_1 cuts off at $k_c t = \pi$. Intersection of the curves $y_1, y_2 > 0$ constitute solutions to the transcendental equation, Equation 7.170.*

where $\varepsilon_d - 1$ represents an 'effective' dielectric constant for the TM modes in the 1-D surface wave.

7.7.3   TM field solutions

Once k_c has been found for a fixed set of parameters k_0, t, ε_d, the TM fields E_z, E_x, H_y can be obtained from Equations 7.163 and 7.164, and the relations in Section 8.5.1:

$$E_z(x, z) = \pm A' \sin(k_c x) e^{i\beta z} \qquad\qquad 0 \le x \le t \qquad\qquad (7.175)$$

$$E_z(x, z) = \pm A' \sin(k_c t) e^{-\kappa(x-t)} e^{i\beta z} \qquad x \ge t \qquad\qquad (7.176)$$

where the constant A' replaces the constant factors in Equation 7.163, $A' = A2i$.

For E_x we find that

$$E_x(x, z) = \pm \frac{i\beta}{k_c} A' \cos(k_c x) e^{i\beta z} \qquad\qquad 0 \le x \le t \qquad\qquad (7.177)$$

$$E_x(x, z) = \mp \frac{i\beta}{\kappa} A' \sin(k_c t) e^{-\kappa(x-t)} e^{i\beta z} \qquad x \ge t \qquad\qquad (7.178)$$

and for H_y that

$$H_y(x, z) = \pm \frac{i\omega\varepsilon_0\varepsilon_d}{k_c} A' \cos(k_c x) e^{i\beta z} \qquad 0 \le x \le t \qquad\qquad (7.179)$$

$$H_y(x, z) = \mp \frac{i\omega\varepsilon_0}{\kappa} A' \sin(k_c t) e^{-\kappa(x-t)} e^{i\beta z} \qquad x \ge t \qquad\qquad (7.180)$$

7.7.4   TE modes

In contrast to the case of plasmon surface waves, the 1-D waveguide between two dielectrics can support TE waves as well as TM waves. The fields specified by TE polarisation are H_z, H_x, E_y, and the scalar wave equation above the dielectric slab is written as

$$\left(\frac{\partial^2 x}{\partial x^2} - \kappa^2 \right) \mathcal{H}_z = 0 \qquad\qquad (7.181)$$

and, since the dielectric slab is non-magnetic, within the slab the wave equation is

$$\left(\frac{\partial^2 x}{\partial x^2} + k_0^2 - \beta^2 \right) \mathcal{H}_z = 0 \qquad\qquad (7.182)$$

The form of the solutions within the slab is

$$\mathcal{H}_z^d(x) = A e^{ik_c x} \pm B e^{-ik_c x} \qquad\qquad (7.183)$$

with $k_c^2 = k_0^2 - \beta^2$. On the air side above the slab we again posit an exponentially decreasing amplitude:

$$\mathcal{H}_z^a(x) = C e^{-\kappa x} \qquad\qquad (7.184)$$

Again, at $x = 0$, the E-field parallel to the ground plane must vanish. Obtaining the E-field from the H-field using the relations in Section 8.5.1, we have

$$\mathcal{E}_y^d(x) = -\frac{i}{k_c^2} \omega\mu_0 \frac{\partial \mathcal{H}_z}{\partial x}$$

$$= \mp \frac{\omega\mu_0}{k_c} A 2i \sin k_c x \qquad\qquad (7.185)$$

and in the air space above the dielectric slab,

$$\mathscr{E}_y^a(x) = \frac{i}{\kappa}\omega\mu_0 C e^{-\kappa x} \qquad (7.186)$$

The expression for $\mathscr{E}_y^d$ in Equation 7.185 implies that the general solution for $\mathscr{H}_z^d(x)$ in Equation 7.183 specialises to

$$\mathscr{H}_z^d(x) = \pm A2 \cos k_c x \qquad (7.187)$$

and that, therefore, the H-field is maximum at $x = 0$, the ground-plane boundary. At first thought this condition might seem perplexing since no field should exist in the ground plane. What the boundary condition really means is that an infinitesimal distance above the boundary, the H-field amplitude is $\pm 2A$, and an infinitesimal distance below the boundary, the H-field is null.

At the boundary $x = t$, $\mathscr{H}_z$ and $\mathscr{E}_y$ must match:

$$\pm A2 \cos k_c t = C e^{-\kappa t} \qquad (7.188)$$

$$\mp \frac{\omega\mu_0}{k_c} A2i \sin k_c t = \frac{i}{\kappa}\omega\mu_0 C e^{-\kappa t} \qquad (7.189)$$

Dividing Equation 7.188 by Equation 7.189 and rearranging results in

$$-\cot k_c t = \frac{\kappa}{k_c} \qquad (7.190)$$

Again, we set

$$y_1 = -\cot k_c t \qquad (7.191)$$

$$y_2 = \frac{\kappa t}{k_c t} = \sqrt{\frac{k_0^2(\varepsilon_d - 1)}{k_c^2} - 1} \qquad (7.192)$$

and plot y_1, y_2 with $k_c t$ the independent variable. Solutions to Equation 7.190 are at the positive intersections of y_1, y_2. The general form of the solutions is shown in Figure 7.16. In the first zone $0 \le k_c t \le \pi/2$, no TE modes are possible because all the y_1 solutions are negative and therefore unphysical. The first TE mode, TE$_1$, appears in the zone $\pi/2 \le k_c t \le \pi$. From Figures 7.15 and 7.16 we see that the energy ordering of the first three modes is TM$_0$, TE$_1$, and TM$_1$. Subsequent TE modes first appear at $k_c t = 3\pi/2, 5\pi/2, 7\pi/2 \ldots$. The cutoff frequencies are then:

$$v_c = \frac{nc}{4t\sqrt{\varepsilon_d - 1}} \qquad n = 1, 3, 5, \ldots \qquad (7.193)$$

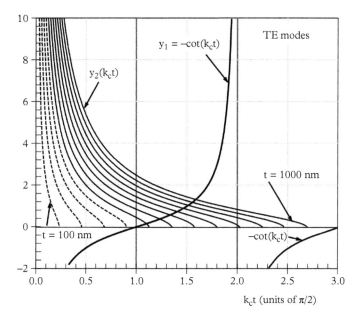

Figure 7.16 *The expressions y_1, y_2 plotted as a function of $k_c t$. The function y_2 is plotted for a range of thicknesses t from 100 nm to 1000 nm. The first four dashed curves indicate representative values of $y_2(k_c t)$ that do not admit solutions. Cutoff for the first allowed TE mode (TE$_1$) occurs at $k_c t = \pi/2$. Intersection of the curves in the y domain where $y_1, y_2 > 0$ constitute solutions to the transcendental equation, Equation 7.190.*

7.7.5 TE field solutions

Once k_c has been found for a fixed set of parameters k_0, t, ε_d, the TE fields H_z, H_x, E_y can be obtained from Equations 7.184 – 7.187 and the relations in Section 8.5.1:

$$H_z(x, z) = \pm 2A \cos(k_c x) e^{i\beta z} \qquad\qquad 0 \le x \le t \qquad\qquad (7.194)$$

$$H_z(x, z) = \pm 2A \cos(k_c t) e^{-\kappa(x-t)} e^{i\beta z} \qquad x \ge t \qquad\qquad (7.195)$$

For $H_x(x, z)$:

$$H_x(x, z) = \mp \frac{\beta}{k_c} i2A \sin(k_c x) e^{i\beta z} \qquad\qquad 0 \le x \le t \qquad\qquad (7.196)$$

$$H_x(x, z) = \pm \frac{i\beta}{\kappa} 2A \cos(k_c t) e^{-\kappa(x-t)} e^{i\beta z} \qquad x \ge t \qquad\qquad (7.197)$$

and for $E_y(x, z)$:

$$E_y(x, z) = \mp \frac{i\omega\mu_0}{k_c} 2A \sin(k_c x) e^{i\beta z} \qquad 0 \le x \le t \qquad (7.198)$$

$$E_y(x, z) = \pm \frac{i\omega\mu_0}{\kappa} 2A \cos(k_c t) e^{-\kappa(x-t)} e^{i\beta z} \qquad x \ge t \qquad (7.199)$$

As we can see from Figures 7.15 and 7.16, the energy ordering of the first three modes is TM_0, TE_1, and TM_1. Figures 7.17 and 7.18 show two illustrative examples of the dielectric guided waves for a structure in which a 200 nm silicon slab is sandwiched

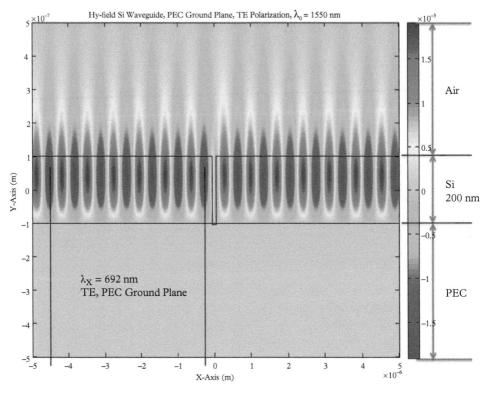

Figure 7.17 *Guided waves propagating along x in a 200 nm silicon slab. The figure shows the H_y component in a TE mode. The top layer is air and the bottom layer is a ground plane perfect electric conductor (PEC). The guided waves are launched from a 100 nm wide slit at the centre of the structure. The source is a total-field-scattered-field (TFSF) source with $\lambda_0 = 1550$ nm impinging on the slit along the y (vertical) axis. The effective wavelength of the guided wave is $\lambda_x = 692$ nm, and therefore the effective refractive index of the guiding structure is $n_{eff} = 2.24$. The two black vertical lines on the left-hand side are reference markers for measuring the effective wavelength. Note that the z- and x-axes are reversed from Figure 7.14.*

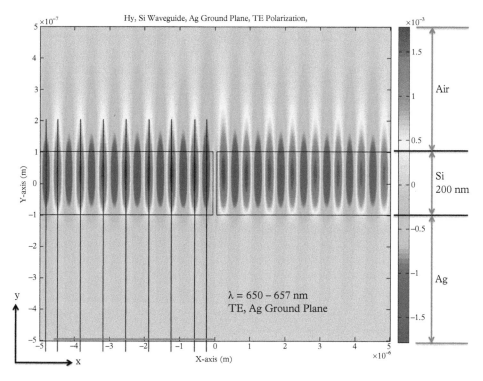

Figure 7.18 *Guided waves propagating along x in a 200 nm silicon slab. The figure shows the H_y component in a TE mode. The top layer is air and the bottom layer is a ground plane of silver (Ag). The guided waves are launched from a 100 nm wide slit at the centre of the structure. The source is a total-field-scattered-field (TFSF) source with $\lambda_0 = 1550$ nm impinging on the slit along the y (vertical) axis. The effective wavelength of the guided wave is $\lambda_x = 654$ nm, and therefore the effective refractive index of the guiding structure is $n_{eff} = 2.37$. The black vertical lines on the left-hand side are reference markers for measuring the effective wavelength. Note that the z- and x-axes are reversed from Figure 7.14.*

between air and a ground plane or 'perfect electric conductor' (PEC). The figures are plots of numerical solutions to Maxwell's equations using the finite-difference, time-domain (FDTD) method.

7.8   Summary

The chapter begins with a brief historical sketch of the importance of surface waves to the early days of radio communication before launching into the development of TE polarised surfaces (that do not exist) and TM polarised surface waves (that do). The dispersion models of the free-electron gas and the Drude metal are then introduced. The component and net Poynting vectors are then presented both for lossless and lossy (real) metals. Surface waves are guided waves, and in anticipation of the chapter on waveguides, the first discussion of boundary wave matching is worked out. The chapter ends with a discussion of waves at the boundary of dielectrics subject to a ground plane. This situation is reminiscent of the 'stripline' geometry used in the electrical engineering of micro-electronics. The waves at the top surface of the stripline are evanescent, and in this sense can be considered 'surface waves' although they are really part of the propagating waves inside the Si waveguide.

7.9 Exercises

1. It is a well-known fact that aluminium (Al), soon after exposure to air, exhibits a transparent aluminium oxide layer (Al_2O_3). This oxide layer is about 100 nm thick and has an index of refraction $n_{Al_2O_3} = 1.77$. What would be the effective index of refraction of a surface plasmon polariton (SPP) wave propagating at the metal/metal oxide interface if the incident light has a wavelength of 400 nm? The index of refraction of Al at $= 400$ nm is $n = 0.381 + i4.883$.

2. A SPP wave propagates at the interface between air and a gold (Au) layer. Calculate the phase and group velocity if the incident wavelength is 780 nm. In the common case of good conductors where $|\varepsilon'_m| \gg \varepsilon''$, the group velocity is, to good approximation,

$$v_g \simeq \mathrm{Re}\left[c \left\{ \sqrt{\frac{\varepsilon_d \varepsilon'_m}{\varepsilon_d + \varepsilon'_m}} + \frac{\omega}{2} \cdot \left[\frac{\varepsilon_d \varepsilon'_m}{\varepsilon_d \varepsilon'_m} \right]^{3/2} \cdot \left[\frac{1}{\varepsilon'^2_m} \left(\frac{d\varepsilon'_m}{d\omega} \right) + \frac{1}{\varepsilon'^2_d} \left(\frac{d\varepsilon'_d}{d\omega} \right) \right] \right\}^{-1} \right]$$

(7.200)

and from the Drude model the complex dielectric constant for Au can be expressed as

$$\varepsilon'_m(\omega) = 1 - \left(\frac{\omega_p}{\omega} \right)^2$$

(7.201)

$$\varepsilon''_m = \left(\frac{\omega_p^2}{\omega^3 \tau} \right)$$

(7.202)

The bulk plasmon resonance frequency is denoted by ω_p, and τ is the characteristic damping time in the metal. For Au the values are $\omega_p = 1.37 \times 10^{16}\,\mathrm{s^{-1}}$ and $\tau = 2.47 \times 10^{-14}\,\mathrm{s}$.

3. Calculate the spatial dependence of the electric and magnetic field amplitudes of an SPP wave propagating at the interface between a silver (Ag) layer and a glass substrate. The incident wave is 780 nm. Use the Drude model of metals to calculate the permittivity of Ag at the given wavelength. The bulk plasma frequency of Ag is $\omega_p = 1.40 \times 10^{16}$ rad/s and the relaxation rate is $\Gamma = 1.0 \times 10^{14}\,\mathrm{s^{-1}}$. The dielectric constant of glass is taken to be $\varepsilon = 2.25$.

4. With reference to the preceding problem, calculate the SPP energy flux along the propagation direction (x) at two points: $x = 0$ and $x = 100$ nm.

5. An SPP wave propagates along the x direction at an air-gold interface. The free-space wavelength of the incident wave exciting the SPP is 632 nm, and the origin of the SPP wave is at $x = 0$. Calculate the spatial extent of the SPP amplitude along the z direction, normal to the interface, at $x = 0$. Determine the 1/e points for the amplitude and intensity on the air side and on the metal side of the interface.

6. Suppose a plane wave (free-space wavelength $\lambda = \lambda_0$) is normally incident on a dielectric-metal interface. Consider the energy density at the interface between the two media: a dielectric with permittivity $\varepsilon_0 \varepsilon_d$ and a metal with permittivity $\varepsilon_0 \varepsilon_m$. Write the relation between the electric and magnetic energy densities on the dielectric side and the metal side of the interface. Assume the dielectric is lossless but $\varepsilon_m = \varepsilon'_m + i\varepsilon''_m$.

7. In Section 7.7 we developed the properties of surface waves propagating at the interface between two dielectrics, one of which was grounded. Follow the same procedure of field matching at the boundaries to find the surface waves at an air-dielectric interface where the dielectric is in contact with a perfect electrical conductor (PEC). Find the transcendental equations for TM polarisation, analogous to Equation 7.168.

7.10 Further reading

1. A. Sommerfeld, *Partial Differential Equation in Physics*, Lectures on Theoretical Physics Vol. VI, Chap. VI, Problems of Radio Academic Press, New York (1964).

2. A. Sommerfeld, *Ueber die fortpflanzung elektrodynamischer wellen längs eines drahte. Ann Phys Chem* vol 67, pp. 233–290 (1899).

3. J. Zenneck, *Fortplfanzung ebener elektromagnetischer Wellen längs einer ebenen Leiterfläche. Ann Phys* vol 23, pp. 846–866 (1907).

4. H. Raether, *Surface plasmons on Smooth and Rough Surfaces and on Gratings*. Springer, Berlin (1988).

5. M. Born and E. Wolf, *Principles of Optics*, 6th edition, Pergamon Press (1993).

6. T-I. Jeon and D. Grischkowsky, *THz Zenneck surface wave (THz surface plasmon) propagation on a metal sheet*. Applied Physics Letters, vol. 88, p. 061113 (2006).

7. D. M. Pozar, *Microwave Engineering*, 3rd edition, John Wiley & Sons (2005).

8. P. B. Johnson and R. W. Christy, *Optical Constants of the Noble Metals*, Phys Rev B vol 6, pp. 4370–4379 (1972).

9. M. A. Ordal, R. J. Bell, R. W. Alexander, L. L. Long, and M. R. Querry, *Optical properties of fourteen metals in the infrared and far infrared: Al, Co, Cu, Au, Fe, Pb, Mo, Ni, Pd, Pt, Ag, Ti, V and W*. Appl Optics vol 24, pp. 4493–4499 (1985).

8

Transmission Lines and Waveguides

8.1 Introduction

In Chapter 2 we summarised the elements of electromagnetics in terms of electric and magnetic vector force fields related by Maxwell's equations. We saw in Section 2.4 that the two time-varying equations, Faraday's law, Equation 2.27 and the Maxwell–Ampère law, Equation 2.28, give rise to electromagnetic wave solutions that propagate through a medium or, as we saw in Chapter 7, at the interface between media. These solutions are generally distributed throughout a spatial extent that is large with respect to the characteristic wavelength.

In contrast, the conventional radio and microwave circuit theory of electrical engineering considers time-varying electromagnetic phenomena from the standpoint of various combinations of *lumped elements* (capacitors, inductors, and resistors), localised in space, and linked by interconnections of negligible impedance. Furthermore, the spatial extent of these circuit elements, together with their sources of voltage and current, oscillating from tens of kilohertz to hundreds of megahertz, is usually much smaller than the characteristic wavelength. The result is that all the elements of the circuits are subject to the same time dependence without retardation. This time behaviour is termed *quasi-static*, and the physical arrangement of the circuit can be said to be *subwavelength*. The subwavelength scaling similarity between dielectric and metallic nanostructures interacting with optical driving fields and electrical circuits driven by conventional voltage and current oscillators suggests that circuit analysis might find useful application in the design of functional plasmonic and photonic devices. The main goal of this chapter, therefore, is to explore this possibility.

8.2 Elements of conventional circuit theory

Circuit theory is essentially the application of Maxwell's equations to problems commonly encountered in electrical engineering. The conditions that validate circuit theory are: (1) that electrical effects happens instantaneously throughout the circuit—the quasi-static approximation; (2) that the *net* charge on every component is null; and (3) that

Light-Matter Interaction. Second Edition. John Weiner and Frederico Nunes.
© John Weiner and Frederico Nunes 2017. Published 2017 by Oxford University Press.

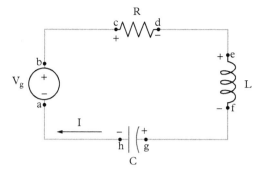

Figure 8.1 *Schematic of an electrical circuit showing lumped elements of resistor, inductor, and capacitor in series with a voltage source. Also shown is the direction of current relative to the polarity of the voltage source.*

magnetic coupling between or among lumped components of the circuit are negligible. A typical circuit comprising a source, resistor, capacitor, and inductor are shown in Figure 8.1.

8.2.1 Kirchhoff's rules

The starting point in circuit analysis are the two rules annunciated by Gustav Kirchhoff in 1845. The rules are as follows. (1) The scalar sum of the electric potential differences V_i around any closed circuit loop is null:

$$\sum_i V_i = 0 \tag{8.1}$$

(2) The scalar sum of charge currents I_i flowing out of a junction node is null:

$$\sum_i I_i = 0 \tag{8.2}$$

8.2.1.1 Kirchhoff's voltage rule

We consider first the voltage rule, which is the more relevant of the two for our purposes. Equation 8.1 derives from the differential form of Faraday's law. By applying Stokes' theorem to Equation 2.27 we have

$$\int_S (\nabla \times \mathbf{E}) \cdot d\mathbf{S} = \oint \mathbf{E} \cdot d\mathbf{l} = - \int_S \frac{\partial \mathbf{B}}{\partial t} \cdot d\mathbf{S} \tag{8.3}$$

where $\mathbf{E}, \mathbf{B}$ are the usual electric and magnetic induction fields, $\mathbf{S}$ is the outward surface normal through which the curl of $\mathbf{E}$ protrudes, and $\mathbf{l}$ is the line around the boundary of $\mathbf{S}$, the positive direction of which is taken by using the right-hand rule. With respect to Figure 8.1, we take the line integral around the circuit and define the voltage difference between two reference points a, b as

$$V_{ba} = - \int_a^b \mathbf{E} \cdot d\mathbf{l} \tag{8.4}$$

Then using Equation 8.3 we have

$$-\int_a^b \mathbf{E} \cdot d\mathbf{l} - \int_c^d \mathbf{E} \cdot d\mathbf{l} - \int_e^f \mathbf{E} \cdot - \int_g^h \mathbf{E} \cdot d\mathbf{l} = \int_S \frac{\partial \mathbf{B}}{\partial t} \cdot d\mathbf{S} = \frac{\partial}{\partial t} \int_S \mathbf{B} \cdot d\mathbf{S} \qquad (8.5)$$

or

$$V_0(t) + V_{dc} + V_{fe} + V_{hg} = \frac{\partial}{\partial t} \int_S \mathbf{B} \cdot d\mathbf{S} \qquad (8.6)$$

where $V_0(t)$ is the voltage source connected to terminals a, b, V_{dc} is the resistive voltage drop, and V_{fe} and V_{hg} the inductive and capacitive voltage drops, respectively. In order to identify Equation 8.6 with the Kirchhoff rule we have to make two assumptions: (1) that there is no changing magnetic flux traversing the plane of the circuit loop. In simple circumstances where we only consider R, L, C lumped circuits, this condition is not hard to realise. Even if there is a changing magnetic field present, it can usually be incorporated into the inductive voltage term to make an 'effective' voltage drop. (2) Voltage drops along the wire connecting the lumped elements in the loop are negligible.

8.2.1.2   Resistance

We know from Ohm's law the current density $\mathbf{J}$ through a conductor is proportional to the E-field along the conductor, $\mathbf{J} = \sigma \mathbf{E}$, where σ is the conductivity. Then the voltage drop across the resistor is

$$V_{dc} = -\int_c^d \mathbf{E} \cdot d\mathbf{l} = -\int_c^d \frac{\mathbf{J}}{\sigma} \cdot d\mathbf{l} \qquad (8.7)$$

We also know that $\int \mathbf{J} \cdot d\mathbf{l}$ can be written in terms of the total current I flowing through the conductor and the resistivity ρ, the inverse of the conductivity:

$$V_{dc} = -\int_c^d \frac{I}{A} \rho \, dl = -IR \qquad (8.8)$$

where the resistance R is defined as

$$R = \int_c^d \frac{\rho}{A} \, dl \qquad (8.9)$$

and A is the cross-sectional area of the conductor. The familiar expression Equation 8.8 is valid at low frequencies where the current flows through the conductor uniformly over A. At high frequencies the current in not uniform across A, passing only near the surface and within the skin depth.

8.2.1.3   *Induction*

Here we apply Faraday's law to the local inductive element labelled L in Figure 8.1 (not the whole circuit loop). The integral has to be over a closed path linking the terminals e, f. The first branch of the loop goes through the current-carrying coil from e to f. The second branch is the return path from f to e along an arbitrary path in the space outside the coil. This second branch contributes nothing to the voltage drop across the coil:

$$\oint \mathbf{E} \cdot d\mathbf{l} = \int_e^f \mathbf{E} \cdot d\mathbf{l} + \left[\int_f^e \mathbf{E} \cdot d\mathbf{l} = 0 \right] \tag{8.10}$$

Then,

$$V_{fe} = \int_e^f \mathbf{E} \cdot d\mathbf{l} = -\frac{\partial}{\partial t} \int_{\mathbf{S}} \mathbf{B} \cdot d\mathbf{S} \tag{8.11}$$

With the definition of the inductance L as the integral of the magnetic induction $\mathbf{B}$ over the surface surrounding the inductive element, per unit current I passing through the coil of the inductor, we have

$$L = \frac{\int \mathbf{B} \cdot d\mathbf{S}}{I} \tag{8.12}$$

Therefore:

$$V_{fe} = -\frac{\partial (LI)}{\partial t} = -L\frac{\partial I}{\partial t} \tag{8.13}$$

8.2.1.4   *Capacitance*

We assume that the capacitive element can be represented by a parallel plate capacitor. From elementary electrostatics we know that the definition of capacitance is the total charge Q accumulated on one plate divided by the voltage V_{hg} across the plate:

$$C = \frac{Q}{V} \tag{8.14}$$

But

$$Q = \int I \, dt \tag{8.15}$$

and thus the voltage drop across the capacitor in terms of the capacitance C and the current looping through the circuit is

$$V_{hg} = -V_{gh} = -\frac{\int I \, dt}{C} \tag{8.16}$$

Putting all three lumped circuit elements together with the positive voltage source $V_{ba} = V_s$ and using the Kirchhoff voltage rule, we have

$$V_s - IR - L\frac{dI}{dt} - \frac{1}{C}\int I\,dt = 0 \qquad (8.17)$$

8.2.1.5 *Kirchhoff's current rule*

Kirchhoff's current rule, that the charge currents I_i flowing into and out of a circuit junction sum to zero, is illustrated in Figure 8.2. The basis for Equation 8.2 is essentially the charge conservation relation that derives from the Maxwell–Ampère equation

$$\nabla \times \mathbf{H} = \mathbf{J} + \frac{\partial \mathbf{D}}{\partial t} \qquad (8.18)$$

Taking the divergence of both sides of this equation, and remembering that div · curl = 0 we find that

$$\nabla \cdot \mathbf{J} + \frac{\partial \nabla \cdot \mathbf{D}}{\partial t} = 0$$

$$\nabla \cdot \left(\mathbf{J} + \frac{\partial \mathbf{D}}{\partial t}\right) = 0 \qquad (8.19)$$

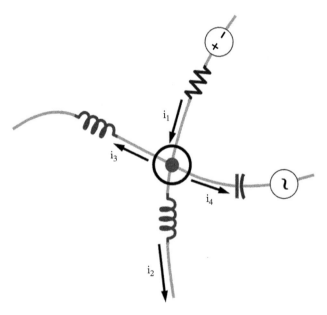

Figure 8.2 *Circuit junction point from which currents enter and leave. The Kirchhoff current rule states that the sum of the currents into and out of the junction must be zero.*

Then using Stokes' theorem we can write

$$\int_A \nabla \cdot \left(\mathbf{J} + \frac{\partial \mathbf{D}}{\partial t} \right) dA = \oint_c \left(\mathbf{J} + \frac{\partial \mathbf{D}}{\partial t} \right) \cdot d\mathbf{l} = 0 \tag{8.20}$$

Equation 8.20 says that the sum of the conduction current density $\mathbf{J}$ and displacement current density $\partial \mathbf{D}/\partial t$ around a closed loop c defining a surface A is null. From a circuit point of view, the closed loop can be considered any loop around a junction point into and out of which currents flow along highly spatially localised wires. The integral over the conduction current can be considered a sum over the current I_i of wire i by writing

$$\oint_c \mathbf{J} \cdot \delta(l_i - l) \, d\mathbf{l} = \sum_i I_i \tag{8.21}$$

For harmonically time-varying circuits the second term on the right in Equation 8.20 can be written as

$$\oint_c \frac{\partial \mathbf{D}}{\partial t} \cdot d\mathbf{l} = \oint_c -i\omega \mathbf{D} \cdot d\mathbf{l} \simeq 0 \tag{8.22}$$

since at radio and microwave frequencies the displacement current is negligible compared to the conduction current. At optical frequencies the displacement current might become significant and in this regime Kirchhoff's current rule has to be checked on a case-by-case basis. We shall assume as a working assumption that Kirchhoff's current rule is valid and write

$$\oint_c \left(\mathbf{J} + \frac{\partial \mathbf{D}}{\partial t} \right) \cdot d\mathbf{l} \simeq \sum_i I_i \simeq 0 \tag{8.23}$$

8.3 Transmission lines

8.3.1 Lumped-element circuit analysis of a transmission line

In conventional circuit analysis, lumped circuit elements such as resistors, inductors, and capacitors are assumed to be dimensionally 'subwavelength', i.e. much smaller than the characteristic wavelength of the driving field. In contrast, a transmission line is a circuit entity transporting voltage and current waves over distances much greater than a wavelength. However, Kirchhoff's two rules can be used to analyse a transmission line as a succession of lumped circuit elements. A two-wire transmission line extending along the z direction can be represented by the diagram in Figure 8.3 that shows one segment of a repeating circuit laid out along Δz. The lumped elements R, L, and C represent the resistance, inductance, and capacitance per unit length, respectively. The fourth quantity, G, represents the 'shunt conductance' per unit length due to dielectric absorption between the two conductors. The series resistance in the two conductors is due to the finite (but very high) conductivity of the metal, while shunt conductance

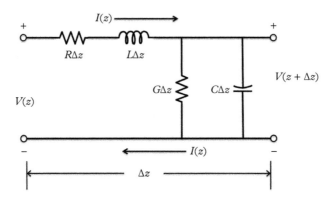

Figure 8.3 *Lumped circuit element representation of a transmission line repeating segment with length Δz. The transmission line extends along the z direction.*

is due to the finite (but very low) conductivity of the insulating dielectric. These four quantities divide into two groups: R and G are dissipative and represent loss, while L and C are reactive and represent stored energy.

We can use the Kirchhoff rules to analyse this circuit segment. Proceeding around the circuit and applying the voltage rule (Equation 8.1) we have

$$V(z, t) - R\Delta z I(z, t) - L\Delta z \frac{\partial I(z, t)}{\partial t} - V(z + \Delta z) = 0 \tag{8.24}$$

while the current rule (Equation 8.2) gives

$$I(z, t) - G\Delta z V(z + \Delta z, t) - C\Delta z \frac{\partial V(z + \Delta z)}{\partial t} - I(z + \Delta z, t) = 0 \tag{8.25}$$

Now divide both equations by Δz and take the limit as $\Delta z \to 0$. The result is

$$\frac{\partial V}{\partial z} = -RI(z, t) - L\frac{\partial I(z, t)}{\partial t} \tag{8.26}$$

$$\frac{\partial I}{\partial z} = -GV(z, t) - C\frac{\partial V(z, t)}{\partial t} \tag{8.27}$$

As usual we assume harmonic time variation so the partial derivative terms with respect to t simplify to

$$\frac{dV}{dz} = -[R - i\omega L]\, I(z) \tag{8.28}$$

$$\frac{dI}{dz} = -[G - i\omega C]\, V(z) \tag{8.29}$$

These two equations can be uncoupled to yield

$$\frac{d^2 V(z)}{d z^2} - \tilde{\beta}^2 V(z) = 0 \tag{8.30}$$

$$\frac{d^2 I(z)}{d z^2} - \tilde{\beta}^2 I(z) = 0 \tag{8.31}$$

with

$$\tilde{\beta}^2 = (R - i\omega L)(G - i\omega C) \tag{8.32}$$

$$\tilde{\beta} = \pm\sqrt{(R - i\omega L)(G - i\omega C)} \tag{8.33}$$

and

$$\tilde{\beta} \equiv i\beta - \gamma \tag{8.34}$$

The solutions are very reminiscent of the plane wave solutions of Chapter 3, Section 3.1. In the positive z space the voltage and current solutions look like damped travelling waves propagating in the forward z direction. The voltage wave forward travelling takes the form

$$V = V_0^+ e^{\tilde{\beta} z} = V_0^+ e^{i(\beta + i\gamma)z} \tag{8.35}$$

where β is the propagation parameter[1] and γ the dissipation term. In the negative z space we have backwards travelling waves:

$$V = V_0^- e^{-\tilde{\beta} z} = V_0^- e^{-i(\beta + i\gamma)z} \tag{8.36}$$

The general solution is some linear combination of forward and backward travelling waves:

$$V = V_0^+ e^{\tilde{\beta} z} + V_0^- e^{-\tilde{\beta} z} \tag{8.37}$$

Usually the amplitude of the backward propagating wave V_0^- will not be equal to the amplitude of the forward wave V_0^+, and if the backward wave is reflected (the usual case), the forward and backward amplitudes will be related by a reflection coefficient. From Equation 8.28 we can write $I(z)$ in terms of V_0^+ and V_0^-:

$$I(z) = I_0^+ + I_0^- = -\frac{dV/dz}{R - i\omega L} = -\frac{i\beta - \gamma}{R - i\omega L} \left[V_0^+ - V_0^- \right] \tag{8.38}$$

---

[1] Common usage in optics and electromagnetics is to denote k for the propagation parameter. In transmission line theory, developed by and for electrical engineering, the term β is more common.

In analogy to Equation 3.28 we define an impedance Z_0 as the ratio of the forward voltage amplitude to the forward current amplitude:

$$Z_0 = \frac{V_0^+}{I_0^+} \tag{8.39}$$

Substituting from Equation 8.38, and using Equations 8.32 and 8.34, we have

$$Z_0 = -\frac{R - i\omega L}{i\beta - \gamma} = \sqrt{\frac{R - i\omega L}{G - i\omega C}} \tag{8.40}$$

where we have used the negative root of $\tilde{\beta}^2$ from Equation 8.33 in the denominator of Equation 8.40. The transmission line current can now be written in terms of the voltage and the characteristic impedance:

$$I(z) = \frac{1}{Z_0} \left[V_0^+ e^{\tilde{\beta}z} + V_0^- e^{-\tilde{\beta}z} \right] \tag{8.41}$$

In most practical transmission lines R and G are very small, and in the limit of a lossless transmission line we find

$$Z_0 \to \sqrt{\frac{L}{C}} \tag{8.42}$$

8.3.2   Lossless plane-parallel transmission line

If we start with a lossless transmission line as the point of departure, we can calculate the inductance per unit length, a function of the transmission line geometry, directly from Faraday's law. Figure 8.4 shows the magnetic flux lines between two parallel plates carrying a constant current I in opposite directions. The separation between the plates is given by s and the width of the plates is w. The ratio of w/s is sufficiently great that the magnetic flux density $\mathbf{B}$ is mostly concentrated between the plates and can be considered constant. The magnetic field between the plates can be related to the current flowing in one of the plate conductors through the integral form of the Maxwell–Ampère law, Equation 2.28. Using Stokes' theorem as we did in Equation 8.3 for Faraday's law, we write

$$\int_S (\nabla \times \mathbf{H}) \cdot dS = \oint \mathbf{H} \cdot dl = \int \mathbf{J} \cdot dS \tag{8.43}$$

Applying this relation to the cross-sectional area ABCD of Figure 8.4 we have

$$H_0 w = I \tag{8.44}$$

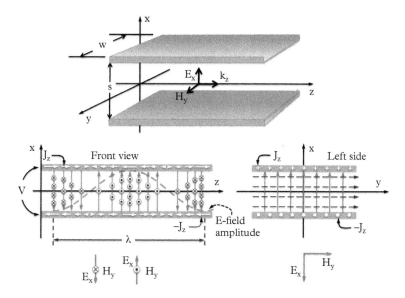

Figure 8.4 *Idealised parallel-plate transmission line with no resistive losses in the current-carrying plates and no shunt conductance between the plates.*

where $I = \int J \cdot dS$, the integral of the current density over ABCD. Now we substitute the definition of inductance L from Equation 8.12 into Equation 8.44, taking the $\int B \cdot dS$ over the surface shown in Figure 8.4. The inductance per unit length along z is then

$$\frac{dL}{dz} = \bar{L} = \mu_0 \frac{s}{w} \tag{8.45}$$

8.3.2.1   *Voltage and current along a transmission line*

Equation 8.45 expresses the constant distributed inductance along the planar waveguide of Figure 8.4. The voltage gradient along the guide is then obtained from Equation 8.13:

$$\frac{\partial V}{\partial z} = -\bar{L}\frac{\partial I}{\partial t} \tag{8.46}$$

and the current gradient from the distributed capacitance, obtained from Equation 8.17:

$$\frac{\partial I}{\partial z} = -\bar{C}\frac{\partial V}{\partial t} \tag{8.47}$$

8.3.3   Correspondence to plane waves

8.3.3.1   *Transmission line voltage and current waves*

Equations 8.46 and 8.47 bear a close resemblance to Equations 3.18–3.21. Just as we found for Equations 8.30 and 8.31, differential 'Helmholtz-like' equations for V and I separately can be obtained for the lossless transmission line following the same procedure that produced Equations 3.23 and 3.24 for plane waves. Differentiation of Equation 8.46 by z and Equation 8.47 by t results in

$$\frac{\partial^2 V}{\partial z^2} - LC\frac{\partial^2 V}{\partial t^2} = 0 \tag{8.48}$$

and eliminating V in favour of an equation for I by reversing the differentiation variables for each equation results in

$$\frac{\partial^2 I}{\partial z^2} - LC\frac{\partial^2 I}{\partial t^2} = 0 \tag{8.49}$$

Just as in the plane-wave case we associate LC with $1/v^2$ where v is the phase velocity of the $V - I$ transmission wave running in the two-slab transmission line of Figure 8.4. Furthermore, if we assume time-harmonic phasor solutions to Equations 8.48 and 8.49 we can write the solution to the voltage equation as

$$V = V_0 e^{i(\beta z - \omega t)} \tag{8.50}$$

Then substituting this solution into Equation 8.48 yields

$$\frac{\partial^2 V}{\partial z^2} + LC\omega^2 V = 0 \tag{8.51}$$

We see therefore that the phase velocity v and propagation parameter β for the $V - I$ excitation of the transmission line is

$$v = \frac{1}{\sqrt{LC}} \quad \text{and} \quad \beta = \frac{\omega}{v} = \sqrt{LC}\,\omega \tag{8.52}$$

Given the voltage solution, Equation 8.50, we can find the current by substituting Equation 8.50 into Equation 8.46 and integrating with respect to time:

$$\frac{\partial I}{\partial t} = -\frac{1}{L}\frac{\partial V}{\partial t} = -\frac{1}{L}i\beta V \tag{8.53}$$

then

$$I = -\frac{1}{L}i\beta \int V\, dt = \frac{1}{L}\frac{\beta}{\omega}V = \sqrt{\frac{C}{L}}\cdot V \tag{8.54}$$

The ratio of the voltage to current amplitudes has units of resistance and is identified with the *characteristic impedance* of the transmission line:

$$Z_0 = \frac{V}{I} = \sqrt{\frac{L}{C}} \tag{8.55}$$

which is, as expected, in agreement with Equation 8.42.

8.3.3.2   *Reflection on a lossless terminated line*

So far we have considered the characteristic impedance Z_0 of the transmission line itself, which is determined essentially by the geometric parameters of the line—the separation and symmetry between the conductors and the nature of the insulating dielectric between them. If now we insert a load impedance Z_L across the line conductors, the 'impedance mismatch' will provoke some fraction of the incident wave to reflect at the load position and the transmission line is said to be 'terminated' by Z_L. Suppose we put the load at $z = 0$. We know that in general V and I on the otherwise lossless line are related by

$$V(z) = V_0^+ e^{i\beta z} + V_0^- e^{-i\beta z} \tag{8.56}$$

$$I(z) = \frac{1}{Z_0}\left[V_0^+ e^{i\beta z} - V_0^- e^{-i\beta z}\right] \tag{8.57}$$

At $z = 0$ we must have

$$Z_L = \frac{V(0)}{I(0)} = \frac{V_0^+ + V_0^-}{V_0^+ - V_0^-} Z_0 \tag{8.58}$$

so the amplitude of the reflected wave in terms of the line and load impedances is

$$V_0^- = \left(\frac{Z_L - Z_0}{Z_L + Z_0}\right) V_0^+ \tag{8.59}$$

Now, in analogy to Equations 3.139 and 3.140 for plane wave reflection, we define[2] a reflection coefficient Γ as

$$\Gamma \equiv -\frac{V_0^-}{V_0^+} = \frac{Z_0 - Z_L}{Z_0 + Z_L} \tag{8.60}$$

---

[2] We choose here the phase of the reflected voltage to be consistent with the convention used in Section 3.2.3 for reflected electromagnetic plane waves. The electrical engineering literature commonly uses $\Gamma = V_0^-/V_0^+$.

The voltage and current on the transmission line are then given by

$$V(z) = V_0^+ \left(e^{i\beta z} - \Gamma e^{-i\beta z}\right) \tag{8.61}$$

$$I(z) = I_0^+ \left(e^{i\beta z} + \Gamma e^{-i\beta z}\right) \tag{8.62}$$

We see that if the load is 'impedance matched' to the characteristic impedance of the transmission line, then reflected waves are suppressed.

Note that Z_L is defined as the ratio of voltage to current amplitude at $z = 0$. Since the superposition of incident and reflected waves set up a standing wave on the line, we can expect the ratio $V(z)/I(z)$ to vary with z. Therefore the impedance along the line varies with z when the load is mismatched. If we seek the impedance Z a distance $-d$ from the load, then from Equations 8.39, 8.61, and 8.62 we have

$$\begin{aligned} Z_{\text{line}} &= Z_0 \cdot \frac{e^{-i\beta d} - \Gamma e^{i\beta d}}{e^{-i\beta d} + \Gamma e^{i\beta d}} \\ &= Z_0 \cdot \frac{(1 - \Gamma) - i(1 + \Gamma)\tan\beta d}{(1 + \Gamma) - i(1 - \Gamma)\tan\beta d} \\ &= Z_0 \cdot \frac{Z_L - iZ_0\tan\beta d}{Z_0 - iZ_L\tan\beta d} \end{aligned} \tag{8.63}$$

Figure 8.5 shows the transmission line terminated in the load Z_L and the impedance Z_{line} a distance $-d$ to the left of the terminating load.

8.3.3.3 Power propagation on a lossless transmission line

The cycle-averaged power on a transmission line is given by the real part of the product of the voltage and current. Using Equations 8.61 and 8.62:

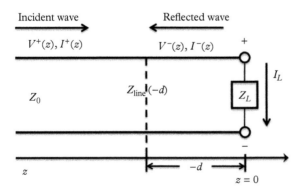

Figure 8.5 *Transmission line impedance $Z_{\text{line}}(-d)$ with the line terminated by a load $Z_L \neq Z_0$. The impedance mismatch sets up a reflection on the line and a standing wave due to the superposition of $V^+(z)$ and $V^-(z)$ waves.*

$$P_T = \frac{1}{2}\text{Re}\left[V(z)I^*(z)\right] \tag{8.64}$$

$$= \frac{1}{2}\frac{|V_0^+|^2}{Z_0}\text{Re}\left[1 + \Gamma^* e^{i2\beta z} - \Gamma e^{-i2\beta z} - |\Gamma|^2\right] \tag{8.65}$$

$$= \frac{1}{2}\frac{|V_0^+|^2}{Z_0}\left[1 - |\Gamma|^2\right] \tag{8.66}$$

This power expression is reminiscent of the power calculated from the Poynting vector of a propagating plane wave, Equations 3.166, 3.167, and 3.170. Clearly if the line is impedance matched so that there is no reflection, $\Gamma = 0$, then all the power propagates down the line. If $\Gamma \neq 0$ then some power is reflected back along the incident direction, setting up a $V - I$ standing wave along the line.

8.3.3.4   *Transmission and reflection at a line junction*

Suppose we have two transmission lines, each with a characteristic impedance Z_1, Z_2, respectively, and they are joined at $z = 0$ with the Z_1 line along $z < 0$ and the Z_2 line along $z > 0$. Assume further that the second line with impedance Z_2 is infinitely long (or terminated with an impedance matching load) so that there are no reflections at the end of it. At the junction the reflection coefficient is

$$\Gamma = \frac{Z_1 - Z_2}{Z_1 + Z_2} \tag{8.67}$$

so that the voltage along the $z < 0$ section is

$$V_1(z) = V_0^+ \left(e^{i\beta_1 z} - \Gamma e^{-i\beta_1 z}\right) \qquad z < 0 \tag{8.68}$$

The transmission junction is illustrated in Figure 8.6. On the $z > 0$ side, the voltage travels only in the $+z$ direction and the transmitted amplitude is some fraction T of V_1^+. The transmitted wave is then represented by

$$V_2(z) = TV_1^+ e^{i\beta_2 z} \qquad z > 0 \tag{8.69}$$

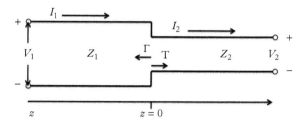

Figure 8.6 *Two transmission lines joined at $z = 0$. The line to the left ($z < 0$) is characterised by line impedance Z_1, and the line to the right ($z > 0$) by Z_2. Reflection coefficient Γ and transmission coefficient T determine the amplitude of the reflected and transmitted waves, respectively.*

Table 8.1 *Equivalent quantities between plane waves and transmission lines.*

Plane Wave	Transmission Line
$E_x(z)$	$V(z)$
$H_y(z)$	$I(z)$
$k = \omega\sqrt{\mu\varepsilon}$	$\beta = \omega\sqrt{LC}$
$Z = \sqrt{\frac{\mu}{\varepsilon}}$	$Z = \sqrt{\frac{L}{C}}$
$\mathcal{R}$	Γ
$\mathcal{T}$	T

At the junction itself the two voltages must be equal:

$$V_1(z = 0) = V_2(z = 0) \tag{8.70}$$

$$V_1^+(1 - \Gamma) = TV_1^+ \tag{8.71}$$

$$\frac{2Z_2}{Z_1 + Z_2} = T \tag{8.72}$$

So we find that in terms of the characteristic impedances in the two lines the reflection and transmission coefficients are analogous to reflection and transmission of plane waves at a material surface.

$$\Gamma = \frac{Z_1 - Z_2}{Z_1 + Z_2} \tag{8.73}$$

$$T = \frac{2Z_2}{Z_1 + Z_2} \tag{8.74}$$

8.3.4   Equivalence of plane waves and transmission lines

The correspondence between electromagnetic plane wave propagation (Section 3.1, Figure 3.1) through media characterised by μ, ε, and voltage/current transport along a lossless transmission line (Section 8.3.2) characterised by L, C, is indicated in Table 8.1. Analogous Helmholtz-like wave equations govern propagation, reflection, and transmission in both cases.

8.4   Special termination cases

In this section we discuss some special cases for Z_L in the lossless terminated transmission line, Figure 8.5 in Section 8.3.3.

8.4.1 Shorted transmission line

If we set $Z_L = 0$ then the reflection coefficient, from Equation 8.60, becomes unity, $\Gamma = 1$, and from Equations 8.61 and 8.62 the transmission voltage and current become

$$V(z) = V_0^+ \left(e^{i\beta z} - e^{-i\beta z} \right) = i2V_0^+ \sin \beta z \qquad (8.75)$$

$$I(z) = I_0^+ \left(e^{i\beta z} + e^{-i\beta z} \right) = 2I_0^+ \cos \beta z \qquad (8.76)$$

Note that at the load point ($z = 0$) the voltage is null and the current is maximum as would be expected from a short circuit. Note also that since the product $V(z)I^*(z)$ is pure imaginary no power is delivered to the load. The impedance along the line, from Equation 8.63 is

$$Z_{\text{line}} = -iZ_0 \tan \beta d \qquad (8.77)$$

8.4.2 Open circuit transmission line

Suppose $Z_L = \infty$ so that in Figure 8.5 the load is completely removed. In this case, from Equation 8.60, $\Gamma = -1$, and there is 100 % reflection at the load. The voltage and current expressions become

$$V(z) = V_0^+ \left(e^{i\beta z} + e^{-i\beta z} \right) = 2V_0^+ \cos \beta z \qquad (8.78)$$

$$I(z) = I_0^+ \left(e^{i\beta z} - e^{-i\beta z} \right) = i2I_0^+ \sin \beta z \qquad (8.79)$$

As expected, the voltage is maximum at the load point and the current is null. The expression for the impedance along the line, from Equation 8.63, is

$$Z_{\text{line}} = iZ_0 \cot \beta d \qquad (8.80)$$

and again no power is transmitted to the load.

8.4.3 Line impedance at $d = -\lambda/2$

At half-wave points along the line, Equation 8.63 shows that

$$Z_{\text{line}} = Z_L \qquad (8.81)$$

At these points the impedance is independent of the transmission line characteristics.

8.4.4   Line impedance at $d = -\lambda/4$

At the quarter-wave points we have, from Equation 8.63,

$$Z_{\text{line}} = Z_1 \frac{Z_L - iZ_1 \tan \beta d}{Z_1 - iZ_L \tan \beta d} \tag{8.82}$$

$$\lim_{\beta d \to \pi/2} Z_{\text{line}} = \frac{Z_1^2}{Z_L} \tag{8.83}$$

This quarter-wave transformer can be used as a $\lambda/4$ length of line, with impedance Z_1 to match an input transmission line of impedance Z_0 to a given load Z_L. Figure 8.7 shows how this matching can be accomplished. In order to suppress reflection at the Z_0/Z_1 junction, Γ must be set equal to zero there. Therefore,

$$Z_{\text{line}} = Z_0 \qquad \text{at} -d = \lambda/4 \tag{8.84}$$

and

$$Z_1 = \sqrt{Z_0 Z_L} \tag{8.85}$$

The line impedance at the quarter-wave point, Equation 8.84, is called a *quarter-wave transformer* because the load impedance is 'transformed' by the square of the characteristic impedance of the transmission line.

8.4.5   Slightly lossy transmission lines

An ideal, lossless transmission line is characterised by L and C, themselves calculated from Maxwell's equations and the geometry of the line. However, a real transmission line always has some resistance in the conducting elements and some leakage

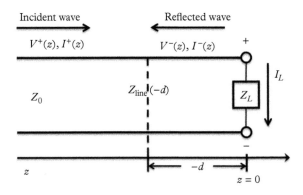

Figure 8.7 *Diagram of how a segment of transmission line with impedance Z_1 and length $\lambda/4$ can be used to match an incoming transmission line Z_0 with a load Z_L. The distance -d is equal to a quarter wavelength.*

current in the dielectric separating them. As shown in Figure 8.3, these departures from ideality are characterised by R in the conductors and G, the 'shunt conductance' through the insulating dielectric material. The complex propagation parameter is given by Equation 8.32,

$$\tilde{\beta}^2 = (R - i\omega L)(G - i\omega C) \tag{8.86}$$

or

$$\tilde{\beta} = \pm\sqrt{(R - i\omega L)(G - i\omega C)} \tag{8.87}$$

Assuming that $R \ll \omega L$ and $G \ll \omega C$ we can write Equation 8.87 as

$$\tilde{\beta} = \pm i\omega\sqrt{LC}\sqrt{1 + i\frac{R}{\omega L}} \cdot \sqrt{1 + i\frac{G}{\omega C}} \tag{8.88}$$

Expanding the square-root expression as the first two terms in a Taylor's series,

$$\tilde{\beta} = \pm i\omega\sqrt{LC}\left(1 + \frac{i}{2}\frac{R}{\omega L}\right) \cdot \left(1 + \frac{i}{2}\frac{G}{\omega C}\right) \tag{8.89}$$

$$= \pm i\omega\sqrt{LC}\left(1 + \frac{i}{2}\frac{R}{\omega L} + \frac{i}{2}\frac{G}{\omega C} - \frac{RG}{(\omega LC)^2}\right) \tag{8.90}$$

Since $R/\omega L$ and $G/\omega C$ are already small compared to unity, the last term on the right can be dropped, and we write

$$\tilde{\beta} \simeq \pm i\omega\sqrt{LC}\left[1 + \frac{i}{2}\left(\frac{G}{\omega C} + \frac{R}{\omega L}\right)\right] \tag{8.91}$$

$$\simeq \pm i\omega\sqrt{LC} \mp \frac{1}{2}\left(G\sqrt{\frac{L}{C}} + R\sqrt{\frac{C}{L}}\right) \tag{8.92}$$

We then identify the real propagation parameter β:

$$\beta = \omega\sqrt{LC} \tag{8.93}$$

and the dissipation term γ:

$$\gamma = \frac{1}{2}\left(G\sqrt{\frac{L}{C}} + R\sqrt{\frac{C}{L}}\right) \tag{8.94}$$

$$= \frac{1}{2}\left(GZ_0 + \frac{R}{Z_0}\right) \tag{8.95}$$

The signs of β and γ have been chosen so that a wave propagating in the positive z direction has a positive argument, $e^{i\beta z}$, and the real exponential term γ dissipates the amplitude to zero as $z \to \infty$:

$$e^{i\tilde{\beta}z} = e^{i(\beta+i\gamma)z} = e^{(i\beta-\gamma)z} \tag{8.96}$$

8.5   Waveguides

Up to now we have studied plane waves in Chapter 3 and transmission lines in the present chapter. Waves propagating within or on guiding structures, which essentially define the transverse field boundary conditions, are related to unbounded plane waves and are even more closely related to transmission lines. But guided waves also exhibit unique characteristics that require extension of our familiar ideas of voltage, current, capacitance, and inductance. Waveguiding at microwave frequencies was developed for radar during WWII, but now integrated optical circuits at the nanoscale use waveguides to transport optical signals onto and around integrated microchips. We discuss in this section the elements of waveguide theory and the relation guided waves bear to plane electromagnetic waves and to transmission lines.

The prototypical transmission line, consisting of two parallel slab conductors separated by a dielectric, is shown in Figure 8.4. In a sense, this structure can be considered a waveguide since the E-field terminates on the two conducting slabs separated by a constant distance in the $y - z$ plane, although the H-field is unbounded along y. It is clear from the front and side views, that the E- and H-field lines are confined to the $x-y$ plane with no field components along the direction of propagation, z. Waves with this property are called transverse electromagnetic (TEM) waves. An ideal plane wave propagating in space also exhibits phase fronts confined to the plane orthogonal to the propagation direction, so they are TEM waves as well. Note that if we added two conducting slabs in the $x - z$ plane so as to enclose a volume with a single conductor of rectangular $x - y$ cross section, no TEM field could exist since the four conducting sides would define an interior space of constant voltage (and hence a null E-field). However, waveguides with rectangular, cylindrical, and even arbitrarily shaped cross sections do exist, but in order to understand them we have to extend our ideas beyond the familiar TEM wave. Waves propagating within structures of a single conductor are classified into two types: transverse electric (TE) waves with an H-field component along the propagation direction z (but E-field only in the transverse $x-y$ plane), and transverse magnetic (TM) waves with an E-field component along the propagation direction (but only with transverse H-field components). Note that TE and TM *waves* (or modes) should not be confused with TE and TM *polarisation* discussed at length in Chapters 3 and 7.

8.5.1   Field solutions for TM and TE waves

We assume field solutions to the Helmholtz equation that will have some spatial variation in the $x - y$ plane, subject to the boundary conditions of a known perfect electric conductor (PEC) guide structure, and propagate along the z direction with $e^{i\beta z}$. For the time being we will take the propagation parameter to be real (lossless medium). The general form of the phasor fields are then

$$E(x, y, z) = \left[\mathscr{E}(x, y) + \mathscr{E}_z(x, y)\hat{z} \right] e^{i\beta z} \tag{8.97}$$

$$H(x, y, z) = \left[\mathscr{H}(x, y) + \mathscr{H}_z(x, y)\hat{z} \right] e^{i\beta z} \tag{8.98}$$

Once again we write the two curl equations of Maxwell with harmonic time dependence, no current sources in non-magnetic material ($\mu = \mu_0$):

$$\nabla \times \mathbf{E} = i\omega\mu_0\mathbf{H}$$

$$\nabla \times \mathbf{H} = -i\omega\varepsilon\mathbf{E}$$

and write out explicitly the various components of $\mathbf{H}$ and $\mathbf{E}$. For the H-field components we have

$$\frac{\partial E_z}{\partial y} - i\beta E_y = i\omega\mu_0 H_x \tag{8.99}$$

$$i\beta E_x - \frac{\partial E_z}{\partial x} = i\omega\mu_0 H_y \tag{8.100}$$

$$\frac{\partial E_y}{\partial x} - \frac{\partial E_x}{\partial y} = i\omega\mu_0 H_z \tag{8.101}$$

and for the E-field components

$$\frac{\partial H_z}{\partial y} - i\beta H_y = -i\omega\varepsilon E_x \tag{8.102}$$

$$i\beta H_x - \frac{\partial H_z}{\partial x} = -i\omega\varepsilon E_y \tag{8.103}$$

$$\frac{\partial H_y}{\partial x} - \frac{\partial H_x}{\partial y} = -i\omega\varepsilon E_z \tag{8.104}$$

We can use Equations 8.99 and 8.103 to write H_x in terms of derivatives of E_z and H_z in the transverse $x - y$ plane. Similarly we can use Equations 8.100 and 8.102 to also write H_y in terms of transverse derivatives of E_z, H_z. Proceeding in the same way for E_x and E_y we have finally four equations,

$$H_x = -\frac{i}{\left(k^2 - \beta^2\right)}\left[\omega\varepsilon\frac{\partial E_z}{\partial y} - \beta\frac{\partial H_z}{\partial x}\right] \tag{8.105}$$

$$H_y = \frac{i}{\left(k^2 - \beta^2\right)}\left[\omega\varepsilon\frac{\partial E_z}{\partial x} + \beta\frac{\partial H_z}{\partial y}\right] \tag{8.106}$$

$$E_x = \frac{i}{\left(k^2 - \beta^2\right)}\left[\omega\mu_0\frac{\partial H_z}{\partial y} + \beta\frac{\partial E_z}{\partial x}\right] \tag{8.107}$$

$$E_y = -\frac{i}{\left(k^2 - \beta^2\right)}\left[\omega\mu_0\frac{\partial H_z}{\partial x} - \beta\frac{\partial E_z}{\partial y}\right] \tag{8.108}$$

where k, as always, is related to the frequency ω and velocity of light through the medium $(v = 1/\sqrt{\varepsilon\mu_0})$ by the expression $k = \omega\sqrt{\varepsilon\mu_0}$. The factor $\left(k^2 - \beta^2\right)$ can be interpreted as the square of a propagation parameter in the x–y plane since β is always along the z-axis. This transverse propagation parameter is called the 'cutoff' parameter k_c:

$$k_c^2 = k^2 - \beta^2 \tag{8.109}$$

8.5.1.1   TE waves

In the case of TE waves, $E_z = 0$ and Equations 8.105–8.108 become

$$H_x = \frac{i}{k_c^2} \cdot \beta\frac{\partial H_z}{\partial x} \tag{8.110}$$

$$H_y = \frac{i}{k_c^2} \cdot \beta\frac{\partial H_z}{\partial y} \tag{8.111}$$

$$E_x = \frac{i}{k_c^2} \cdot \omega\mu_0\frac{\partial H_z}{\partial y} \tag{8.112}$$

$$E_y = -\frac{i}{k_c^2} \cdot \omega\mu_0\frac{\partial H_z}{\partial x} \tag{8.113}$$

In order to find the transverse fields we have to use the Helmholtz equation to find the permitted values of H_z. We write the expression

$$\left(\nabla^2 + k^2\right) H_z = \left(\frac{\partial^2}{\partial x^2} + \frac{\partial^2}{\partial y^2} + \frac{\partial^2}{\partial z^2} + k^2\right) H_z = 0 \tag{8.114}$$

But since

$$\frac{\partial^2 H_z}{\partial z^2} = -\beta^2 H_z \tag{8.115}$$

the Helmholtz equation reduces to a differential equation for the transverse H_z field gradients:

$$\left(\frac{\partial^2}{\partial x^2} + \frac{\partial^2}{\partial y^2} + k_c^2\right)\mathcal{H}_z = 0 \tag{8.116}$$

Allowed solutions will be a function of the transverse boundary conditions.

8.5.1.2 *TM waves*

For TM waves we follow the same procedure except we set $H_z = 0$ and write Equations 8.105–8.108 as

$$H_x = -\frac{i}{k_c^2}\cdot\omega\varepsilon\frac{\partial E_z}{\partial y} \tag{8.117}$$

$$H_y = \frac{i}{k_c^2}\cdot\omega\varepsilon\frac{\partial E_z}{\partial x} \tag{8.118}$$

$$E_x = \frac{i}{k_c^2}\cdot\beta\frac{\partial E_z}{\partial x} \tag{8.119}$$

$$E_y = \frac{i}{k_c^2}\cdot\beta\frac{\partial E_z}{\partial y} \tag{8.120}$$

The Helmholtz equation for E_z reduces to

$$\left(\frac{\partial^2}{\partial x^2} + \frac{\partial^2}{\partial y^2} + k_c^2\right)\mathcal{E}_z = 0 \tag{8.121}$$

and allowed solutions will again be a function of the transverse boundary conditions.

8.5.2 Parallel plate waveguide

Here we apply the formal development of the last section to a simple but very useful example, the parallel plate waveguide. This structure is essentially the same as the parallel plate transmission line studied in Section 8.3.2, but here we will see that the transmission line TEM waves (with no field components in the direction of propagation) can be complemented with TE modes and TM modes. The two conducting plates of the waveguide are separated by s between the plates (along the x direction) and have a width w (along the y direction) that is much greater than the separation s. We can therefore assume that the solutions we seek will not be functions of y.

8.5.2.1 *TE modes*

In order to find the allowed TE modes we start with the reduced Helmholtz equation for the TE case, Equation 8.116, and write the solutions for $\mathcal{H}_z$. By inspection, we write the general solution in terms of the real sin and cos functions and refer to the coord-inate system of Figure 8.4. The amplitudes A and B are determined by the boundary conditions:

$$\mathcal{H}_z(x, y) = A \sin k_c x + B \cos k_c x \tag{8.122}$$

Since we posit that the plates are perfect conductors (and the intervening dielectric lossless), we have for boundary conditions:

$$E_y(0, y) = 0 \quad \text{and} \quad E_y(s, y) = 0 \tag{8.123}$$

Then from Equation 8.113:

$$E_y(x, y) = -\frac{i}{k_c^2} \cdot \omega \mu_0 \frac{\partial H_z}{\partial x} = -\frac{i}{k_c^2} \cdot \omega \mu_0 \frac{\partial \mathcal{H}_z}{\partial x} e^{i\beta z} \tag{8.124}$$

$$= -\frac{i\omega\mu_0}{k_c} [A \cos k_c x - B \sin k_c x] \, e^{i\beta z} \tag{8.125}$$

The boundary conditions, Equation 8.123, require that $A = 0$ and puts a condition on the argument of the sin term, that at $x = s$ the E-field must be null, so

$$k_c s = n\pi \quad n = 1, 2, 3 \ldots \quad \text{or} \quad k_c = \frac{n\pi}{s} \quad n = 1, 2, 3 \ldots \tag{8.126}$$

The boundary conditions impose that only discrete values of k_c are permitted (depend-ing on the plate separation s), and the expression for the propagation parameter along the z direction becomes

$$\beta = \pm\sqrt{k^2 - k_c^2} \quad \text{or} \quad \beta = \pm\sqrt{k^2 - \left(\frac{n\pi}{s}\right)^2} \tag{8.127}$$

The solutions for $\mathcal{H}_z(x, y)$ are

$$\mathcal{H}_z(x, y) = B_n \cos \frac{n\pi}{s} x \tag{8.128}$$

or

$$H_z(x, y) = B_n \cos \left(\frac{n\pi}{s} x\right) e^{i\beta z} \tag{8.129}$$

Note that when $k > k_c$, β is positive and the wave propagates along z. When $k = k_c$, then $\beta = 0$ and the wave is stationary with zero phase velocity. If k falls below k_c, then β becomes pure imaginary. The wave no longer propagates in the waveguide but decays exponentially to $1/e$ of its initial amplitude at a characteristic distance of $x = 1/\beta$. The propagating wave becomes an *evanescent* wave. This threshold behaviour for propagation of mode n within the waveguide is the reason that k_c is called the 'cutoff' parameter. Below cutoff, the wave does not propagate. The cutoff parameter k_c is related in the usual way to wavelength and frequency:

$$k_c = \frac{2\pi}{\lambda_c} \quad \text{and} \quad \omega_c = k_c v \tag{8.130}$$

where v is the propagation velocity in the gap dielectric of the waveguide. If the dielectric is air or vacuum, then is the velocity of light, c. For a given waveguide 'mode' n the cutoff frequency must be greater than $k_{cn} v$.

Once we have the solutions for $H_z(x, y)$ we can get the solutions for all the transverse field components from Equations 8.110–8.113. Since there is no transverse dependence in the y direction, Equations 8.111 and 8.112 show that $H_y = E_x = 0$ and

$$H_x = -i\frac{\beta}{k_c} B_n \sin\left(\frac{n\pi}{s}x\right) e^{i\beta z} \tag{8.131}$$

$$E_y = i\frac{\omega\mu_0}{k_c} B_n \sin\left(\frac{n\pi}{s}x\right) e^{i\beta z} \tag{8.132}$$

Note the difference in sign between H_x and E_y. This sign difference is consistent with the direction of energy propagation, determined by the Poynting vector, along the positive z direction, $\mathbf{S} = \mathbf{E} \times \mathbf{H}$. The wave impedance is given by the ratio of the E-field amplitude to the H-field amplitude,

$$Z_{TE} = \frac{|E_y|}{|H_x|} = \frac{\omega\mu_0}{\beta} = \frac{k}{\beta}Z_0 \tag{8.133}$$

For the parallel plate waveguide the TE modes consist of a 'triplet' of components, H_z, H_x, E_y and the modes are labelled TE_n, $n = 1, 2, 3 \ldots$.

8.5.2.2 TM modes

To find the field components for the TM modes we follow a parallel procedure to the TE case. The first step is to find the $\mathcal{E}_z$ solutions from the reduced Helmholtz equation, Equation 8.121, subject to the boundary conditions on the parallel plate waveguide at $x = 0$ and $x = s$. Once again the general solution is

$$\mathcal{E}_z(x, y) = A \sin k_c x + B \cos k_c x \tag{8.134}$$

with boundary conditions that $\mathcal{E}_z(0, y) = \mathcal{E}_z(s, y) = 0$. Therefore $B = 0$ and $\sin k_c x = 0$ when $x = 0$ or when $x = s$. The solutions to Equation 8.121 that respect the boundary conditions are therefore

$$\mathcal{E}_z = A_n \sin\left(\frac{n\pi}{s}\right) x \quad n = 0, 1, 2, 3 \ldots \tag{8.135}$$

and

$$E_z = A_n \sin\left(\frac{n\pi}{s}\right) x e^{i\beta z} \tag{8.136}$$

with

$$\beta = \pm\sqrt{k^2 - k_c^2} = \pm\sqrt{k^2 - \left(\frac{n\pi}{s}\right)^2} \tag{8.137}$$

From Equations 8.117–8.120 we find the transverse field components:

$$H_y = i\frac{\omega\varepsilon}{k_c} A_n \cos\left(\frac{n\pi}{s}x\right) e^{i\beta z} \tag{8.138}$$

$$E_x = i\frac{\beta}{k_c} A_n \cos\left(\frac{n\pi}{s}x\right) e^{i\beta z} \tag{8.139}$$

The impedance for the TM waves is

$$Z_{\text{TM}} = \frac{E_x}{H_y} = \frac{\beta}{\omega\varepsilon} = \frac{\beta}{k} Z_0 \tag{8.140}$$

and the TM modes consist of a triplet of three field components: E_z, E_x, H_y. Note that for the lowest $n = 0$ mode, there is no cut-off frequency, $E_z = 0$, and the H_y, E_x components have constant amplitudes. In fact the TM_0 mode is identical to the TEM_0 mode, represented in Figure 8.4.

8.6   Rectangular waveguides

The parallel plate waveguide is essentially a one-dimensional problem since the transverse spatial variation of the TE and TM modes depend only on x. Here we examine the more realistic case of a rectangular waveguide with finite dimensions in both x and y. We assume again boundary conditions on a perfect electrical conductor (PEC) and apply those E-field conditions at the waveguide edges along x and y. By convention we denote the long and short sides of the rectangular cross section as l, and h, respectively, and align l along x and h along y so that the lower left corner of the waveguide cross section is at the coordinate origin. The guide is filled with some non-magnetic, lossless dielectric material characterised by permeability μ_0 and permittivity ε. Figure 8.8 shows the layout for the rectangular waveguide.

8.6.1   TE modes

The transverse electric (TE) modes are defined as those waves that propagate in the guide with E-fields only transverse to the z (propagation) direction. Therefore, we seek H_z solutions to the transverse Helmholtz equation:

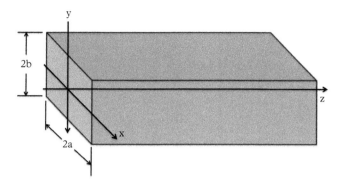

Figure 8.8 *Rectangular waveguide with sides x = 2a and y = 2b.*
Inside the conducting boundaries the waveguide is filled with a
lossless dielectric characterised by μ_0, ε.

$$\left(\frac{\partial^2}{\partial x^2} + \frac{\partial^2}{\partial y^2} + k_c^2 \right) \mathcal{H}_z(x, y) = 0 \tag{8.141}$$

The solutions will be functions of both coordinates, but we can expect them to be uncoupled. Therefore we can posit solutions of the form $\mathcal{H}_z(x, y) = X(x)Y(y)$. Substitution into Equation 8.141 and division by $X(x)Y(y)$ results in

$$\frac{1}{X} \frac{\partial^2 X}{\partial x^2} + \frac{1}{Y} \frac{\partial^2 Y}{\partial y^2} = -k_c^2 \tag{8.142}$$

Since this result must be true for all values of x, y, the terms on the left must themselves be equal to constants that we write

$$\frac{1}{X} \frac{\partial^2 X}{\partial x^2} = -k_x^2 \qquad\qquad \frac{1}{Y} \frac{\partial^2 Y}{\partial y^2} = -k_y^2 \tag{8.143}$$

and

$$k_x^2 + k_y^2 = k_c^2 \tag{8.144}$$

The solution for $\mathcal{H}_z(x, y)$ is a product of the general solution in x and the general solution in y:

$$\mathcal{H}_z(x, y) = (A \cos k_x x + B \sin k_x x) \cdot (C \cos k_y y + D \sin k_y y) \tag{8.145}$$

The boundary conditions at $x, y = 0$ and $x = a, y = b$ are

$$E_x(x, 0), E_x(x, b) = 0 \quad \text{and} \quad E_y(0, y), E_y(a, y) = 0 \qquad (8.146)$$

The transverse E-fields are obtained from the longitudinal H-field $\mathcal{H}_z$ by using Equations 8.112 and 8.113:

$$E_x = i\frac{\omega\mu_0}{k_c^2} k_y (A \cos k_x x + B \sin k_x x) \cdot (-C \sin k_y y + D \cos k_y y) \qquad (8.147)$$

$$E_y = i\frac{\omega\mu_0}{k_c^2} k_x (-A \sin k_x x + B \cos k_x x) \cdot (C \cos k_y y + D \sin k_y y) \qquad (8.148)$$

The condition on $E_x(x, y = 0)$ implies that $D = 0$ and the condition on $E_x(x, y = b)$ implies that $k_y = n\pi/b$ with $n = 0, 1, 2 \ldots$. Similarly, the condition on $E_y(x = 0, y)$ requires that $B = 0$ and the condition on $E_y(x = a, y)$ requires $k_x = m\pi/a$ with $m = 0, 1, 2 \ldots$. Therefore, the general solution for $\mathcal{H}_z(x, y)$, Equation 8.145, becomes

$$\mathcal{H}_z = A_{mn} \cos \frac{m\pi x}{a} \cos \frac{n\pi y}{b} \qquad (8.149)$$

and

$$H_z(x, y, z) = A_{mn} \cos \frac{m\pi x}{a} \cos \frac{n\pi y}{b} e^{i\beta z} \qquad (8.150)$$

The arbitrary amplitude constant A_{mn} is a product of the constants A, C in Equation 8.145. Now we get the transverse fields using Equations 8.110–8.113:

$$H_x = -i\frac{\beta m\pi}{k_c^2 a} A_{mn} \sin \frac{m\pi x}{a} \cos \frac{n\pi y}{b} e^{i\beta z} \qquad (8.151)$$

$$H_y = -i\frac{\beta n\pi}{k_c^2 b} A_{mn} \cos \frac{m\pi x}{a} \sin \frac{n\pi y}{b} e^{i\beta z} \qquad (8.152)$$

$$E_x = -i\frac{\omega\mu_0 n\pi}{k_c^2 b} A_{mn} \cos \frac{m\pi x}{a} \sin \frac{n\pi y}{b} e^{i\beta z} \qquad (8.153)$$

$$E_y = i\frac{\omega\mu_0 m\pi}{k_c^2 a} A_{mn} \sin \frac{m\pi x}{a} \cos \frac{n\pi y}{b} e^{i\beta z} \qquad (8.154)$$

As always we get the cutoff propagation parameters and frequencies from

$$\beta = \sqrt{k^2 - k_c^2} = \sqrt{k^2 - (k_x^2 + k_y^2)} = \sqrt{k^2 - \left(\frac{m\pi}{a}\right)^2 - \left(\frac{n\pi}{b}\right)^2} \qquad (8.155)$$

The propagation parameter along z becomes real when

$$k > k_c = \sqrt{\left(\frac{m\pi}{a}\right)^2 - \left(\frac{n\pi}{b}\right)^2} \tag{8.156}$$

The TE wave impedance is denoted by Z_{TE} and is independent of the mode labelling n, m:

$$Z_{TE} = \frac{E_x}{H_y} = \frac{k}{\beta}Z_0 \tag{8.157}$$

Because we have posited that $a > b$, Equation 8.156 shows that the TE mode with the lowest cutoff frequency is $TE_{mn} = TE_{10}$.

8.6.2 TM modes

The transverse magnetic modes are defined as those modes with H-fields only in the transverse $(x - y)$ plane. Therefore, the field component in the z direction must be an E-field. The waves propagate along z and E_z is written as the product $\mathscr{E}_z(x, y)e^{i\beta z}$. Therefore, we seek $\mathscr{E}_z$ solutions to the same Helmholtz wave equation as for the TE modes:

$$\left(\frac{\partial^2}{\partial x^2} + \frac{\partial^2}{\partial y^2} + \frac{\partial^2}{\partial z^2} + k_c^2\right)\mathscr{E}_z = 0 \tag{8.158}$$

The general solution for $\mathscr{E}_z$ has the same form as that for $\mathscr{H}_z$, Equation 8.145, and subject to boundary conditions,

$$E_z(x, 0), E_z(x, b) = 0 \quad \text{and} \quad E_z(0, y), E_z(a, y) = 0 \tag{8.159}$$

The expression for E_z is therefore

$$E_z(x, y, z) = \mathscr{E}_z(x, y)e^{i\beta z} = B_{mn}\sin\frac{m\pi x}{a}\sin\frac{n\pi y}{b}e^{i\beta z} \tag{8.160}$$

with $m, n = 1, 2, 3 \ldots$. As with the TE modes the cutoff propagation parameter is given by

$$k_c^2 = k_x^2 + k_y^2 \tag{8.161}$$

and the boundary conditions dictate that

$$k_x = \frac{m\pi}{a} \quad \text{and} \quad k_y = \frac{n\pi}{b} \tag{8.162}$$

The transverse field components are obtained from E_z by using Equations 8.117–8.120:

$$H_x = -i\frac{\omega\varepsilon n\pi}{bk_c^2} B_{mn} \sin\frac{m\pi x}{a} \cos\frac{n\pi y}{b} e^{i\beta z} \tag{8.163}$$

$$H_y = i\frac{\omega\varepsilon m\pi}{ak_c^2} B_{mn} \cos\frac{m\pi x}{a} \sin\frac{n\pi y}{b} e^{i\beta z} \tag{8.164}$$

$$E_x = i\frac{\beta m\pi}{ak_c^2} B_{mn} \cos\frac{m\pi x}{a} \sin\frac{n\pi y}{b} e^{i\beta z} \tag{8.165}$$

$$E_y = i\frac{\beta n\pi}{bk_c^2} B_{mn} \sin\frac{m\pi x}{a} \cos\frac{n\pi y}{b} e^{i\beta z} \tag{8.166}$$

The propagation parameter is given, as usual, by

$$\beta = \pm\sqrt{k^2 - k_c^2} = \pm\sqrt{k^2 - \left(\frac{m\pi}{a}\right)^2 - \left(\frac{n\pi}{b}\right)^2} \tag{8.167}$$

We can see from Equations 8.163–8.166 that TM_{01} or TM_{10} vanish (in fact all TM modes with a zero index vanish) so the lowest TM mode is TM_{11} with a cutoff propagation parameter given by

$$k_c = \sqrt{\left(\frac{\pi}{a}\right)^2 + \left(\frac{\pi}{b}\right)^2} \tag{8.168}$$

If one of the sides becomes much longer than the other (say, $a \gg b$), then

$$k_c \to \frac{\pi}{b} \tag{8.169}$$

and the cutoff k_c becomes equal to the parallel plate waveguide case (Section 8.5). Finally, the TM mode impedance is given by

$$Z_{TM} = \frac{E_x}{H_y} = \frac{\beta}{k} Z_0 \tag{8.170}$$

8.7   Cylindrical waveguides

The cylindrical geometry is important not only for waveguides and transmission lines in the conventional microwave domain but also for nanoscale holes and hole arrays in the visible and near infrared regions. A sensible point of departure for understanding light transmission through these cylindrically symmetric structures is an analysis of the field modes they can support. The procedure is really no different than what we have done for the rectangular waveguides. We choose a coordinate system in which the solutions to the Helmholtz equation will separate, and we will be able to write the solutions to E_z and

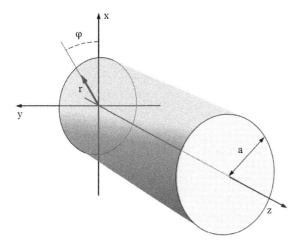

Figure 8.9 *Schematic of cylindrical waveguide. The cylinder surface is assumed to be a perfect conductor. Cylindrical coordinates are r, φ and z as shown. The radius of the cylinder is a.*

H_z as a product of three factors $R(r)\Phi(\varphi)e^{i\beta z}$ where $R(r)$ is the radial field dependence and $\Phi(\varphi)$ is the azimuthal angular dependence around the z-axis. A schematic of the guide is shown in Figure 8.9.

8.7.1 TE modes

As always, TE modes are defined by the presence of electric fields only in the transverse plane. The field function along the z-axis must be an H-field, and we write the Helmholtz equation.

$$\nabla^2 H_z + k^2 H_z = 0 \tag{8.171}$$

Then we write the ∇^2 or Laplacian operator in cylindrical coordinates and factor $H_z(r, \varphi, z) = \mathcal{H}_z(r, \varphi)e^{i\beta z}$. We take the expression for the Laplacian operator from Equation D.32 of Appendix D:

$$\left(\frac{\partial^2}{\partial r^2} + \frac{1}{r}\frac{\partial}{\partial r} + \frac{1}{r^2}\frac{\partial^2}{\partial \varphi^2} + k_c^2 \right) \mathcal{H}_z(r, \varphi) = 0 \tag{8.172}$$

Now we take the posited product solution

$$\mathcal{H}_z(r, \varphi) = R(r)\Phi(\varphi) \tag{8.173}$$

and substitute it into Equation 8.172. After multiplication by r^2 we have

$$\frac{r^2}{R}\frac{d^2 R}{dr^2} + \frac{r}{R}\frac{dR}{dr} + r^2 k_c^2 = -\frac{1}{\Phi}\frac{d^2 \Phi}{d\varphi^2} \tag{8.174}$$

The left-hand side is a function only of r and the right-hand side only of φ. In order for the equation to be valid for the entire range of r, φ the two sides must be equal to a constant. We set the separation constant equal to l_φ^2 and write

$$-\frac{1}{\Phi}\frac{d^2\Phi}{d\varphi^2} = l_\varphi^2 \tag{8.175}$$

or rearranging:

$$\frac{d^2\Phi}{d\varphi^2} + l_\varphi^2\Phi = 0 \tag{8.176}$$

From the left-hand side of Equation 8.174 we have

$$r^2\frac{d^2R}{dr^2} + r\frac{dR}{dr} + \left(r^2k_c^2 - l_\varphi^2\right)R = 0 \tag{8.177}$$

Note that l_φ is unitless. From inspection we can write down a solution to Equation 8.176:

$$\Phi = e^{il_\varphi\varphi} \tag{8.178}$$

Now in order to represent a physical entity Φ must be a single-valued function, meaning that $\Phi[l_\varphi(\varphi + 2\pi)]$ must have the same value as $\Phi(l_\varphi\varphi)$ or

$$\Phi[(l_\varphi(\varphi + 2\pi)] = e^{il_\varphi\varphi}e^{il_\varphi 2\pi} = \Phi(l_\varphi\varphi) \tag{8.179}$$

Therefore, the separation constant l_φ must be an integer, $l_\varphi = n = 1, 2, 3 \ldots$. The general solution to Equation 8.176 can also be written in terms of real sin and cos functions:

$$\Phi(\varphi) = A\sin n\varphi + B\cos n\varphi \tag{8.180}$$

The radial equation becomes

$$r^2\frac{d^2R}{dr^2} + r\frac{dR}{dr} + \left(r^2k_c^2 - n^2\right)R = 0 \tag{8.181}$$

This expression is one of the forms of Bessel's differential equation. There exists a family of functions that are solutions to this equation. In fact there are two independent families, Bessel functions of the first kind, $J_n(x)$, and of the second kind, $N_n(x)$. The functions of the second kind, the Neumann functions, tend to $-\infty$ as $x \to 0$ so they are not physically acceptable solutions. The Bessel functions of the first kind are finite at the origin and exhibit physically acceptable behaviour. The first few Bessel functions

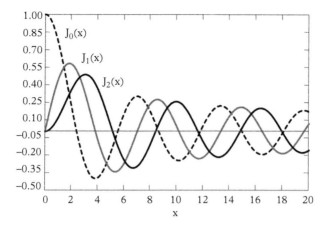

Figure 8.10 *First three Bessel functions $J_n(x)$.*

$J_n(x)$ are plotted in Figure 8.10. Putting together the angular and radial solutions to the reduced Helmholtz equation, we have

$$\mathcal{H}_z(r, \varphi) = (A \sin n\varphi + B \cos n\varphi) J_n(k_c r) \tag{8.182}$$

The next step is to apply boundary conditions at $r = 0$ and $r = a$, where a is the radius of the hole. But in order to do so we must write the E-field and H-field components in the r, φ directions in terms of the z-axis components. These components are found from writing down the six Maxwell curl equations in cylindrical coordinates and reducing them to four relations of r, φ in terms of z as was done for Equations 8.105–8.108. The result is

$$E_r = i\frac{1}{k_c^2} \left(\beta \frac{\partial E_z}{\partial r} + \frac{\omega \mu_0}{r} \frac{\partial H_z}{\partial \varphi} \right) \tag{8.183}$$

$$E_\varphi = i\frac{1}{k_c^2} \left(\frac{\beta}{r} \frac{\partial E_z}{\partial \varphi} - \omega \mu_0 \frac{\partial H_z}{\partial r} \right) \tag{8.184}$$

$$H_r = -i\frac{1}{k_c^2} \left(\frac{\omega \varepsilon}{r} \frac{\partial E_z}{\partial \varphi} - \beta \frac{\partial H_z}{\partial r} \right) \tag{8.185}$$

$$H_\varphi = i\frac{i}{k_c^2} \left(\omega \varepsilon \frac{\partial E_z}{\partial r} + \frac{\beta}{r} \frac{\partial H_z}{\partial \varphi} \right) \tag{8.186}$$

On the perfectly conducting wall of the cylinder, the E-field tangent to the wall must vanish. Therefore, $E_\varphi(r = a, \varphi) = 0$. From Equation 8.184 we have

$$E_\varphi(r, \varphi, z) = -i\frac{\omega \mu_0}{k_c} (A \sin n\varphi + B \cos n\varphi) J_n'(k_c r) e^{i\beta z} \tag{8.187}$$

where $J'_n(k_c r)$ is the radial derivative of the Bessel function. Applying the boundary condition at $r = a$:

$$E_\varphi(a, \varphi, z) = -i\frac{\omega\mu_0}{k_c}(A\sin n\varphi + B\cos n\varphi)J'_n(k_c a)e^{i\beta z} = 0 \qquad (8.188)$$

Since the r dependence is only in the Bessel term, we must have

$$J'_n(k_c a) = 0 \qquad (8.189)$$

But this requirement fixes k_c to the zeros of $J'_n(k_c a)$. Each function J'_n, labelled by n, will have a series of zeros where $J'_n[(k_c a)_{nm}]$ corresponds to the mth zero of the nth Bessel function derivative. The arguments of the zeros (or roots) of J'_n are not simple multiples of π as in the case of sin and cos functions, they have to be found numerically and Table 8.2 provides values for the first few arguments $(k_c a)_{nm}$ corresponding to the zeros of the J'_n functions. The first three Bessel function derivatives are plotted in Figure 8.11. Evidently we obtain k_c for the nmth mode from

$$k_{c nm} = \frac{k_c a_{nm}}{a} \qquad (8.190)$$

We obtain the propagation parameter along z for the nmth mode in the usual way,

$$\beta_{nm} = \sqrt{k^2 - k_{c nm}^2} \qquad (8.191)$$

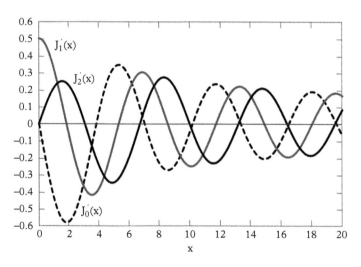

Figure 8.11 *Derivatives of the first three Bessel functions, showing the positions of the first few zero-crossings.*

Table 8.2 *Some values of* $(k_c a)_{nm}$ *for TE in a cylindrical waveguide.*

n	$(k_c a)_{n1}$	$(k_c a)_{n2}$	$(k_c a)_{n3}$
0	3.832	7.016	10.174
1	1.841	5.331	8.536
2	3.054	6.706	9.970

and with β_{nm} in hand we construct $H_z(r, \varphi, z) = \mathcal{H}_z(r, \varphi)e^{i\beta_{nm}z}$. Then from Equations 8.183–8.186 we calculate the transverse fields for the nmth mode. The notation nm has been omitted from $k_{c_{nm}}$ and β_{nm} so as not to encumber the expressions more than necessary:

$$E_r = i\frac{\omega\mu_0 n}{k_c^2 r} (A\cos n\varphi - B\sin n\varphi) J_n(k_c r)e^{i\beta z} \tag{8.192}$$

$$E_\varphi = -i\frac{\omega\mu_0}{k_c} (A\sin n\varphi + B\cos n\varphi) J_n'(k_c r)e^{i\beta z} \tag{8.193}$$

$$H_r = i\frac{\beta}{k_c} (A\sin n\varphi + B\cos n\varphi) J_n'(k_c r)e^{i\beta z} \tag{8.194}$$

$$H_\varphi = i\frac{\beta n}{k_c^2 r} (A\cos n\varphi - B\sin n\varphi) J_n(k_c r)e^{i\beta z} \tag{8.195}$$

In fact, the A and B amplitude factors in the angular part of the transverse field expressions are not independent. We can easily see this from Equation 8.178. The 'normalisation integral' of the angular solution in complex form is

$$\int_0^{2\pi} \Phi^* \Phi \, d\varphi = 2\pi = \text{const.} \tag{8.196}$$

and writing the normalisation of the angular solution in terms of real sin and cos functions,

$$\int_0^{2\pi} (A\sin n\varphi + B\cos n\varphi)^2 \, d\varphi = \frac{1}{2}(A^2 + B^2) = \text{const.} \tag{8.197}$$

or in other words the only restriction on the amplitudes of the sin and cos terms is that the sum of the squares of A and B must always be equal to some constant. Therefore, we can set $A = 0$ and let $B^2 = $ const. Then the transverse fields can be written in a somewhat simpler form:

$$E_r = -i\frac{\omega\mu_0 n}{k_c^2 r} (B\sin n\varphi) J_n(k_c r)e^{i\beta z} \tag{8.198}$$

$$E_\varphi = -i\frac{\omega\mu_0}{k_c} (B\cos n\varphi) J_n'(k_c r)e^{i\beta z} \tag{8.199}$$

$$H_r = i\frac{\beta}{k_c} (B\cos n\varphi) J_n'(k_c r)e^{i\beta z} \tag{8.200}$$

$$H_\varphi = -i\frac{\beta n}{k_c^2 r} (B\sin n\varphi) J_n(k_c r)e^{i\beta z} \tag{8.201}$$

The TE mode impedance is given by

$$Z_{TE} = \frac{E_r}{H_\varphi} = \frac{k}{\beta}Z_0 \tag{8.202}$$

8.7.2   TM modes

In the case of TM modes, there is no component H_z along the axis of symmetry, and we seek solutions to the Helmholtz equation for $E_z(r, \varphi, z) = \mathscr{E}_z(r, \varphi)e^{i\beta z}$. As with the TE modes we have a reduced Helmholtz equation for the transverse coordinates:

$$\left(\frac{\partial^2}{\partial r^2} + \frac{1}{r}\frac{\partial}{\partial r} + \frac{1}{r^2}\frac{\partial^2}{\partial\varphi^2} + k_c^2\right)\mathscr{E}_z(r, \varphi) = 0 \tag{8.203}$$

This expression is the same as Equation 8.172 so the general solutions are the same:

$$\mathscr{E}_z(r, \varphi) = (A\sin n\varphi + B\cos n\varphi) J_n(k_c r) \tag{8.204}$$

But in this case the boundary conditions on the E-field can be applied directly to the solutions for the Helmholtz equation:

$$\mathscr{E}_z(r = a, \varphi) = 0 \tag{8.205}$$

Therefore, from Equation 8.204 we have

$$J_n(k_c a) = 0 \tag{8.206}$$

and we get the cutoff parameter for the TM_{nm} mode from the zeros of J_n Bessel function. Table 8.3 shows some values for the roots of J_n. Once again, the cutoff parameters are obtained from

$$k_{c nm} = \frac{k_c a_{nm}}{a} \tag{8.207}$$

Table 8.3 *Some values of $(k_c a)_{nm}$ for TM in a cylindrical waveguide.*

n	$(k_c a)_{n1}$	$(k_c a)_{n2}$	$(k_c a)_{n3}$
0	2.405	5.520	8.654
1	3.832	7.016	10.174
2	5.135	8.417	11.620

The propagation parameter along z is

$$\beta_{nm} = \sqrt{k^2 - k_c^2} \qquad (8.208)$$

The transverse fields from Equations 8.183–8.186 and setting $A = 0$ are

$$E_r = i\frac{\beta}{k_c}\,(B\cos n\varphi)\,J_n'(k_c r)e^{i\beta z} \qquad (8.209)$$

$$E_\varphi = -i\frac{\beta n}{k_c^2 r}\,(B\sin n\varphi)\,J_n(k_c r)e^{i\beta z} \qquad (8.210)$$

$$H_r = i\frac{\omega\varepsilon n}{k_c^2 r}\,(B\sin n\varphi)\,J_n(k_c r)e^{i\beta z} \qquad (8.211)$$

$$H_\varphi = i\frac{\omega\varepsilon}{k_c}\,(B\cos n\varphi)\,J_n'(k_c r)e^{i\beta z} \qquad (8.212)$$

Finally, the mode impedance is given by

$$Z_{\text{TM}} = \frac{E_r}{H_\varphi} = \frac{\beta}{k}Z_0 \qquad (8.213)$$

8.8 Networks of transmission lines and waveguides

Earlier in this chapter we discussed the properties of lumped circuit elements and how these properties, such as capacitance, inductance, and resistance, can be adapted to distributed structures like transmission lines and waveguides. The purpose of the present section is to show how transmission lines and waveguides themselves can be grouped together into a network to achieve some predetermined design result without having to solve Maxwell's equations directly for elaborate source and boundary conditions. The ultimate goal is then to adopt these ideas to nanoscale structures in order to build networks of 2-D surface or gap plasmon waveguides, or 1D interfacial waveguides. The discussion here will be somewhat brief since an extended discussion of networked-transmission-line theory is really a digression from the subject of light–matter interaction. The idea is to introduce key concepts and notation so that readers not trained in electrical engineering can grasp the significance of these analysis tools.

8.8.1 Impedance and admittance matrices

Figure 8.12 shows a generalised waveguide network where we see propagation into and away from some arrangement of couplers, dividers, and amplifiers, etc. that constitute the network. Often our interest in focused on the inputs and outputs at the network 'ports', labelled in the figure as ports $S_1 - S_4$. The ports are assumed to be single-mode waveguides, and we can in principle, know the transverse wave properties and the propagation parameter of the active input and output ports. We can write the voltage and current passing the nth port entrance-exit plane:

$$V_n = V_n^+ + V_n^-$$
$$I_n = I_n^+ - I_n^-$$

and relating the voltages and currents by the impedance Z for each port of the four-port network of Figure 8.12,

$$\begin{bmatrix} V_1 \\ V_2 \\ V_3 \\ V_4 \end{bmatrix} = \begin{bmatrix} Z_{11} & Z_{12} & Z_{13} & Z_{14} \\ Z_{21} & Z_{22} & Z_{23} & Z_{24} \\ Z_{31} & Z_{32} & Z_{33} & Z_{34} \\ Z_{41} & Z_{42} & Z_{43} & Z_{44} \end{bmatrix} \cdot \begin{bmatrix} I_1 \\ I_2 \\ I_3 \\ I_4 \end{bmatrix} \tag{8.214}$$

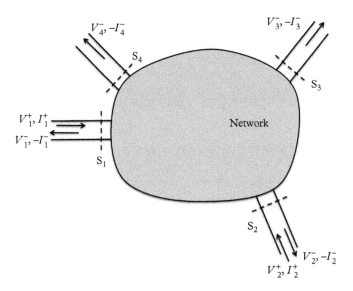

Figure 8.12 *A schematic waveguide network with two ports S_1 and S_2 showing inputs and outputs and two ports S_3 and S_4 showing only outputs. The outputs at S_1 and S_2 can be thought of as reflected waves.*

or in short-hand matrix notation,

$$[V] = [Z] \cdot [I] \tag{8.215}$$

The inverse of the impedance is called the admittance Y, and the inverse of the impedance matrix $[Z]$ is the admittance matrix $[Y]$:

$$[Y] = [Z]^{-1} \tag{8.216}$$

or written out explicitly for the four-port case, Equation 8.216 becomes

$$\begin{bmatrix} I_1 \\ I_2 \\ I_3 \\ I_4 \end{bmatrix} = \begin{bmatrix} Y_{11} & Y_{12} & Y_{13} & Y_{14} \\ Y_{21} & Y_{22} & Y_{23} & Y_{24} \\ Y_{31} & Y_{32} & Y_{33} & Y_{34} \\ Y_{41} & Y_{42} & Y_{43} & Y_{44} \end{bmatrix} \cdot \begin{bmatrix} V_1 \\ V_2 \\ V_3 \\ V_4 \end{bmatrix} \tag{8.217}$$

The matrix equation, Equation 8.215, means that for each element Z_{ij} we have

$$Z_{ij} = \frac{V_i}{I_j} \tag{8.218}$$

where all I_k with $k \neq j$ have been set to zero. Equation 8.218 means that we can determine the impedance Z_{ij} if we inject current I at port j and measure the resulting voltage V at port i with all the other ports closed or 'open-circuited' in engineering parlance. For the admittance matrix we have the analogous situation:

$$Y_{ij} = \frac{I_i}{V_j} \tag{8.219}$$

which means that in order to determine the admittance matrix element Y_{ij}, we apply a voltage V to port j and measure the current I at port i with all other ports disconnected from the network (open-circuited). It is important to remember that the network can be lossy, in which case the matrix elements Z_{ij} or Y_{ij} are complex. It can be shown that lossless networks have all Z_{ij} and Y_{ij} pure imaginary and that *reciprocal* networks are represented by $[Z]$ and $[Y]$ matrices that are symmetric with respect to an interchange of indices. That is, for example, $Z_{ij} = Z_{ji}$. Physically, the reciprocal property means that if a current injected at port m produces a voltage at port n, then injecting the same current in port n will produce the same voltage at port m. Networks of linear, passive components, such as resistors, capacitors, and inductors, are reciprocal. Networks with active non-linear components with gain, such as amplifiers, are not reciprocal.

8.8.2   Illustrative example: two-port voltage divider

We illustrate these ideas with a simple two-port voltage divider network depicted in Figure 8.13. The first step is to shut off port S_2 and inject current I_1 at port S_1. Then:

$$Z_{11} = \frac{V_1}{I_1} = Z_A + Z_C \tag{8.220}$$

because no current runs through Z_B. The 'transfer impedance' Z_{12} is obtained by shutting off port S_1 and injecting current into port S_2. No current flows through Z_A so all the injected current I_2 flows through Z_B and Z_C. Then $V_2 = I_2(Z_B + Z_C)$. The voltage V_1 appearing at the port S_1 is just the current I_2 flowing through the 'pull-up' resistor Z_C. Therefore, Z_{12} is Z_C. It is easy to verify that this network is reciprocal and that, therefore, $Z_{22} = Z_{11}$ and $Z_{12} = Z_{21}$.

8.8.3   The scattering matrix

For high-frequency networks in the microwave and optical regime, it is difficult to characterise voltage and current points because the transverse amplitudes across wavefronts need not be constant. Therefore, instead of using impedance and admittance, the scattering matrix (or S-matrix) is often employed instead. An S-matrix relates voltage *waves* incident and reflected at the various network ports. Incident and reflected power from travelling and standing waves lend themselves to easier measurement in high-frequency networks. Consider port 1 in Figure 8.12. The S-matrix relating the incident voltage wave amplitude V_1^+ to the reflected wave amplitude V_1^- is given by

$$\begin{bmatrix} V_1^- \\ V_2^- \\ V_3^- \\ V_4^- \end{bmatrix} = \begin{bmatrix} S_{11} & S_{12} & S_{13} & S_{14} \\ S_{21} & S_{22} & S_{23} & S_{24} \\ S_{31} & S_{32} & S_{33} & S_{34} \\ S_{41} & S_{42} & S_{43} & S_{44} \end{bmatrix} \cdot \begin{bmatrix} V_1^+ \\ V_2^+ \\ V_3^+ \\ V_4^+ \end{bmatrix} \tag{8.221}$$

(a)

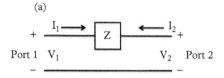

(b)

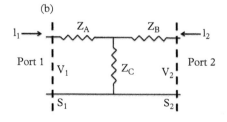

Figure 8.13 *(a) Two-port impedance network. (b) Two-port impedance network implementing a voltage divider. Note that the directions of I_1 and I_2 point into the ports towards the network.*

The usual rules of matrix multiplication obtain here. Thus, for a given network, the voltage amplitude of a reflected wave at port i is related to the voltage amplitude of an incident wave at port j by

$$V_i^- = S_{ij} V_j^+ \tag{8.222}$$

with the implicit assumption that all other ports are quiescent with voltages at the port inputs set to zero. The diagonal matrix elements express the ratio of reflected amplitude to incident amplitude at a given port, assuming all other ports are quiescent (i.e. matched impedances at all other ports). Thus, S_{ii} is a reflection coefficient at port i. Since the S-matrix is essentially intended for waveguide networks, the ports are assumed to have some characteristic impedance (for example, $Z_0 = 50\Omega$). The S_{ij} matrix element expresses the transmission coefficient from port j to port i.

8.8.4 Relation of S-matrix to Z-matrix

We know that the voltage and current at port n is the net result of transmission and reflection:

$$V_n = V_n^+ + V_n^- \tag{8.223}$$
$$I_n = I_n^+ - I_n^- \tag{8.224}$$

Now suppose that the characteristic impedance at port n is unity, $Z_{0n} = 1$. Then we can also write

$$I_n = I_n^+ - I_n^- = V_n^+ - V_n^- \tag{8.225}$$

and the voltage, current, and impedance matrices are related by

$$[V] = [Z] \cdot [I] = [Z] \cdot \left([V^+] - [V^-]\right) = [V^+] + [V^-]$$

Rearranging by grouping matrices $[V^+]$ and $[V^-]$ we have

$$[V^+] - [V^+][Z] = -[V^-] - [Z][V^-] = -\left([V^-] + [Z][V^-]\right) \tag{8.226}$$

We can matrix-factor this last matrix equation if we define a unit matrix $[U]$ with only elements of unity on the diagonal and zero elements elsewhere. Then we have

$$[V^+] \left([U] - [Z]\right) = -[V^-] \left([U] + [Z]\right) \tag{8.227}$$

or

$$[V^+]\,([Z]-[U]) = [V^-]\,([U]+[Z])$$
$$\frac{[V^-]}{[V^+]} = \frac{[Z]-[U]}{[Z]+[U]} \tag{8.228}$$

Now division by a matrix really means multiplication by the matrix inverse so

$$\frac{[V^-]}{[V^+]} = ([Z]-[U]) \cdot ([Z]+[U])^{-1} \tag{8.229}$$

and from the definition of the S-matrix, Equation 8.221, we can identify the S-matrix with the right-hand side of Equation 8.229:

$$[S] = \frac{[V^-]}{[V^+]} = ([Z]-[U]) \cdot ([Z]+[U])^{-1} \tag{8.230}$$

8.8.5   The ABCD matrix

Although the preceding sections have assumed that a network has n ports, in most practical cases circuits are designed with many two-port networks in series (in 'cascade' in engineering parlance). A given two-port network is described by the matrix equation

$$\begin{bmatrix} V_1 \\ I_1 \end{bmatrix} = \begin{bmatrix} A & B \\ C & D \end{bmatrix} \cdot \begin{bmatrix} V_2 \\ I_2 \end{bmatrix} \tag{8.231}$$

In equation form:

$$V_1 = AV_2 + BI_2 \tag{8.232}$$
$$I_1 = CV_2 + DI_2 \tag{8.233}$$

The diagram in Figure 8.14 shows how the ABCD matrix functions. If we have two networks in series, the overall result is obtained by matrix multiplication:

$$\begin{bmatrix} V_1 \\ I_1 \end{bmatrix} = \begin{bmatrix} A_1 & B_1 \\ C_1 & D_1 \end{bmatrix} \cdot \begin{bmatrix} A_2 & B_2 \\ C_2 & D_2 \end{bmatrix} \cdot \begin{bmatrix} V_3 \\ I_3 \end{bmatrix} \tag{8.234}$$

Figure 8.14 *Schematic ABCD network. Note that the positive current I_2 is pointing away from the port entrance. This ABCD-matrix convention differs from the S-matrix where all positive quantities point into the port.*

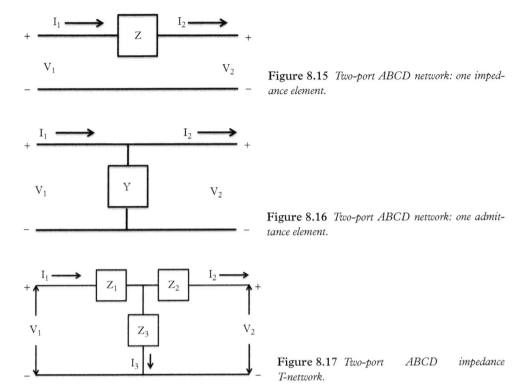

Figure 8.15 *Two-port ABCD network: one imped-ance element.*

Figure 8.16 *Two-port ABCD network: one admit-tance element.*

Figure 8.17 *Two-port   ABCD   impedance T-network.*

The ABCD matrices can be used to construct elaborate networks from a few simple elements. For example, suppose we have an elementary two-port network consisting of one impedance element, as shown in Figure 8.15. By applying Ohm's law we can identify the ABCD matrix elements. By inspection, it is easy to determine that $A = 1$, $B = Z$, $C = 0$, and $D = 1$. Another example is a two-port network with a single admittance as shown in Figure 8.16. Again, by Ohm's law and remembering the $Y = 1/Z$, a simple inspection of the coupled equations reveals that $A = 1, B = 0, C = Y$, and $D = 1$.

Figure 8.17 shows an impedance T-network. The ABCD matrix for this network can be obtained by a more systematic approach. First consider the A element and Equation 8.232. We see that if no current flows at the output (open circuit with $I_2 = 0$), then $A = V_1/V_2$. But in that case, from Ohm's law $V_1 = I_1 (Z_1 + Z_3)$ and $V_2 = I_1 Z_1$. Therefore, $A = (Z_1 + Z_3)/Z_1$. To determine the B element we set $V_2 = 0$ (output short circuit). Then $B = V_1/I_2$. The result of applying Ohm's law (or the Kirchhoff voltage and current rules) is that $B = Z_1 + Z_2 + Z_1 Z_2/Z_3$. Similarly, we consider Equation 8.233 and set successively $I_2, V_2 = 0$ to find the matrix elements C and D. The result is $C = 1/Z_3$ and $D = 1 + Z_2/Z_3$.

Another common network is the admittance network shown in Figure 8.18. This arrangement is often called the Pi-network. By methodically 'open-circuiting' and 'shorting' the output terminals, the elements of the ABCD matrix can be determined.

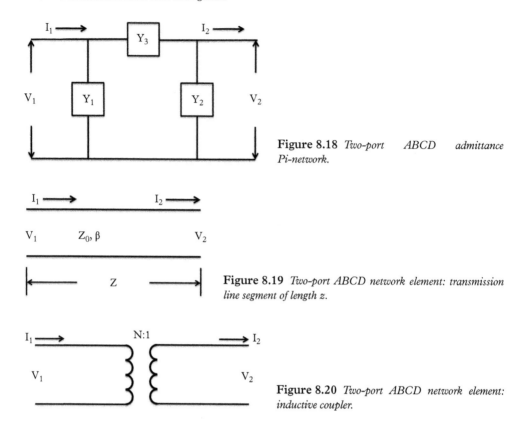

Figure 8.18 *Two-port   ABCD   admittance Pi-network.*

Figure 8.19 *Two-port ABCD network element: transmission line segment of length z.*

Figure 8.20 *Two-port ABCD network element: inductive coupler.*

Two more elementary functions useful for constructing more elaborate networks are the transmission line segment and the inductive transformer. Figures 8.19 and 8.20 sketch out these two circuit functions, and the ABCD matrix elements for all these elementary circuit functions are entered in Table 8.4.

Table 8.4 *ABCD elements for frequently used network functions.*

Network	A	B	C	D
Single Z	1	Z	0	1
Single Y	1	0	Y	1
T-Impedance	$1 + \frac{Z_1}{Z_3}$	$Z_1 + Z_2 + \frac{Z_1 Z_2}{Z_3}$	$\frac{1}{Z_3}$	$1 + \frac{Z_2}{Z_3}$
Pi-Admittance	$1 + \frac{Y_2}{Y_3}$	$\frac{1}{Y_3}$	$Y_1 + Y_2 + \frac{Y_1 Y_2}{Y_3}$	$1 + \frac{Y_1}{Y_3}$
Ideal Transmission Line	$\cos \beta z$	$-i Z_0 \sin \beta z$	$-\frac{i}{Z_0} \sin \beta z$	$\cos \beta z$
Real Transmission Line	$\cosh \beta z$	$Z_0 \sinh \beta z$	$\frac{1}{Z_0} \sinh \beta z$	$\cosh \beta z$
Inductive Coupler	N	0	0	$\frac{1}{N}$

8.8.6   Reciprocal and lossless networks

Most common network elements such as capacitors, inductors, and resistors are linear, and passive in the sense that they do not contribute to voltage gain. The matrix elements representing these networks are in general complex quantities. Therefore, in general, a Z, Y, S, or ABCD matrix will contain $2N^2$ independent elements, where N is the number of ports in the network. For components with linear, passive response, however, the matrices are always *symmetric* so that $X_{ij} = X_{ji}$. All the network matrix representations considered here are symmetric. Furthermore, it can also be shown that if the network is lossless, containing only reactive elements that store power but do not dissipate it, the network elements will be pure imaginary quantities. Coupling networks that contain capacitance and induction but no significant resistance fall into this category.

8.8.7   Comparison of impedance matrix and ABCD matrix

We can specialise the impedance matrix, discussed in Section 8.8.1, to the two-port network,

$$\begin{bmatrix} V_1 \\ V_2 \end{bmatrix} = \begin{bmatrix} Z_{11} & Z_{12} \\ Z_{21} & Z_{22} \end{bmatrix} \cdot \begin{bmatrix} I_1 \\ -I_2 \end{bmatrix} \tag{8.235}$$

Note that the sign of I_2 has been reversed. The reason is that we seek to compare the ABCD network scheme with the Z scheme. But the impedance network shown in Figure 8.13 assumes that positive current I_2 points *into* port 2, whereas in the ABCD network I_2 points outwards from the port. Written out in equation form we have

$$V_1 = Z_{11}I_1 - Z_{12}I_2 \tag{8.236}$$

$$V_2 = Z_{21}I_1 - Z_{22}I_2 \tag{8.237}$$

The set of Z-coupled equations, Equations 8.236 and 8.237, must be compared to the set of ABCD-coupled equations, Equations 8.232 and 8.233. Suppose we take I_2 output current to zero (open-circuit port 2). Then from Equation 8.232 $A = V_1/V_2$ and from Equations 8.236 and 8.237 $Z_{11} = V_1/I_1$ and $Z_{21} = V_2/I_1$. From these relations we find that the A element of the ABCD matrix is related to the impedance matrix elements by $A = Z_{11}/Z_{21}$. If we short-circuit port 2 ($V_2 = 0$) we find $B = Z_{11}(Z_{22}/Z_{21}) - Z_{12}$. Proceeding similarly for C and D, we find $C = 1/Z_{21}$ and $D = Z_{22}/Z_{21}$.

The equivalence between the ABCD matrix elements, the impedance matrix elements, and the impedance-equivalent circuit for the T-network is summarised in Table 8.5. The T-network, implemented with impedance elements, is shown in Figure 8.21.

Table 8.5 *Equivalence between ABCD matrix and Z-matrix elements in two-port networks.*

ABCD Matrix	Z-Matrix	T-Network
A	$\dfrac{Z_{11}}{Z_{21}}$	$1 + \dfrac{Z_1}{Z_3}$
B	$\dfrac{Z_{11}Z_{22} - Z_{12}Z_{21}}{Z_{21}}$	$Z_1 + Z_2 + \dfrac{Z_1 Z_2}{Z_3}$
C	$\dfrac{1}{Z_{21}}$	$\dfrac{1}{Z_3}$
D	$\dfrac{Z_{22}}{Z_{21}}$	$1 + \dfrac{Z_2}{Z_3}$

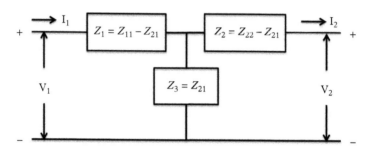

Figure 8.21 *T-Network equivalent circuit showing Z_1, Z_2, Z_3 impedances in terms of Z-matrix elements.*

8.9 Nanostructures and equivalent circuits

8.9.1 Nanosphere driven by a harmonic electric field

We discuss how the response of nanoscale objects to light in the visible-near IR range can be considered as an 'equivalent circuit'. Our discussion follows the treatment of N. Engheta and his research group (Reference 3. listed in Section 8.12). We begin by revisiting the sphere subject to an electric field that we considered in Chapter 2, Section 2.6. Let us consider a homogenous sphere of nanoscale dimension, radius r_0, immersed in a harmonically varying E-field, $E = E_0 e^{-i\omega t}$. We posit the permittivity of the material as ε without specifying, for the present, whether the sphere is dielectric or conductive. Since the sphere is subwavelength in size, we can use the quasistatic approximation for the E-fields in the space within and outside the sphere. We have already obtained these fields in Equations 2.121 and 2.122, but we rewrite them here for convenience. The permittivity of the sphere itself is labelled ε_{in} and outside the sphere the permittivity is ε_{out}:

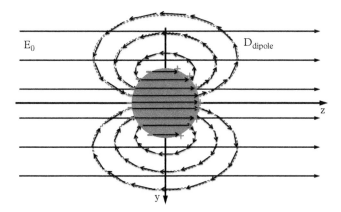

Figure 8.22 *Polarised sphere subject to an external E-field*
$E_z = E_0 e^{-i\omega t}$. E-field lines are sketched in the vicinity of the
sphere, and displacement field (D-field) lines are shown within the
sphere and looping outside the sphere in a dipole pattern. Note that
the D-field lines normal to the surface are continuous through the
interface.

$$E_{in}(r, \theta) = \left[\frac{3\varepsilon_{out}}{\varepsilon_{in} + 2\varepsilon_{out}}\right] E_0 \hat{e}_z \tag{8.238}$$

$$E_{out}(r, \theta) = E_0 \hat{e}_z + E_0 \left[\frac{\varepsilon_{in} - \varepsilon_{out}}{\varepsilon_{in} + 2\varepsilon_{out}}\right] \frac{r_0^3}{r^3} \left(2 \cos \theta \, \hat{e}_r + \sin \theta \, \hat{e}_\theta\right) \tag{8.239}$$

Figure 8.22 shows the shape and direction of the applied E-field lines (along the z direction) in the vicinity of the sphere, and the displacement field inside the sphere, $\mathbf{D}_{in} = \varepsilon_{in} \mathbf{E}_{in}$, and outside the sphere, $\mathbf{D}_{out} = \varepsilon_{out} \mathbf{E}_{out}$. At the surface of the sphere, matching conditions impose that the normal component of the displacement field be continuous across the surface:

$$\varepsilon_{in} \mathbf{E}_{in} \cdot \hat{n} = \varepsilon_{out} \left[E_0 \hat{e}_z \cdot \hat{n} + E_0 \left[\frac{\varepsilon_{in} - \varepsilon_{out}}{\varepsilon_{in} + 2\varepsilon_{out}}\right] (2 \cos \theta \hat{e}_r \cdot \hat{n} + \sin \theta \hat{e}_\theta \cdot \hat{n})\right]$$

$$\varepsilon_{in} \left[\frac{3\varepsilon_{out}}{\varepsilon_{in} + 2\varepsilon_{out}}\right] E_0 \hat{e}_z \cdot \hat{n} = \varepsilon_{out} \left[E_0 \hat{e}_z \cdot \hat{n} + E_0 \left[\frac{\varepsilon_{in} - \varepsilon_{out}}{\varepsilon_{in} + 2\varepsilon_{out}}\right] 2 \hat{e}_z \cdot \hat{n}\right] \tag{8.240}$$

The applied displacement field, parallel to z, *inside the sphere* is given by

$$\mathbf{D}_{in_{appl}} = \varepsilon_{in} \mathbf{E}_{in} = \varepsilon_{in} E_0 \hat{e}_z \tag{8.241}$$

The *net* or residual displacement field within the sphere is the difference between the *total* displacement field (left-hand side of Equation 8.240) and the *applied* displacement field:

$$\mathbf{D}_{\text{net}} = \varepsilon_{\text{in}} \left[\frac{\varepsilon_{\text{out}} - \varepsilon_{\text{in}}}{\varepsilon_{\text{in}} + 2\varepsilon_{\text{out}}} \right] E_0 \hat{e}_z \qquad (8.242)$$

The component normal to the sphere surface of this net displacement field is

$$\mathbf{D}_{\text{net}} \cdot \hat{n} = \varepsilon_{\text{in}} \left[\frac{\varepsilon_{\text{out}} - \varepsilon_{\text{in}}}{\varepsilon_{\text{in}} + 2\varepsilon_{\text{out}}} \right] E_0 \hat{e}_z \cdot \hat{n} \qquad (8.243)$$

Now we subtract Equation 8.243 from both sides of Equation 8.240 and rearrange the terms to put the applied fields inside and outside the sphere on the left and everything else on the right:

$$(\varepsilon_{\text{in}} - \varepsilon_{\text{out}}) E_0 \hat{e}_z \cdot \hat{n} = \varepsilon_{\text{in}} \left[\frac{\varepsilon_{\text{in}} - \varepsilon_{\text{out}}}{\varepsilon_{\text{in}} + 2\varepsilon_{\text{out}}} \right] E_0 \hat{e}_z \cdot \hat{n} + \varepsilon_{\text{out}} \left[\frac{\varepsilon_{\text{in}} - \varepsilon_{\text{out}}}{\varepsilon_{\text{in}} + 2\varepsilon_{\text{out}}} \right] E_0 2 \hat{e}_z \cdot \hat{n} \quad (8.244)$$

The displacement current is given by the time derivative of the displacement field. Taking the time derivative of Equation 8.244 and integrating over the surface of the half-sphere with positive charge in Figure 8.22 yields

$$-i\omega (\varepsilon_{\text{in}} - \varepsilon_{\text{out}}) E_0 \cdot \pi r_0^2 = -i\omega \varepsilon_{\text{in}} \left[\frac{\varepsilon_{\text{in}} - \varepsilon_{\text{out}}}{\varepsilon_{\text{in}} + 2\varepsilon_{\text{out}}} \right] E_0 \cdot \pi r_0^2$$

$$- i\omega \varepsilon_{\text{out}} \left[\frac{\varepsilon_{\text{in}} - \varepsilon_{\text{out}}}{\varepsilon_{\text{in}} + 2\varepsilon_{\text{out}}} \right] E_0 \cdot 2\pi r_0^2 \qquad (8.245)$$

The integration over the surface is carried out by noting that the $\hat{e}_z \cdot \hat{n}$ factor in Equation 8.244 is $\cos \theta$ and that therefore the integration over an infinitesimal of half the sphere surface dS is

$$r_0^2 \int \hat{e}_z \cdot \hat{n} \, dS = r_0^2 \int_0^{\pi/2} \cos \theta \sin \theta \, d\theta \, d\varphi$$

$$= r_0^2 2\pi \left[\frac{1}{2} \sin^2 \theta \right]_0^{\pi/2} = \pi r_0^2$$

The term on the left-hand side of Equation 8.245 is a bound current source arising from the polarisation of the sphere due to the external applied electric field. The first term on the right is the bound current passing through the sphere surface and the second term is the external 'dipole current' looping back to the source (Figure 8.22). The three terms of Equation 8.245 express the Kirchhoff current law (Section 8.2.1): that all currents entering and leaving a node in a circuit must sum to zero.

8.9.1.1   Dielectric sphere

We suppose for the moment that ε_{in} represents an (almost) lossless dielectric. In that case the permittivity of the sphere is complex but with a negligible imaginary term and $\varepsilon_{\text{in}} > \varepsilon_0$. The circuit of the displacement current can then be represented as in

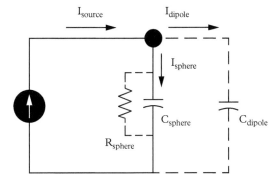

Figure 8.23. Now from the electric field at the surface of the sphere we can find the potential difference across the sphere. The potential is given by

$$V_{\text{sphere}} = -\int_{\sigma} \mathbf{E}_{\text{surf}} \cdot d\mathbf{l}$$

$$= -2 \int_{-\pi/2}^{0} \left[\frac{\varepsilon_{\text{out}} - \varepsilon_{\text{in}}}{\varepsilon_{\text{in}} + 2\varepsilon_{\text{out}}} \right] E_0 \cos\theta \, r_0 \sin\theta \, d\theta$$

$$= r_0 \left[\frac{\varepsilon_{\text{out}} - \varepsilon_{\text{in}}}{\varepsilon_{\text{in}} + 2\varepsilon_{\text{out}}} \right] E_0 \tag{8.246}$$

Now from V_{sphere} and I_{sphere} we can identify the characteristic impedance of the sphere as

$$Z_{\text{sphere}} = \frac{V_{\text{sphere}}}{I_{\text{sphere}}}$$

$$= \frac{i}{\omega \varepsilon_{\text{in}} \pi \, r_0} \tag{8.247}$$

and the impedance of the dipolar field is

$$Z_{\text{dipole}} = \frac{V_{\text{sphere}}}{I_{\text{dipole}}}$$

$$= \frac{i}{\omega \varepsilon_{\text{out}} 2\pi \, r_0} \tag{8.248}$$

From the phasor form of the impedance (Section E.2 in Appendix E) we have

$$Z = R + i\,(X_L + X_C) \tag{8.249}$$

$$X_L = -\omega L \qquad \text{inductive reactance} \tag{8.250}$$

$$X_C = \frac{1}{\omega C} \qquad \text{capacitive reactance} \tag{8.251}$$

Comparing Equation 8.251 with Equations 8.247 and 8.248 we see that

$$C_{\text{sphere}} = \varepsilon_{\text{in}} \pi r_0 \tag{8.252}$$
$$C_{\text{dipole}} = \varepsilon_{\text{out}} 2\pi r_0 \tag{8.253}$$

8.9.1.2 Metallic sphere

In the case of a metallic sphere, the real part of the permittivity can be strongly negative and the resistive loss non-negligible. In that case, the equivalent circuit can be construed as shown in Figure 8.24. The sphere impedance becomes

$$Z_{\text{sphere}}^{\text{metal}} = \frac{i}{\omega \text{Re}[\varepsilon_{\text{in}}] \pi r_0} \tag{8.254}$$

and comparing Equation 8.254 with Equation 8.250 we see that

$$L_{\text{sphere}}^{\text{metal}} = -\frac{i}{\omega^2 \text{Re}[\varepsilon_{\text{in}}] \pi r_0} \tag{8.255}$$

Thus, we see that the nanosphere can act as a capacitive element or an inductive element by choosing the material property: insulating or conducting.

8.9.2   Equivalent circuit for plasmon surface waves

In this section we reconsider surface waves at the interface between a dielectric and metal. We apply what we have learned about planar transmission lines, their relation to plane waves (Section 8.3.4), and the equivalent circuits that can represent a transmission line. The basic idea is to treat a plane wave incident normally on the interface as a planar transmission line and the interface itself as a junction between two transmission lines: one in the dielectric and the other in the metal.

8.9.2.1 Transmission line equivalent circuit

A plane wave impinging on a dielectric-metal interface can be modelled as a real transmission line. Figure 8.25 shows in the left panel the plane wave incident normal to the

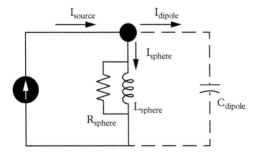

Figure 8.24 *Equivalent circuit of a subwavelength metallic sphere subject to an external oscillating electric field. The source displacement current drives the sphere and Kirchhoff's current rule applies to the branch point node. Current through the sphere exhibits inductive reactance in parallel with a non-negligible resistive loss. The dipole branch exhibits capacitive reactance as in the case of the dielectric sphere.*

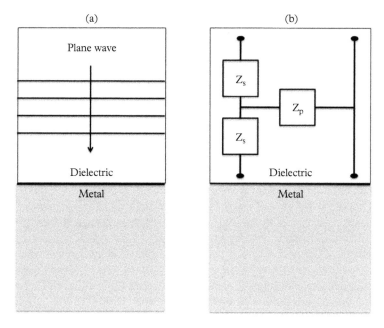

Figure 8.25 *Comparison between plane wave and equivalent impedance T-network. Panel (a) Plane wave incident on a dielectric-metal interface. Panel (b) Equivalent circuit for a transmission line representing the plane wave.*

interface and the right panel the equivalent impedance T-network representing a transmission line. The expressions for the impedances are (see Reference 4 in Section 8.12),

$$Z_s = Z_c \tanh\left(\frac{\beta l}{2}\right) \tag{8.256}$$

$$Z_p = \frac{Z_c}{\sinh(\beta l)} \tag{8.257}$$

where

$$Z_c = \frac{\gamma_c}{i\omega\varepsilon_c} \tag{8.258}$$

and

$$\gamma_c^2 = \beta^2 - \varepsilon_c k_0^2 = -k_c^2 \tag{8.259}$$

where β is the complex propagation constant, l the length of the transmission line, Z_c the characteristic impedance, and ε_c the characteristic dielectric constant of the medium. Using these expressions for the series impedance Z_s and parallel impedance Z_p, and applying the rules for finding the ABCD matrix elements of the T-circuit we developed in Section 8.8.5, we find that for long transmission lines ($l \gg 1/\beta$):

$$A = \cosh \beta l \tag{8.260}$$

$$B = Z_0 \sinh \beta l \tag{8.261}$$

$$C = \frac{1}{Z_0} \sinh \beta l \tag{8.262}$$

$$D = \cosh \beta l \tag{8.263}$$

in accordance with the expressions listed in Table 8.4. Now as the length of the transmission line tends to infinity, $Z \to 0$, and in the limit of very long lines we have the situation shown in Figure 8.26. In fact, the plane wave propagates through the dielectric and the metal, although the propagation length in the metal penetrates only as far as the skin depth. Therefore, we can characterise the light incident on the interface as two transmission lines connected at the interface with two characteristic impedances, Z_0^d and Z_0^m, for the dielectric and metal, respectively. Figure 8.27 shows the correspondence between plane wave propagation in the dielectric and metal, and the equivalent circuit of the interface. Now the transmission line impedances can be written in a particularly useful form if we consider Maxwell's equations in a Laplace-transformed space. The Laplace transform is an integral operator of the form

$$\hat{L} = \int_0^\infty e^{-\gamma z}\, dz \tag{8.264}$$

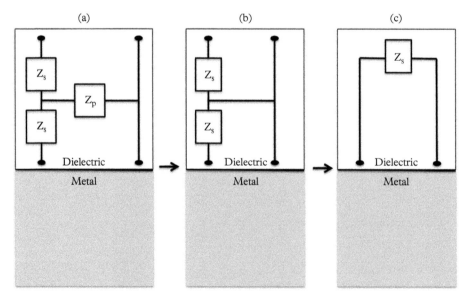

Figure 8.26 *The limiting form of the equivalent circuit of a long transmission line.*

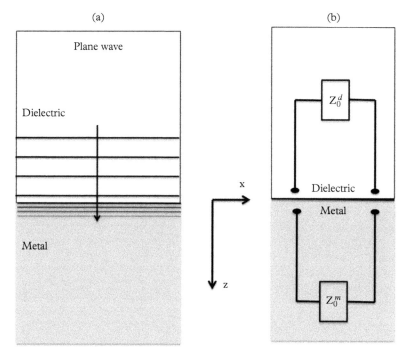

Figure 8.27 *Correspondence between plane waves propagating in dielectric and metal, Panel (a), and two equivalent circuit transmission lines joined at the dielectric-metal interface, Panel (b).*

and transformation of field $\mathbf{A}(x, z)$ is expressed by,

$$\tilde{\mathbf{A}}(\gamma_0) = \hat{L}\mathbf{A}(x, z) = \int_0^\infty \mathbf{A}(x, z)e^{-\gamma z}\, dz \qquad (8.265)$$

with

$$\gamma^2 = -k_z^2 \qquad (8.266)$$

where γ may be complex to take into account transmission line losses and k_z the propagation parameter in the z direction. Details of this transformation and its consequences are discussed in Reference 6 in Section 8.12. Here, we simply state one of the key results for the Maxwell–Ampère law in Laplace space for incident waves with TM polarisation:

$$\gamma \tilde{H}_y = i\omega\varepsilon\tilde{E}_x \qquad (8.267)$$

where $\tilde{H}_y$ and $\tilde{E}_x$ are the Laplace-transformed y and x components of the H- and E-field, respectively; ω and ε have their usual meanings of frequency and permittivity. We identify the characteristic impedance of the transmission line with the corresponding impedance of the Laplace-transformed wave:

$$Z_0 = \frac{\tilde{E}_x}{\tilde{H}_y} = \frac{\gamma}{i\omega\varepsilon} \tag{8.268}$$

Any surface waves travelling in the x direction must have the same propagation parameter β in the dielectric and metal materials. With k_0, the wave propagation vector in free space, we have from a k-vector addition on each side of the interface:

$$\gamma_d = \sqrt{\beta^2 - \varepsilon_d k_0^2} = ik_z^d \tag{8.269}$$

$$\gamma_m = \sqrt{\beta^2 - \varepsilon_m k_0^2} = ik_z^m \tag{8.270}$$

The picture we have now is of two transmission lines along z that join at the dielectric-metal interface. The characteristic impedances for the dielectric and metal transmission lines are

$$Z_0^d = \frac{\gamma_d}{i\omega\varepsilon_d} \tag{8.271}$$

$$Z_0^m = \frac{\gamma_m}{i\omega\varepsilon_m} \tag{8.272}$$

8.9.2.2   Transmission line transverse resonance

When a transmission line is populated by a wave that is just at the cutoff condition, k_c, a standing wave is present in the line, and no net energy propagates along the line. A transmission line in this state is said to be in 'resonance'. The two transmission lines, along z in our problem, are transverse to the interface x. In order for stable surface waves to propagate along the interface, they must be subject to the same restriction that no energy propagates along z. Therefore, a necessary condition for stable, propagating surface waves along x is that our transmission lines along z fulfil the resonance condition. Since no net wave propagates along z, the net propagation parameter along z must be null. Therefore, according to Equations 8.271 and 8.272, a necessary condition for transverse resonance in our two joined transmission lines is

$$Z_0^d = Z_0^m \qquad \text{for all } z \tag{8.273}$$

Therefore:

$$\frac{\gamma_d}{i\omega\varepsilon_d} = \frac{\gamma_m}{i\omega\varepsilon_m} \tag{8.274}$$

or

$$\frac{k_z^d}{\varepsilon_d} = \frac{k_z^m}{\varepsilon_m} \tag{8.275}$$

This 'impedance matching' condition at the interface, arising from the transmission line at resonance, is equivalent to the field-matching condition that we found in Section 7.3, Equation 7.32. Thus, we see that a transmission line point of view gives us a supplementary insight into the physical conditions required for stable surface waves.

8.9.3 Equivalent circuit of a dielectric slit in a metal layer

In this section we consider the equivalent circuit of a dielectric slit milled in a metal layer. We suppose that both the thickness of the metal layer and the slit width are sub-wavelength. Figure 8.28 shows a schematic of the physical slit and the corresponding impedance equivalent circuit. The equivalent circuit consists of the T-circuit, which we considered earlier in Section 8.9.2 for the slit and two symmetrical voltage sources, and impedances that represent voltage and current running in the metal near the surfaces. We posit that a plane wave propagating along z is incident on the top surface and sets up a standing wave there. We saw in Section 3.2.3 that the magnetic field component of the wave is adjacent to the surface and penetrates to the skin depth within a real metal. This time-harmonically oscillating magnetic field, via Faraday's law, induces currents and voltages within the skin depth on the top surface and on the slit walls. A detailed discussion of the currents circulating in the skin depth near the slit is given in Reference 7. in Section 8.12.

The first step in the analysis is to assume that the circuit is symmetrical with the metal wall impedances $Z_1 = Z_2 = Z_m$. Then we consider the circuit with only one source, say the one on the left associated with Z_1 in Figure 8.28, and effectively 'short-circuit' the other source on the right. The resulting circuit consists of a single voltage source and a series-parallel network of impedances that can be combined into a single impedance Z_T. The expression for Z_T is

$$Z_T = \left[\frac{(Z_m + Z_{sd})Z_{pd} + (Z_m + Z_{sd})(Z_m + Z_{sd} + Z_{pd})}{Z_m + Z_{sd} + Z_{pd}} \right] \tag{8.276}$$

Now we impose the 'resonant transmission line' condition to ensure that no power is propagated perpendicular to the slit walls (along the x direction). As we saw in Section 7.6, power is propagated only along z with exponential fall-off of the fields penetrating the metal walls of the slit. This condition is that $Z_T = 0$, which results in the following impedance expression:

$$\left(Z_{sd} + Z_{pd}\right)^2 + (2Z_m)\left(Z_{sd} + Z_{pd}\right) + Z_m^2 - Z_m + Z_{pd}^2 = 0 \tag{8.277}$$

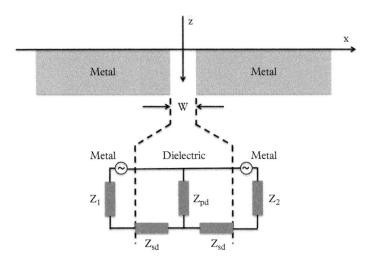

Figure 8.28 *Schematic of a subwavelength slit of width w milled into a subwavelength metal layer. The dielectric in the slit is the same as the dielectric above and below the metal surface. In the usual case the metal is the same on both sides of the slit so $Z_1 = Z_2$. The equivalent circuits of the two dielectric-metal interfaces and the T-circuit of the slit itself are shown in the impedance equivalent circuit.*

Adapting the definitions in Equations 8.256–8.259, we have

$$Z_{sd} = Z_d \tanh \frac{\gamma_d W}{2} \tag{8.278}$$

$$Z_{pd} = \frac{Z_d}{\sinh \gamma_d W} \tag{8.279}$$

where

$$Z_d = \pm \frac{\gamma_d}{i\omega\varepsilon_d} \tag{8.280}$$

and

$$\gamma_d^2 = \beta^2 - \varepsilon_d k_0^2 = -k_d^2 \tag{8.281}$$

with W the width of the slit. Substituting these definitions into Equation 8.277 we arrive, after some algebra, at

$$\tanh \gamma_d W = \pm \frac{2 Z_d Z_m}{Z_d^2 + Z_m^2} = \pm \frac{2 \, (Z_m/Z_d)}{1 + (Z_m/Z_d)^2} \tag{8.282}$$

This expression is the transcendental equation that specifies the allowed modes in the slit that propagate in z and are stable against radiation along x. Now recalling the identity,

$$\tanh x = \frac{2 \tanh (x/2)}{1 + [\tanh (x/2)]^2} \qquad (8.283)$$

and comparing with Equation 8.282 we can identify that

$$\tanh \frac{\gamma_d W}{2} = \pm \frac{Z_m}{Z_d} \qquad (8.284)$$

From the definition of the hyperbolic tangent in terms of exponential functions, this last expression can be further simplified to

$$e^{\gamma_d W} \left[\frac{Z_d + Z_m}{Z_d - Z_m} \right] = \pm 1 \qquad (8.285)$$

This is the final expression for the slit transcendental function in terms of the equivalent circuit quantities. It is consistent with the transcendental equations found for the slit in Chapter 7, obtained from a wave treatment.

8.10 Summary

The chapter begins with a review of Kirchhoff's rules for voltage and current that form the basis of conventional circuit theory. Lumped circuit analysis of transmission lines is next followed by the prototypical lossless, plane-parallel transmission line. The close correspondence between plane waves and transmission lines is emphasised. Special termination cases and line impedances close the discussion on transmission lines, which is really a precursor to waveguides. Waveguides are divided into two types: TE modes and TM modes (not to be confused with TE and TM polarisation). Propagating solutions to the two common geometries, flat slab and cylindrical waveguides, are worked out in detail. The next section treats networks of waveguides. The discussion then shifts to nanostructures and equivalent circuits, including plasmonic waveguiding in a slit geometry.

8.11 Exercises

1. Suppose two lossless transmission lines with different characteristic impedances are coupled. If $Z_1 = 50 \, \Omega$ and $Z_2 = 75 \, \Omega$, calculate the fractional power reflected and transmitted at the interface between the two lines.

2. For a transmission line of length $\lambda/4$, calculate the impedance of the line if it is: (a) shorted, (b) open-circuited.

3. For a rectangular waveguide, calculate the cutoff wave vector k_c for TM modes if the guide is square with side $s = 800$ nm and if the guide is very thin with $s_1 = 800$ nm and $s_2 = 100$ nm. Assume the guide is excited by a plane wave of $\lambda = 500$ nm impinging normally on the front face of the guide. Calculate the characteristic impedance of this transmission line.

4. Consider the interface between silver and glass. Assume that 1 mW power is coupled to the interface and converted to surface plasmon polariton (SPP) waves. From what you learned in Chapter 7, calculate the stable surface wave propagation parameter assuming the exciting field wavelength is 832 nm. Use the following dielectric constants: glass $\varepsilon' = 2.40$, silver $\varepsilon = -32.8 + i0.46$.

5. Using the same data as in Exercise 4., calculate the impedance Z_p of the equivalent T-circuit and the characteristic impedance Z_0 of the corresponding transmission line.

6. Calculate the TE and TM modes that can propagate in a planar metal-dielectric-metal (MDM) waveguide with water between the two metal walls. The distance between the two metal slabs is 0.25 μm and the free-space wavelength is 1.55 μm. The dielectric constant of water at $\lambda_0 = 1.55\mu$m can be considered real and equal to 1.77.

7. Calculate (a) the capacitance of a glass nanosphere with radius equal to 50 nm and (b) the capacitive reactance of the sphere at an excitation wavelength of 832 nm. The dielectric constant of glass is 2.25.

8. Suppose we have a sphere with the same dimensions as in Exercise 7 but fabricated from silver metal. Using the dielectric constant data of Exercise 4, calculate (a) the inductance of the sphere and (b) the inductive reactance.

8.12   Further reading

1. D. M. Pozar, *Microwave Engineering*, 3rd edition, John Wiley & Sons, Hoboken, NJ (2005).

2. S. Ramo, J. R. Whinnery, and T. Van Duzer, *Fields and Waves in Communication Electronics*, 3rd edition, John Wiley & Sons, New York (1994).

3. N. Engheta, A. Saladrino, and A. Alù, *Circuit elements at optical frequencies: nano-inductors, nano-capacitors, and nano-resistors. Phys Rev Lett* vol 95, p. 095504 (2005).

4. C. D. Papageorgiou and J. D. Kanellopoulos, *Equivalent circuits in Fourier space for the study of electromagnetic field. J Phys A: Math Gen* vol 15, pp. 2569–2580 (1982).

5. S. E. Kocabaş, G. Veronis, D. A. B. Miller, and S. Fan, *Transmission Line and Equivalent Circuit Models for Plasmonic Waveguide Components. IEEE J Sel Topics Quant Elec* vol 14, pp. 1462–1472 (2008).

6. F. Nunes and J. Weiner, *Equivalent Circuits and Nanoplasmonics. IEEE T. Nano* vol 8, pp. 298–302 (2009).

7. J. Weiner, *The electromagnetics of light transmission through subwavelength slits in metallic films. Opt Express* vol 19, pp. 16139–16153 (2011).

9

Metamaterials

9.1 Introduction

The term 'metamaterials' describes objects composed of conventional materials but with at least one length dimension well below an optical wavelength. The subwavelength scale in the structural composite gives rise to optical properties unique to the structure and distinct from the bulk properties of the individual components. This topic is an active, rapidly developing research area, and therefore, this chapter must be considered more a status report than an exposition of canonical received wisdom. We discuss the distinguishing features of light interacting with metamaterials with emphasis on their simplest realisations: slab waveguides and periodic layered material in which strong permittivity modulation from positive (dielectrics) to negative (metals) gives rise to uncommon optical behaviour. We begin with a brief discussion of two of these unconventional phenomena that have figured importantly in the motivation for metamaterials development: left-handed materials and negative refractive index [1]. The propagation of light through stacked layers is then treated first at conventional length scales [2] and then again in the subwavelength domain.

9.2 Left-handed materials

An elementary expression in physical optics relates the phase velocity of an electromagnetic wave propagating through a material to the permeability μ and permittivity ε of that material:

$$v = \frac{1}{\sqrt{\mu\varepsilon}} \tag{9.1}$$

When light propagates in vacuum the relation is

$$c = \frac{1}{\sqrt{\mu_0\varepsilon_0}} \tag{9.2}$$

Light-Matter Interaction. Second Edition. John Weiner and Frederico Nunes.
© John Weiner and Frederico Nunes 2017. Published 2017 by Oxford University Press.

Normally one only considers the positive solution of the square root, but in fact if μ and ε are both negative, then their product is still positive but the *negative root* can (and should) be assigned to the phase velocity. This possibility was discussed early on in a seminal paper by Veselago [3]. For plane harmonic waves in an isotropic material the wave vector k is related to v through the frequency ω by

$$k = \frac{\omega}{v} \tag{9.3}$$

and consequently if $v < 0$ then $k < 0$. Since $k = nk_0$, where n is the refractive index of the material and $k_0 = \omega/c$, the refractive index of a material with negative permeability and permittivity is also negative. Suppose we have a plane wave propagating in an isotropic material such that the E, H fields are oriented along the x, y axes in the positive direction:

$$E_x = E_0 e^{ik_z z} e^{-i\omega t} \tag{9.4}$$
$$H_y = H_0 e^{ik_z z} e^{-i\omega t} \tag{9.5}$$

Then according to Maxwell's curl equations

$$k_z E_x = \omega \mu H_y \tag{9.6}$$
$$k_z H_y = \omega \varepsilon E_x \tag{9.7}$$

Clearly if $\mu, \varepsilon > 0$ we have a conventional 'right-handed' relation between E_x, H_y, k_z. If μ, ε are both negative, however, k_z points along $-z$ when E_x, H_y point along x, y; and the relation among the three vectors is said to be 'left-handed'. Figure 9.1 shows the relationship between E, H, k in right-handed and left-handed systems. A 'left-handed' material is therefore one with *both* $\mu, \varepsilon < 0$. In another seminal article, Pendry [4] pointed out that a slab of left-handed material would focus propagating *and* evanescent waves emanating from a source, and therefore would be the realisation of a 'perfect' lens, unfettered by the usual imaging diffraction limit.

Up to this point analysis of the properties of left-handedness in optics was entirely speculative because no known natural material possessed the necessary property of both negative permeability and permittivity. However, if the spatial gradients of the E, H fields are very small (i.e. the characteristic spatial dimensions are 'subwavelength')

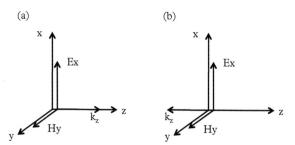

(a)   (b)

Figure 9.1 *(a) Right-handed material. (b) Left-handed material.*

then we get an important simplification in the equations governing propagation. We remember from Equations 7.3–7.8 of Chapter 7 that

$$\frac{\partial E_y}{\partial z} = -(i\omega\mu_0\mu)H_x$$

$$\frac{\partial H_x}{\partial z} - \frac{\partial H_z}{\partial x} = -i(\omega\varepsilon_0\varepsilon)E_y \qquad \text{TE polarisation}$$

$$\frac{\partial E_y}{\partial x} = (i\omega\mu_0\mu)H_z$$

and

$$\frac{\partial H_y}{\partial z} = (i\omega\varepsilon_0\varepsilon)E_x$$

$$\frac{\partial E_x}{\partial z} - \frac{\partial E_z}{\partial x} = (i\omega\mu_0\mu)H_y \qquad \text{TM polarisation}$$

$$\frac{\partial H_y}{\partial x} = -(i\omega\varepsilon_0\varepsilon)E_z$$

In the subwavelength regime the second equation in each set of TE and TM polarisation will be much smaller than the other two and can be dropped. We see, therefore, that waves TE polarised depend only on the permeability and waves TM polarised depend only on the permittivity. Negative permittivity does exist in available materials in the optical frequency range, notably good conductors such as gold and silver. Therefore, in order to demonstrate a negative refractive index we only need to study metamaterial subwavelength structures with overall negative permittivity in some frequency range, even if the permeability remains essentially equal to μ_0.

9.3   Negative index metamaterials and waveguides

The simplest structures that fulfil these requirements are two-dimensional waveguides consisting of a dielectric core and metal cladding, termed metal-insulator-metal (MIM) waveguides, or the opposite arrangement of a metal core and dielectric cladding, (IMI) waveguides. A schematic diagram of a typical MIM waveguide is shown in Figure 9.2. The waveguide consists of a high-index, low-loss core, in this case gallium phosphide (GaP), sandwiched between two silver (Ag) claddings. A plane-wave or modal-wave light source emits in the z direction and excites waveguide modes propagating along z. These waveguide modes are actually surface plasmons excited at the two metal-dielectric interfaces and coupled to form symmetric and antisymmetric waveguide modes with respect to the z centreline. The optical properties of these waveguides have been studied in a series of articles [5–8], and Figure 9.3 summarises the dispersion relations for Ag-GaP-Ag MIM structures for three different core thicknesses. Ignoring for the moment the fact that these modes are lossy when implemented with real metals, we can determine the regions of negative index behaviour from inspecting the real dispersion curve, Figure 9.3a. A well-known result from physical optics identifies the phase velocity of a harmonic wave

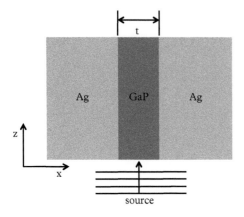

Figure 9.2 *Diagram of a typical MIM waveguide. The core thickness t is subwavelength (usually some tens of nanometres) and the metal cladding width is a factor of ten or more greater than the core. The source can be a plane wave or an amplitude modulated 'modal' wave with appropriate symmetry to excite a desired waveguide mode, symmetric or antisymmetric.*

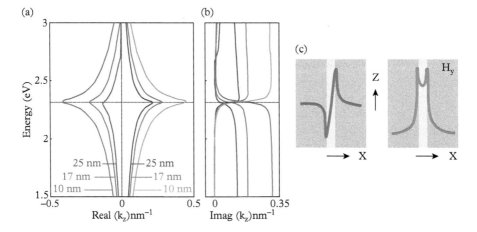

Figure 9.3 *(a) Dispersion diagram (Re(k_x) vs. mode energy) for waveguide modes supported by waveguide structure shown in Figure 9.2. (b) Same as (a) for Im(k_x). Blue curves correspond to symmetric modes; red curves to antisymmetric modes with three different widths, t. The horizontal dotted line is the energy of the surface plasmon resonance. (c) Waveforms for the two lowest-energy antisymmetric and symmetric modes. Figure adapted from Figure 2 of [8] and used with permission.*

travelling in the z direction as $v_\varphi = \omega/k_z$ and the group velocity (equivalent to the energy velocity for lossless, dispersionless media) as $v_\gamma = d\omega/dk_z$. When the phase velocity and group velocity have opposite signs, a lossless medium is said to exhibit negative index behaviour. Thus, we see from Figure 9.3a that the lower right quadrant of the diagram indicates positive phase velocity and positive group velocity for the symmetric modes; while the upper left quadrant shows negative phase velocity and positive group velocity for antisymmetric modes. The antisymmetric modes in this quadrant exhibit negative index behaviour.

In the more general case of lossy media, the propagation vectors become complex and the association of group velocity with energy or flux velocity becomes ambiguous. However, we can always associate the energy flux of any electromagnetic wave with the Poynting vector $S = E \times H$. Therefore, the more general criterion, including lossy materials, is that if the real part of S and v_φ propagate in opposite directions, the waveguide mode exhibits negative index behaviour. In order for the modes to have any practical significance, the imaginary part of k_z must be small compared to the real part. Assuming that propagation along z is represented by a factor $e^{ik_z z}$ it is clear that $\mathrm{Im}(k_z) > 0$, $\mathrm{Re}(k_z) < 0$, and $\mathrm{Im}(k_z)/\mathrm{Re}(k_z) \ll 1$ are equivalent criteria for significant negative index modes. Inspection of Figure 9.3a and b shows that these criteria are satisfied for asymmetric modes above the plasmon resonance energy (upper left-hand quadrant, Figure 9.3a), consistent with our finding for the equivalent lossless case.

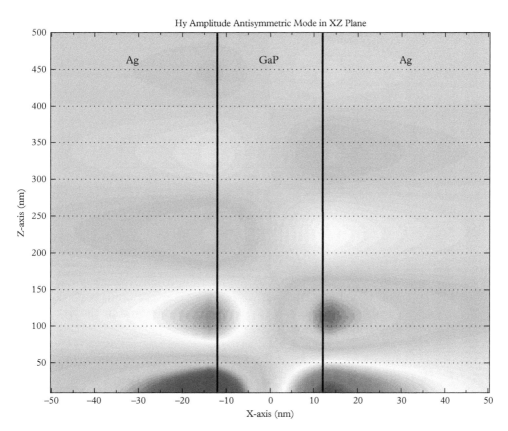

Figure 9.4 *Real part of H_y amplitude of the lowest energy antisymmetric waveguide mode in the $X - Z$ plane for a Ag-GaP-Ag structure with a core width of 25 nm. Red indicates positive amplitude and blue indicates negative amplitude. The figure illustrates the H_y amplitude antisymmetry along z with respect to the node at $x = 0$.*

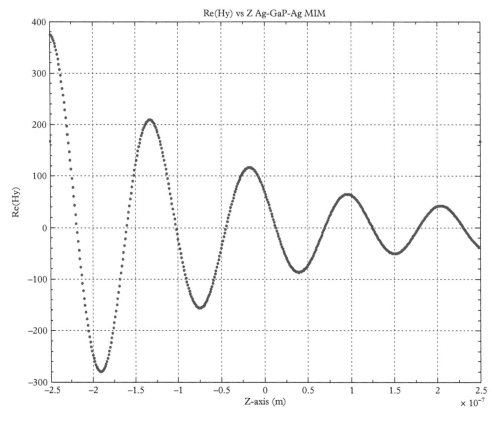

Figure 9.5 *Plot of Re(H_y) as a function of z along the waveguide at x = 12.5 nm. The decrease in amplitude along z is due to dissipation from the lossy real metal.*

Figures 9.4 and 9.5 show the results of an FDTD (Finite Difference Time Domain) numerical simulation of the lowest-energy antisymmetric mode propagating in a Ag-GaP-Ag waveguide with a core width of 25 nm. The real part of the amplitude of H_y in the X–Z plane is shown in Figure 9.4 and the propagation along z is shown in Figure 9.5.

9.4   Reflection and transmission in stacked layers

9.4.1   Matrix formulation

In Section 8.8.5 of Chapter 8 we discussed a matrix approach to relate input and output voltages and currents in linear circuits. A similar approach can be used to relate transmission and reflection, first treated at a single interface in Chapter 3 Section 3.2, across an optical 'network' consisting of material layers with differing indices of refraction.

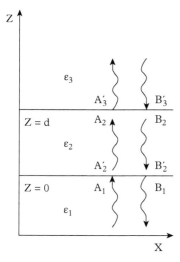

Figure 9.6 *Diagram of the three-layer stratified material with permittivities $\varepsilon_1, \varepsilon_2, \varepsilon_3$. Wavy arrows indicate plane waves propagating, reflecting, and transmitting at the $z = 0$ and $z = d$ interfaces. The A, B coefficients indicate the wave amplitudes at the interfaces.*

Figure 9.6 shows a plane wave entering a three-layer stratified medium from the region $z < 0$ with permittivity ε_1, partially transmitting and reflecting at the $z = 0$ boundary, propagating from $z = 0$ to $z = d$ in the layer with permittivity ε_2, and then once again undergoing reflection and transmission at the $z = d$ boundary between layer 2 and layer 3 with ε_3. Consider the electric field of the wave propagating in the XZ plane:

$$E = E(z)e^{i(nk_0 x - \omega t)} \tag{9.8}$$

with n the index of refraction and $n = \sqrt{\varepsilon}$. The properties of the layers are invariant along the axes orthogonal to z, and we assume the wave polarised TM (H_y, E_x, E_z) or TE (E_y, H_x, H_z). In the following discussion we will concentrate on TM polarisation because it gives rise to surface plasmons that play an important role in many important phenomena involving metamaterials. Expressions for TE polarisation are left to the reader as an exercise. The propagation angles $\mathbf{k} \cdot \mathbf{r} = k_x x + k_z z$ are not necessarily normal to the boundaries and may be reflected and transmitted at angles according to the Fresnel laws (see Section 3.2). The components k_z and k_x are related to k:

$$k_x^2 + k_z^2 = k^2 \tag{9.9}$$

In the electrical and optical engineering literature [9] the component of the wave vector transverse to the propagation direction is often denoted by β:

$$k_x = \beta = \sqrt{k^2 - k_z^2} \tag{9.10}$$

Dropping the time dependence for the moment, we can write the travelling wave moving up and down along the z direction,

$$E(z) = Ue^{(ik_z z)} + De^{-(ik_z z)} = A(z) + B(z) \tag{9.11}$$

where A, B are the complex amplitudes of the waves. On each side of the $z = 0$ and $z = d$ boundaries there are two amplitude pairs. Specifically at the $z = 0$ boundary we have A_1, A_2' and B_1, B_2'. These two amplitude pairs must be related by the Fresnel laws of reflection and transmission and can be expressed as two-element column vectors transformed by a 2×2 matrix D_{12}:

$$\begin{pmatrix} A_1 \\ B_1 \end{pmatrix} = D_{12} \begin{pmatrix} A_2' \\ B_2' \end{pmatrix} \tag{9.12}$$

where D_{12} itself consists of a product of two other matrices, the inverse of a matrix D_1 and another matrix D_2. Thus:

$$D_{12} = D_1^{-1} D_2 \tag{9.13}$$

where D_1, D_2 express the field continuity conditions at the $z = 0$ boundary. Figure 9.7 shows in detail, reflection and transmission at the $z = 0$ boundary from a plane wave incident at angle θ_1 from $z < 0$. Referring to Figure 9.7 and restricting the discussion to TM polarisation we write the continuity conditions for the E_x and H_y field components:

$$-E_1 \cos \theta_1 + E_1' \cos \theta_1 = -E_2 \cos \theta_2 + E_2' \cos \theta_2 \tag{9.14}$$

$$\sqrt{\varepsilon_1} E_1 + \sqrt{\varepsilon_1} E_1' = \sqrt{\varepsilon_2} E_2 + \sqrt{\varepsilon_2} E_2' \tag{9.15}$$

where we have used $H_y = \sqrt{\varepsilon/\mu} E_x$ and have assumed non-magnetic material with $\mu = 1$. In matrix form the continuity relations are

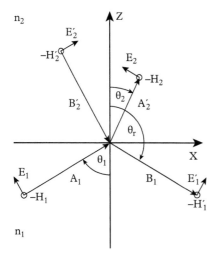

Figure 9.7 *Detail of the incident, reflected, and transmitted field components for TM polarisation at the $z = 0$. Note the phase convention that E_x changes sign on reflection. Note also that the H_y components are aligned so that the Poynting vectors point in the directions of propagation.*

$$D_1 \begin{pmatrix} E_1 \\ E_1' \end{pmatrix} = D_2 \begin{pmatrix} E_2 \\ E_2' \end{pmatrix} \tag{9.16}$$

$$\begin{pmatrix} -\cos\theta_1 & \cos\theta_1 \\ \sqrt{\varepsilon_1} & \sqrt{\varepsilon_1} \end{pmatrix} \begin{pmatrix} E_1 \\ E_1' \end{pmatrix} = \begin{pmatrix} -\cos\theta_2 & \cos\theta_2 \\ \sqrt{\varepsilon_2} & \sqrt{\varepsilon_2} \end{pmatrix} \begin{pmatrix} E_2 \\ E_2' \end{pmatrix} \tag{9.17}$$

Multiplying both sides of Equation 9.16 by D_1^{-1} results in Equation 9.13. Writing out the D_{12} matrix explicitly, we have,

$$\begin{pmatrix} E_1 \\ E_1' \end{pmatrix} = D_1^{-1} D_2 = D_{12} = \frac{1}{2} \begin{bmatrix} \dfrac{\cos\theta_2}{\cos\theta_1} + \sqrt{\dfrac{\varepsilon_2}{\varepsilon_1}} & -\dfrac{\cos\theta_2}{\cos\theta_1} + \sqrt{\dfrac{\varepsilon_2}{\varepsilon_1}} \\ -\dfrac{\cos\theta_2}{\cos\theta_1} + \sqrt{\dfrac{\varepsilon_2}{\varepsilon_1}} & \dfrac{\cos\theta_2}{\cos\theta_1} + \sqrt{\dfrac{\varepsilon_2}{\varepsilon_1}} \end{bmatrix} \begin{pmatrix} E_2 \\ E_2' \end{pmatrix} \tag{9.18}$$

The incident and transmission angles $\cos\theta_1, \cos\theta_2$ can be expressed in terms of the propagation vector components k_{z1}, k_{z2}, the refractive indices n_1, n_2, or permittivities $\varepsilon_1, \varepsilon_2$ on each side of the boundary:

$$k_{z1} = n_1 k_0 \cos\theta_1 \quad \text{and} \quad k_{z2} = n_2 k_0 \cos\theta_2 \tag{9.19}$$

and

$$\frac{\cos\theta_2}{\cos\theta_1} = \frac{n_1 k_{z2}}{n_2 k_{z1}} = \sqrt{\frac{\varepsilon_1}{\varepsilon_2}} \frac{k_{z2}}{k_{z1}} \tag{9.20}$$

Then D_{12} can be expressed as

$$D_{12} = \frac{1}{2} \sqrt{\frac{\varepsilon_2}{\varepsilon_1}} \begin{bmatrix} \dfrac{k_{z2}}{k_{z1}} \dfrac{\varepsilon_1}{\varepsilon_2} + 1 & -\dfrac{k_{z2}}{k_{z1}} \dfrac{\varepsilon_1}{\varepsilon_2} + 1 \\ -\dfrac{k_{z2}}{k_{z1}} \dfrac{\varepsilon_1}{\varepsilon_2} + 1 & \dfrac{k_{z2}}{k_{z1}} \dfrac{\varepsilon_1}{\varepsilon_2} + 1 \end{bmatrix} \tag{9.21}$$

and evidently

$$D_{23} = \frac{1}{2} \sqrt{\frac{\varepsilon_3}{\varepsilon_2}} \begin{bmatrix} \dfrac{k_{z3}}{k_{z2}} \dfrac{\varepsilon_2}{\varepsilon_3} + 1 & -\dfrac{k_{z3}}{k_{z2}} \dfrac{\varepsilon_2}{\varepsilon_3} + 1 \\ -\dfrac{k_{z3}}{k_{z2}} \dfrac{\varepsilon_2}{\varepsilon_3} + 1 & \dfrac{k_{z3}}{k_{z2}} \dfrac{\varepsilon_2}{\varepsilon_3} + 1 \end{bmatrix} \tag{9.22}$$

Again referring to Figure 9.7, and after setting $E_2' = 0$, we can express the Fresnel reflection and transmission coefficients as

$$r_{12} = \frac{-\dfrac{k_{z2}}{k_{z1}} \dfrac{\varepsilon_1}{\varepsilon_2} + 1}{\dfrac{k_{z2}}{k_{z1}} \dfrac{\varepsilon_1}{\varepsilon_2} + 1} \quad \text{and} \quad t_{12} = \frac{2}{\sqrt{\dfrac{\varepsilon_2}{\varepsilon_1}} \left(\dfrac{k_{z2}}{k_{z1}} \dfrac{\varepsilon_1}{\varepsilon_2} + 1 \right)}$$

$$r_{23} = \frac{-\dfrac{k_{z3}}{k_{z2}} \dfrac{\varepsilon_2}{\varepsilon_3} + 1}{\dfrac{k_{z3}}{k_{z2}} \dfrac{\varepsilon_2}{\varepsilon_3} + 1} \quad \text{and} \quad t_{23} = \frac{2}{\sqrt{\dfrac{\varepsilon_3}{\varepsilon_2}} \left(\dfrac{k_{z3}}{k_{z2}} \dfrac{\varepsilon_2}{\varepsilon_3} + 1 \right)} \tag{9.23}$$

From which we can write D_{12} as

$$D_{12} = \frac{1}{t_{12}} \begin{bmatrix} 1 & r_{12} \\ r_{12} & 1 \end{bmatrix} \tag{9.24}$$

and similarly

$$D_{23} = \frac{1}{t_{23}} \begin{bmatrix} 1 & r_{23} \\ r_{23} & 1 \end{bmatrix} \tag{9.25}$$

In the region $0 < z < d$ the plane wave propagates freely with no change in amplitude or direction. From Figure 9.6 we can write the propagation matrix P as

$$\begin{pmatrix} A_2 \\ B_2 \end{pmatrix} = P_2 \begin{pmatrix} A_2' \\ B_2' \end{pmatrix} = \begin{pmatrix} e^{i\varphi} & 0 \\ 0 & e^{-i\varphi} \end{pmatrix} \begin{pmatrix} A_2' \\ B_2' \end{pmatrix} \tag{9.26}$$

The propagation phase $\varphi = k_{z2}d$, where d is the thickness (along z) of the second layer. Similarly to the boundary at $z = 0$, we have at the boundary $z = d$:

$$\begin{pmatrix} A_2 \\ B_2 \end{pmatrix} = D_2^{-1}D_3 \begin{pmatrix} A_3' \\ B_3' \end{pmatrix} = D_{23} \begin{pmatrix} A_3' \\ B_3' \end{pmatrix} \tag{9.27}$$

We see from Equations 9.12, 9.26, and 9.27 that the amplitudes just above $z = d$ can be related to those just below $z = 0$ by a sequence of matrix multiplications:

$$\begin{pmatrix} A_1 \\ B_1 \end{pmatrix} = D_{12}P_2^{-1}D_{23} \begin{pmatrix} A_3' \\ B_3' \end{pmatrix} = D_1^{-1}D_2P_2^{-1}D_2^{-2}D_3 \begin{pmatrix} A_3' \\ B_3' \end{pmatrix} \tag{9.28}$$

This sequential scheme can easily be generalised to a multilayer stack starting with amplitudes A_i, B_i and ending with amplitudes A_f', B_f':

$$\begin{pmatrix} A_i \\ B_i \end{pmatrix} = D_i^{-1} \left[\prod_{j=1}^{N} D_j^{-1}P_j^{-1}D_j \right] D_f \begin{pmatrix} A_f' \\ B_f' \end{pmatrix} \tag{9.29}$$

and we define the matrix M as

$$M = D_i^{-1} \left[\prod_{j=1}^{N} D_j^{-1}P_j^{-1}D_j \right] D_f \tag{9.30}$$

and

$$\begin{pmatrix} A_i \\ B_i \end{pmatrix} = \begin{pmatrix} M_{11} & M_{12} \\ M_{21} & M_{22} \end{pmatrix} \begin{pmatrix} A_f' \\ B_f' \end{pmatrix} \tag{9.31}$$

Taking into account that after the light exits the final interface, $B'_f = 0$, we can express the overall reflection and transmission coefficients r, t as

$$r = \frac{B_i}{A_i} \tag{9.32}$$

$$t = \frac{A'_f}{A_i} \tag{9.33}$$

and using Equation 9.31:

$$r = \frac{M_{21}}{M_{11}} \quad \text{and} \quad t = \frac{1}{M_{11}} \tag{9.34}$$

In the relatively simple case of a three-layer stack we can write out explicitly the elements of the M matrix. The matrix multiplications are most easily carried out using the expressions for D_{12} and D_{23} in terms of the reflection and transmission coefficients $r_{12}, r_{23}, t_{12}, t_{23}$, in Equations 9.24 and 9.25:

$$M = \frac{1}{t_{12}t_{23}} \begin{bmatrix} e^{-i\varphi_2} + r_{12}r_{23}e^{i\varphi_2} & r_{23}e^{-i\varphi_2} + r_{12}e^{i\varphi_2} \\ r_{12}e^{-i\varphi_2} + r_{23}e^{i\varphi_2} & r_{12}r_{23}e^{-i\varphi_2} + e^{i\varphi_2} \end{bmatrix} \tag{9.35}$$

Then using Equation 9.34:

$$r = \frac{r_{12}e^{-i\varphi_2} + r_{23}e^{i\varphi_2}}{e^{-i\varphi_2} + r_{12}r_{23}e^{i\varphi_2}} \tag{9.36}$$

$$t = \frac{t_{12}t_{23}}{e^{-i\varphi_2} + r_{12}r_{23}e^{i\varphi_2}} \tag{9.37}$$

Substituting for r_{12}, t_{12} and r_{23}, t_{23} into Equation 9.23:

$$r = \frac{\left(1 - \frac{k_{z3}}{k_{z1}}\frac{\varepsilon_1}{\varepsilon_3}\right)\cos\varphi_2 + \left(\frac{k_{z2}}{k_{z1}}\frac{\varepsilon_1}{\varepsilon_2} - \frac{k_{z3}}{k_{z2}}\frac{\varepsilon_2}{\varepsilon_3}\right)i\sin\varphi_2}{\left(\frac{k_{z3}}{k_{z1}}\frac{\varepsilon_1}{\varepsilon_3} + 1\right)\cos\varphi_2 - \left(\frac{k_{z2}}{k_{z1}}\frac{\varepsilon_1}{\varepsilon_2} + \frac{k_{z3}}{k_{z2}}\frac{\varepsilon_2}{\varepsilon_3}\right)i\sin\varphi_2} \tag{9.38}$$

and

$$t = \frac{2}{\sqrt{\frac{\varepsilon_3}{\varepsilon_1}}\left[\left(\frac{k_{z3}}{k_{z1}}\frac{\varepsilon_1}{\varepsilon_3} + 1\right)\cos\varphi_2 - \left(\frac{k_{z2}}{k_{z1}}\frac{\varepsilon_1}{\varepsilon_2} + \frac{k_{z3}}{k_{z2}}\frac{\varepsilon_2}{\varepsilon_3}\right)i\sin\varphi_2\right]} \tag{9.39}$$

If the three-layer stack is symmetric with $\varepsilon_1 = \varepsilon_3$ then the expressions for r and t simplify somewhat

$$r = \frac{\left(\frac{k_{z2}}{k_{z1}}\frac{\varepsilon_1}{\varepsilon_2} - \frac{k_{z1}}{k_{z2}}\frac{\varepsilon_2}{\varepsilon_1}\right) i \sin \varphi_2}{2 \cos \varphi_2 - \left(\frac{k_{z2}}{k_{z1}}\frac{\varepsilon_1}{\varepsilon_2} + \frac{k_{z1}}{k_{z2}}\frac{\varepsilon_2}{\varepsilon_1}\right) i \sin \varphi_2} \tag{9.40}$$

and

$$t = \frac{2}{2 \cos \varphi_2 - \left(\frac{k_{z2}}{k_{z1}}\frac{\varepsilon_1}{\varepsilon_2} + \frac{k_{z1}}{k_{z2}}\frac{\varepsilon_2}{\varepsilon_1}\right) i \sin \varphi_2} \tag{9.41}$$

9.4.2 Periodic stacked layers

We consider here the case of alternating *periodic* layers of two permittivities. The analysis of periodic stacked layers has much in common with condensed matter theory of crystalline materials. The periodicity and Fresnel scattering give rise to transmission windows, stopbands, and 'photonic band gaps' analogous to the physics of metals and doped semiconductors [10],[11]. The development builds on the matrix approach introduced in Section 9.4.1. Figure 9.8 illustrates the situation. The permittivity varies periodically along z:

$$\varepsilon(z) = \varepsilon_1 \qquad 0 < z < d$$
$$\varepsilon(z) = \varepsilon_2 \qquad d < z < \Lambda \tag{9.42}$$

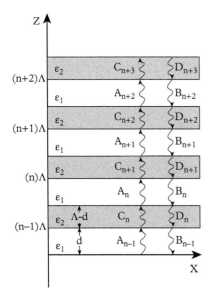

Figure 9.8 *Periodic stacked layers of two different permittivities, $\varepsilon_1, \varepsilon_2$. The repeating unit has a length of Λ. Matrices corresponding to material 1 are denoted A, B, and matrices corresponding to material 2 are denoted C, D.*

where Λ is the period of the alternating layers. We can write the E-field travelling wave along the stack similarly to Equation 9.11:

$$E(z) = A_n e^{ik_{z1}[(z-n\Lambda)]} + B_n e^{-ik_{z1}[(z-n\Lambda)]} \qquad n\Lambda - d < z < n\Lambda \qquad (9.43)$$

$$E(z) = C_n e^{ik_{z2}[(z-(n\Lambda-d)]} + D_n e^{-ik_{z2}[(z-(n\Lambda-d)]} \qquad (n-1)\Lambda < z < n\Lambda - d \qquad (9.44)$$

where k_1, k_2 are related in the usual way to their longitudinal and transverse components propagating in material slabs with permittivities $\varepsilon_1, \varepsilon_2$:

$$k_{z1} = \sqrt{k_1^2 - \beta^2} = k_1 \cos\theta_1$$

$$k_{z2} = \sqrt{k_2^2 - \beta^2} = k_2 \cos\theta_2 \qquad (9.45)$$

The β term is the transverse component defined in Equation 9.10.

Just as in Equations 9.12, 9.13 9.16, and 9.26, the matrix form the E-field coefficients are related by

$$\begin{pmatrix} A_{n-1} \\ B_{n-1} \end{pmatrix} = D_1^{-1} D_2 P_2^{-1} \begin{pmatrix} C_n \\ D_n \end{pmatrix} \qquad (9.46)$$

and

$$\begin{pmatrix} C_n \\ D_n \end{pmatrix} = D_2^{-1} D_1 P_1^{-1} \begin{pmatrix} A_n \\ B_n \end{pmatrix} \qquad (9.47)$$

where the propagation matrices within the slabs ε_1 and ε_2 are written as

$$P_1 = \begin{pmatrix} e^{ik_{z1}d} & 0 \\ 0 & k^{-ik_{z1}d} \end{pmatrix} \qquad (9.48)$$

and

$$P_2 = \begin{pmatrix} e^{ik_{z2}(\Lambda-d)} & 0 \\ 0 & k^{-ik_{z2}(\Lambda-d)} \end{pmatrix} \qquad (9.49)$$

Substituting Equation 9.47 into Equation 9.46, we obtain a matrix that relates the incident and reflected amplitudes from one unit cell to the next:

$$\begin{pmatrix} A_{n-1} \\ B_{n-1} \end{pmatrix} = D_1^{-1} D_2 P_2^{-1} D_2^{-1} D_1 P_1^{-1} \begin{pmatrix} A_n \\ B_n \end{pmatrix} \qquad (9.50)$$

$$\begin{pmatrix} A_{n-1} \\ B_{n-1} \end{pmatrix} = \begin{bmatrix} A & B \\ C & D \end{bmatrix} \begin{pmatrix} A_n \\ B_n \end{pmatrix} \qquad (9.51)$$

Carrying out the explicit matrix multiplications for TM polarisation, we have for the elements of the *ABCD* matrix,

$$A = e^{-i\varphi_2}\left[\cos\varphi_1 - \frac{1}{2}i\sin\varphi_1\left(\frac{k_{z2}\,\varepsilon_1}{k_{z1}\,\varepsilon_2} + \frac{k_{z1}\,\varepsilon_2}{k_{z2}\,\varepsilon_1}\right)\right] \tag{9.52}$$

$$B = e^{i\varphi_2}\left[-\frac{1}{2}\left(\frac{k_{z2}\,\varepsilon_1}{k_{z1}\,\varepsilon_2} - \frac{k_{z1}\,\varepsilon_2}{k_{z2}\,\varepsilon_1}\right)i\sin\varphi_1\right] \tag{9.53}$$

$$C = e^{-i\varphi_2}\left[\frac{1}{2}\left(\frac{k_{z2}\,\varepsilon_1}{k_{z1}\,\varepsilon_2} - \frac{k_{z1}\,\varepsilon_2}{k_{z2}\,\varepsilon_1}\right)i\sin\varphi_1\right] \tag{9.54}$$

$$D = e^{i\varphi_2}\left[\cos\varphi_1 + \frac{1}{2}i\sin\varphi_1\left(\frac{k_{z2}\,\varepsilon_1}{k_{z1}\,\varepsilon_2} + \frac{k_{z1}\,\varepsilon_2}{k_{z2}\,\varepsilon_1}\right)\right] \tag{9.55}$$

where $\varphi_1 = k_{z1}d$ and $\varphi_2 = k_{z2}(\Lambda - d)$.

The reflection and transmission of light through periodic stacked layers is our primary interest, and therefore we develop expressions for Bragg reflection and 'resonant tunnelling' transmission. The latter is particularly important when the stack consists of alternating layers of metal and dielectric because plasmon surface waves form at the metal-dielectric interfaces. These surface waves propagate along X, Y and are evanescent along Z. We will see that, despite this evanescence, under the right conditions light can be transmitted through a periodic stack by resonant tunnelling.

Rewriting Equation 9.51 as a relation between the zeroth layer of an ε_1 slab and the next highest ε_1 layer,

$$\begin{pmatrix} A_1 \\ B_1 \end{pmatrix} = \begin{bmatrix} A & B \\ C & D \end{bmatrix}^{-1} \begin{pmatrix} A_0 \\ B_0 \end{pmatrix} \tag{9.56}$$

we can interpret the inverse of the *ABCD* matrix as the operator that transfers the wave properties from layer 0 to layer 1. In fact, once A_0, B_0 are specified, the inverse of the *ABCD* matrix can transfer these properties to the 'nth' layer of ε_1:

$$\begin{pmatrix} A_n \\ B_n \end{pmatrix} = \begin{bmatrix} A & B \\ C & D \end{bmatrix}^{-n} \begin{pmatrix} A_0 \\ B_0 \end{pmatrix} \tag{9.57}$$

The inverse of the *ABCD* matrix is given by

$$\begin{pmatrix} A & B \\ C & D \end{pmatrix}^{-1} = \frac{1}{\det(ABCD)}\begin{pmatrix} D & -B \\ -C & A \end{pmatrix} \tag{9.58}$$

and it can be verified by direct substitution from Equations 9.52–9.55 that

$$AD - BC = 1 \tag{9.59}$$

Therefore once the properties of the zeroth layer are fixed, they can be transferred to the 'nth' layer by successive matrix multiplications of the inverse $ABCD$ matrix:

$$\begin{pmatrix} A_n \\ B_n \end{pmatrix} = \begin{bmatrix} D & -B \\ -C & A \end{bmatrix}^n \begin{pmatrix} A_0 \\ B_0 \end{pmatrix} \tag{9.60}$$

The $ABCD$ matrix (and its inverse) also sometimes called the 'characteristic' matrix or the 'transfer' matrix, and the theory of field displacement through stacked or stratified layers is often referred to as 'transfer matrix theory' or TMT [12], [13].

9.4.3 Bloch waves

A harmonic plane wave, $E = E_0 e^{i(kz - \omega t)}$, travelling in vacuum or any uniform, isotropic, lossless medium, is periodic in time and space according to the wave angular frequency ω and speed of propagation, v. The simple dispersion relation $k = \omega/v$ specifies the spatial periodicity. If the medium is the vacuum, $k = \omega/c$. A wave travelling in a *periodic* medium must exhibit the symmetry this added periodicity provides as well. The symmetry is translational such that

$$E_K(z) = E_K(z + \Lambda) \tag{9.61}$$

Such a wave is called a Bloch wave [14] and has the form

$$E_K(z) = E_K e^{iKz} \tag{9.62}$$

where $K = 2\pi/\Lambda$ or, taking into account the propagation in the transverse direction x as well,

$$E_K(x, z) = E_K e^{i\beta x} e^{iKz} \tag{9.63}$$

In column vector form we have

$$\begin{pmatrix} A_n \\ B_n \end{pmatrix} = \left(e^{iK\Lambda} \right)^n \begin{pmatrix} A_0 \\ B_0 \end{pmatrix} \tag{9.64}$$

and comparing Equation 9.57 to Equation 9.64:

$$\begin{bmatrix} A & B \\ C & D \end{bmatrix}^{-n} \begin{pmatrix} A_0 \\ B_0 \end{pmatrix} = \left(e^{iK\Lambda} \right)^n \begin{pmatrix} A_0 \\ B_0 \end{pmatrix} \tag{9.65}$$

we see that $e^{iK\Lambda}$ is an eigenvalue of the *inverse* $ABCD$ translation operator matrix. For a one-step (unit periodic cell) $ABCD$ translation in the $-z$ direction:

$$\begin{bmatrix} A & B \\ C & D \end{bmatrix} \begin{pmatrix} A_0 \\ B_0 \end{pmatrix} = e^{-iK\Lambda} \begin{pmatrix} A_0 \\ B_0 \end{pmatrix} \tag{9.66}$$

For a one-step translation in the $+z$ direction,

$$\begin{bmatrix} A & B \\ C & D \end{bmatrix}^{-1} \begin{pmatrix} A_0 \\ B_0 \end{pmatrix} = e^{iK\Lambda} \begin{pmatrix} A_0 \\ B_0 \end{pmatrix} \tag{9.67}$$

The eigenvalues of the matrix in Equation 9.66 can be written in terms of the matrix elements by solving the usual determinant equation:

$$\det \begin{bmatrix} A-\lambda & B \\ C & D-\lambda \end{bmatrix} = 0 \tag{9.68}$$

so

$$e^{-iK\Lambda} = \frac{1}{2}(A + D) \pm \sqrt{\left[\frac{1}{2}(A + D)\right]^2 - 1} \tag{9.69}$$

where we have used $AD - BC = 1$. In order for the Bloch wave on the left to be propagating, the factor $K\Lambda$ in the argument must be real. The wave consists of a real part $\cos K\Lambda$ and an imaginary part $-\sin K\Lambda$. The expression on the right must be also be complex to satisfy the equation, which implies that $(A + D)/2$ must be less than unity for the Bloch wave to propagate through the periodic medium. If $(A + D)/2 > 1$, then the right-hand side is real and the Bloch vector K must then be pure imaginary so that the exponential argument becomes real and the wave evanescent. The sign of K is chosen to ensure this evanescent behaviour. We can use Equations 9.52 and 9.55 to write out $A + D$ in terms of the plane waves of which the Bloch wave is composed. For periodic structures we set $d = d_1, \Lambda - d_1 = d_2$:

$$A+D = 2\cos(k_{z1}d_1)\cos(k_{z2}d_2) - \left(\frac{k_{z2}\,\varepsilon_1}{k_{z1}\,\varepsilon_2} + \frac{k_{z1}\,\varepsilon_2}{k_{z2}\,\varepsilon_1}\right)\sin(k_{z1}d_1)\sin(k_{z2}d_2) \qquad d_1 + d_2 = \Lambda \tag{9.70}$$

For a fixed stack geometry (d_1, d_2) and choice of materials $(\varepsilon_1, \varepsilon_2)$, as ω increases, there will be zones of propagation and zones of evanescence. The non-propagating zones are the 'stopbands' or 'photonic band gaps' and the propagating zones are analogous to the conduction bands in condensed matter theory. They are sometimes termed 'photonic band windows'. The photonic band gaps are truly non-propagating only for periodic structures extending infinitely in $\pm z$. For practical, finite structures the evanescent waves can ultimately tunnel through and emerge propagating at the exit boundary. We shall see in Section 9.4.7 that resonant tunnelling in metamaterials is very important.

We write Equation 9.69 in the following form:

$$\cos K\Lambda = \cos(k_{z1}d_1)\cos(k_{z2}d_2) - \frac{1}{2}\left(\frac{k_{z2}\,\varepsilon_1}{k_{z1}\,\varepsilon_2} + \frac{k_{z1}\,\varepsilon_2}{k_{z2}\,\varepsilon_1}\right)\sin(k_{z1}d_1)\sin(k_{z2}d_2) \tag{9.71}$$

or

$$K = \frac{1}{\Lambda} \cos^{-1} \left[\cos(k_{z1} d_1) \cos(k_{z2} d_2) - \frac{1}{2} \left(\frac{k_{z2} \, \varepsilon_1}{k_{z1} \, \varepsilon_2} + \frac{k_{z1} \, \varepsilon_2}{k_{z2} \, \varepsilon_1} \right) \sin(k_{z1} d_1) \sin(k_{z2} d_2) \right]$$

(9.72)

where we remember that $k_1 = k_0 n_1 = (\omega/c)\sqrt{\varepsilon_1}$ and $k_2 = k_0 n_2 = (\omega/c)\sqrt{\varepsilon_2}$ are the two wave vectors resulting from a single-frequency wave $k_0 = \omega/c$ incident on the periodic structure. Equation 9.72 is the Bloch wave dispersion relation $K(\omega)$. The $\cos K\Lambda$ term on the left-hand side of Equation 9.71 may contain real or complex K. A $\cos$ function with imaginary argument becomes a hyperbolic cosine (cosh) function. In the propagating region at the band gap edge, $K\Lambda = \pi$, and at the centre frequency of the band gap itself

$$K\Lambda = \pi \pm ix$$

(9.73)

where x is yet to be determined. Figure 9.9 plots the plane-wave frequency ω vs $K\Lambda$ for the case where the optical thickness of the two layers is equal $(n_1 d_1 = n_2 d_2)$. Now we *assume* that, given our choice of geometry d_1, d_2 and materials $\varepsilon_1, \varepsilon_2$, a frequency ω_0 can be found such that the plane wave phase accumulation in the two slabs is

$$k_{1z} d_1 = k_{2z} d_2 = \frac{\pi}{2}$$

(9.74)

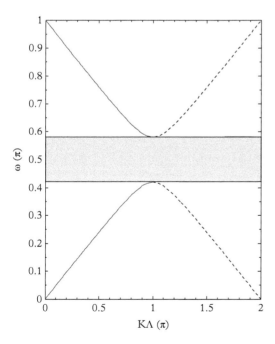

Figure 9.9 *Plot of $K\Lambda$ (π units) as a function of a 'normalised' plane-wave frequency $\omega_N = (n_1 d_1/c)\omega = (n_2 d_2/c)\omega$ (π units). The refractive indices are $n_1 = 1.5$ and $n_2 = 2.5$ for slabs 1 and 2. The optical thicknesses of the two slabs have been chosen to be equal $(n_1 d_1 = n_2 d_2)$. The solid (dashed) trace is for the Bloch wave propagating in the $+z$ ($-z$) direction. The shaded area shows the band gap where the Bloch wave is evanescent.*

To keep the development as simple as possible, we only consider normal incidence so that $k_{z1}, k_{z2} = k_1, k_2 = n_1 k_0, n_2 k_0$. Then, substituting Equation 9.73 into Equation 9.71:

$$\cosh x = \frac{1}{2}\left(\frac{n_1}{n_2} + \frac{n_2}{n_1}\right) \tag{9.75}$$

Taking the inverse hyperbolic cosine of both sides, we find that

$$x \simeq \ln\frac{n_2}{n_1} \simeq 2 \cdot \frac{n_2 - n_1}{n_2 + n_2} \tag{9.76}$$

where we have continued to assume, for the time being, that $\varepsilon_1, \varepsilon_2$ represent lossless dielectrics and that the periodic permittivity modulation in the structure is not too great.

According to Equation 9.73, x is the imaginary part of $K\Lambda$ in the centre of the band gap. At the edge of the band gap, at the boundary between the propagating and non-propagating zones, $x = 0$. Let $\omega - \omega_0$ be the deviation from the plane-wave frequency corresponding to the centre of the band gap and η be the plane-wave phase angle corresponding to this frequency deviation. We then write

$$\eta = \frac{n_1 d_1}{c}\omega - \frac{\pi}{2} = \frac{n_2 d_2}{c}\omega - \frac{\pi}{2} \tag{9.77}$$

or

$$\eta + \frac{\pi}{2} = \frac{n_1 d_1}{c}\omega = \frac{n_2 d_2}{c}\omega \tag{9.78}$$

Now substitute Equations 9.78, 9.75, and 9.73 into Equation 9.71. The result is

$$\cosh[x(\omega)] = \frac{1}{2}\left(\frac{n_2}{n_1} + \frac{n_1}{n_2}\right)\cos^2\eta - \sin^2\eta \tag{9.79}$$

which is a relation expressing x as a function of ω. The plane-wave frequency at the band edge ω_{be} corresponds to $x = 0$ and is given by

$$\sin\eta_{be} = \pm\frac{n_2 - n_1}{n_2 + n_1} \tag{9.80}$$

and assuming that the difference between the indices of refraction is much smaller than their sum,

$$\eta_{be} \simeq \frac{|n_2 - n_1|}{n_2 + n_1} \tag{9.81}$$

The fractional band gap width is

$$\frac{\Delta\omega}{\omega_0} = \frac{2\left(\frac{|n_2-n_1|}{n_2+n_1}\right)}{\pi/2} = \frac{4}{\pi}\left(\frac{|n_2-n_1|}{n_2+n_1}\right) \tag{9.82}$$

9.4.4 Transfer matrix of periodic stacked layers

We are interested in finding the reflection and transmission through a periodic layered stack consisting of N layers. Owing to the periodicity, the transfer matrices of the successive N steps must be identical so the overall transfer matrix must be equivalent to the matrix of a unit-step transfer raised to the power of N. Denote the overall $ABCD$ matrix as $M(N\Lambda)$ and the individual unit-cell displacement matrices as $M_n(\Lambda)$. Then

$$M(N\Lambda) = M_1(\Lambda) \cdot M_2(\Lambda) \cdot M_3(\Lambda) \dots M_N(\Lambda) = M(\Lambda)^N \tag{9.83}$$

Now we invoke a result from the theory of matrices to write an expression for $M(\Lambda)^N$:

$$M(\Lambda)^N = \begin{bmatrix} AU_{N-1}(a) - U_{N-2}(a) & BU_{N-1}(a) \\ CU_{N-1}(a) & DU_{N-1}(a) - U_{N-2}(a) \end{bmatrix} \tag{9.84}$$

where

$$a = \frac{1}{2}(A + D) \tag{9.85}$$

and the functions $U(a)$ are the Chebyshev polynomials of the second kind:

$$U_N(a) = \frac{\sin\left[(N+1)\cos^{-1}(a)\right]}{\sqrt{1-a^2}} \tag{9.86}$$

But from Equations 9.70 and 9.72 we find that

$$\cos^{-1}(a) = K\Lambda \tag{9.87}$$

and manifestly

$$\sqrt{1-a^2} = \sin K\Lambda \tag{9.88}$$

so

$$U_N(a) = \frac{\sin(N+1)K\Lambda}{\sin K\Lambda} \tag{9.89}$$

9.4.5   Bragg reflection

Inspection of Figure 9.10 shows that the reflection coefficient from a plane wave incident normal to the stack must be

$$r_N = \frac{M_{21}}{M_{11}} = \frac{CU_{n-1}}{AU_{N-1} - U_{N-2}} \tag{9.90}$$

and the reflectivity $|r|^2$ is

$$|r_N|^2 = \frac{|C|^2}{|C|^2 + \left(\frac{\sin K\Lambda}{\sin NK\Lambda}\right)^2} \tag{9.91}$$

In writing Equation 9.91 we have used the fact that $D^* = A$, $C^* = B$, and $AD - BC = 1$. At the centre of the band gap $K\Lambda = \pi + ix$ so that the stack reflectivity becomes

$$|r_N|^2 = \frac{|C|^2}{|C|^2 + \left(\frac{\sinh x}{\sinh Nx}\right)^2} \tag{9.92}$$

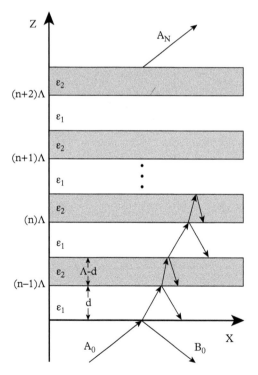

Figure 9.10 *Diagram of a periodic stack of dielectric slabs with permittivities $\varepsilon_1, \varepsilon_2$ and thicknesses (along z) d_1, d_2 with period $\Lambda = d_1 + d_2$. The various paths of transmission and reflection at the successive interfaces is indicated by the arrows.*

As $N \rightarrow \infty$ the second term in the denominator goes to zero and the reflectivity approaches unity.

At the band edge $K\Lambda = \pi$. The second term in the denominator of Equation 9.91 becomes indeterminate and must be evaluated using L'Hôpital's rule. The result is

$$|r_N|^2 = \frac{|C|^2}{|C|^2 + \left(\frac{1}{N}\right)^2} \tag{9.93}$$

Thus, we see that both in the middle of the band gap and at the gap edge, the reflectivity approaches unity with increasing number of periods.

So far we have determined the reflectivity as if the incident plane wave originated in the first layer of the first period. As a practical matter we are usually more interested in the reflectivity of a periodic stack for a plane wave incident from the air or vacuum. We can find this expression from Equations 9.24, 9.25, and 9.26 by supposing that incident plane wave, originating at $-\infty$ in the vacuum, impinges on a very thin layer of material ε_1 before entering the stack. This procedure effectively allows us to use Equation 9.36 while setting the phase shift φ_2 to zero:

$$r = \frac{r_{01} + r_N}{1 + r_{01}r_N} \tag{9.94}$$

where r_{01} is the vacuum-ε_1 reflection coefficient.

9.4.6   Resonant tunnelling

We can write the transmittance or transmissivity from energy conservation as

$$|t_N|^2 = 1 - |r_n|^2 = \frac{1}{1 + |C|^2 \left(\frac{\sin NK\Lambda}{\sin K\Lambda}\right)^2} \tag{9.95}$$

We see that in the band gap, where $\sin NK\Lambda$ and $\sin K\Lambda$ are replaced by their $\sinh x$ counterparts, the transmittance goes to zero as $N \rightarrow \infty$. At the band edge, however, where $K\Lambda = \pi$,

$$|t_N|^2 = \frac{1}{1 + |C|^2 N^2} \tag{9.96}$$

where, from Equation 9.54:

$$|C| = \left[\frac{1}{2}\left(\frac{k_{z2}\,\varepsilon_1}{k_{z1}\,\varepsilon_2} - \frac{k_{z1}\,\varepsilon_2}{k_{z2}\,\varepsilon_1}\right)\sin\varphi_1\right] \tag{9.97}$$

Since $|C|^2$ varies as $\sin^2\varphi_1$ we see that the transmittance can be unity as $\varphi_1 = k_{z1}d_1$ goes through integral multiples of π. For lossless dielectric periodic stacks we, therefore, must conclude that within the bandgap there can be some 'leakage' transmission when the number of unit cells is not too great, but resonant transmission is confined to the adjacent band windows. We shall see later that this conclusion does not necessarily hold when one of the stack components exhibits a real, negative permittivity. Figure 9.11 shows the transmittance, calculated from Equation 9.95 for a periodic stack consisting of 300 nm thick alternating slabs of TiO_2 and SiO_2. Light is incident normal to the surface. A band gap is clearly evident with adjacent rapidly oscillating transmittance maxima. The results of the TMT calculation of the transmittance can be compared to direct numerical simulation using a finite difference time domain (FDTD) technique to solve Maxwell's equations. The numerical simulation result for the same structure is shown in Figure 9.12.

In the band gap the incident light cannot propagate and must therefore decay exponentially. Figure 9.13 plots the intensity of the E_x field component, determined from the FDTD numerical simulation, as a function of distance into the stack along z.

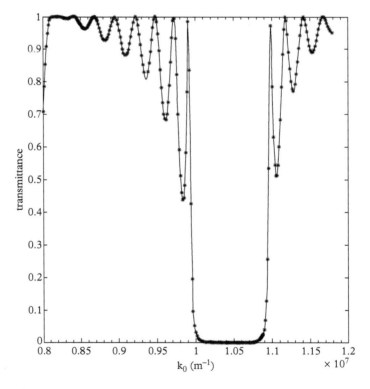

Figure 9.11 *Transmittance through a periodic stack, calculated from Equation 9.95, and composed of 9 periods of alternating TiO_2 and SiO_2 slabs 300 nm thick. The bandgap is centred $k_0 \simeq 1.04 \times 10^7 \ m^{-1}$ with $\Delta k_0 \simeq 0.1 \times 10^7 \ m^{-1}$. Rapid oscillations adjacent to the abrupt band gap edges converge to near-unity transmittance in the band windows on either side of the gap.*

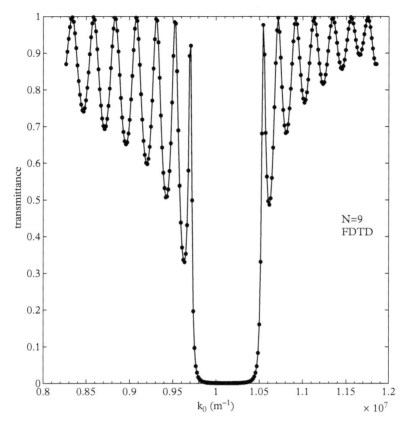

Figure 9.12 *Transmittance through the same periodic stack as in Figure 9.11 calculated by direct numerical solution of Maxwell's equations. The finite difference time domain (FDTD) method is used to calculate the transmittance. Slight dispersion in the indices of refraction over the frequency range is taken into account in this result.*

The intensity decreases in steps with each unit period and exhibits an exponential envelope consistent with Equation 9.73. Figure 9.14 plots the propagating Bloch wave in the stack at the band edge. As expected from Equation 9.76, the envelope indicates a half-wave oscillation.

9.4.7 Metal-dielectric periodic layers

So far in Section 9.4.2 we have considered only lossless dielectric periodic layers. In this section we extend the discussion to alternating layers of dielectric and metal, or more generally between layers of positive and negative permittivity. We will see that surface plasmon modes at the dielectric-metal interfaces play a crucial role in resonant

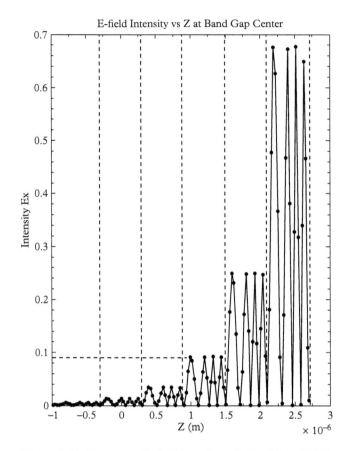

Figure 9.13 *Evanescent E_x field intensity, calculated from FDTD, in the TiO_2-SiO_2 periodic stack at the frequency centre of the band gap. The vertical dashed lines indicated the slab period boundaries. The horizontal dashed line indicates $1/2e$ of the evanescent intensity decay from the point of incidence at $z = +2.7\ \mu m$. The $1/2e$ penetration distance is consistent with Equations 9.73 and 9.76.*

tunnelling of light through this type of periodic stack. We will continue to restrict the discussion to lossless materials in order to keep the mathematical expressions manageable and the physical ideas lucid. Real metals, of course, are always somewhat lossy, but the results obtained here will closely approximate reality as long as the real part of the permittivity is much greater than the imaginary part. Figure 9.15 shows a schematic of alternating layers of metal and dielectric slabs. In this schematic the top and bottom layers are metal followed by alternating layers of dielectric and metal. The dielectric layers (d_1) are usually thicker than the metal layers (d_2) because the absolute value of the permittivities of metals in the optical regime is usually greater than those of dielectrics.

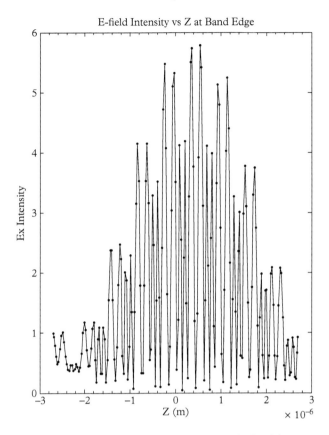

Figure 9.14 *Propagating E_x field intensity, calculated from FDTD, in the TiO_2-SiO_2 periodic stack at the band edge. The amplitude envelope indicates a half period of the $e^{iK\Lambda}$ Bloch wave propagation as expected from Equation 9.73 at the band edge.*

The top and bottom metal layers can be one half the thickness of the interior metal layers to ensure periodic translational symmetry from top to bottom. Often a high index 'coupling' layer is added to the top and bottom of the stack to ensure efficient source coupling at Brewster's angle to the surface plasmon index.

The key difference between dielectric periodic layers and dielectric-metal periodic layers is the appearance of evanescent rather than propagating field components perpendicular to the dielectric-metal interfaces. Applying matching boundary conditions results in surface waves that do propagate along x at the interfaces but are evanescent along z. These waves are called 'surface plasmon' or 'surface plasmon polariton' SPP waves. Their properties for semi-infinite slabs and a single interface have been developed in Chapter 7. Here we recall the three expressions for the surface wave vector components: $k_x^s, k_z^{s,d}, k_z^{s,m}$ from Equations 7.34–7.36,

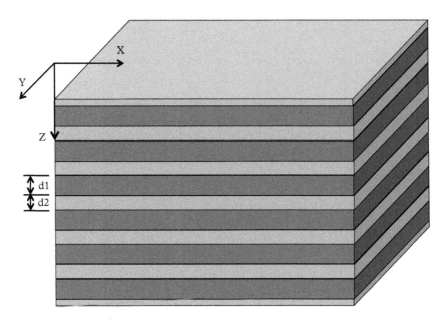

Figure 9.15 *Periodic stack of alternating metal and dielectric layers. The optical entrance and exit (top and bottom) layers are metal with thickness $d_2/2$. Light incident at the top, with E-field vector in the $X - Z$ plane (TM polarisation) couples to surface plasmon modes evanescent in z but propagating in the $x - y$ plane at the metal-dielectric interfaces.*

$$k_x^s = \sqrt{\frac{\varepsilon_d \varepsilon_m}{\varepsilon_d + \varepsilon_m}} k_0 \tag{9.98}$$

$$k_z^{s,d} = \pm \frac{\varepsilon_d}{\sqrt{\varepsilon_d + \varepsilon_m}} k_0 \tag{9.99}$$

$$k_z^{s,m} = \pm \frac{\varepsilon_m}{\sqrt{\varepsilon_d + \varepsilon_m}} k_0 \tag{9.100}$$

In these expressions k_x^s is the x component of the surface wave propagating at the interface, $k_z^{s,d}$ is the z component in the dielectric, and $k_z^{s,m}$ is the z component in the metal. The terms ε_m, ε_d are the real metal and dielectric permittivities, respectively. We remember that the metal permittivities are negative, and therefore if $|\varepsilon_m| > \varepsilon_d$, the k_z components will be pure imaginary. We write k_z components as in Equations 7.39 and 7.40,

$$k_z^{s,d} \to \pm i \kappa_s^d = \pm i \left| \frac{\varepsilon_d}{\sqrt{\varepsilon_d + \varepsilon_m}} \right| k_0 \tag{9.101}$$

$$k_z^{s,m} \to \pm i \kappa_s^m = \pm i \left| \frac{\varepsilon_m}{\sqrt{\varepsilon_d + \varepsilon_m}} \right| k_0 \tag{9.102}$$

The choice of sign ($\pm$) is made so that the wave amplitude decreases exponentially with increasing distance from the interface ($\pm z$). Specifically, if there is a metal-dielectric interface at $z = 0$, and the dielectric is on the $+z$ side, then

$$k_z^{s,d} \rightarrow i\kappa_s^d \qquad z > 0 \tag{9.103}$$

$$k_z^{s,m} \rightarrow -i\kappa_s^m \qquad z < 0 \tag{9.104}$$

and the components of the surface wave projecting onto the dielectric side take on the following forms,

$$H_y^{s,d}(x, z, t) = H_{0y} e^{-\kappa_s^d z} e^{i(k_x^s x - \omega t)} \tag{9.105}$$

$$E_x^{s,d}(x, z, t) = -\frac{i\kappa_s^d}{\omega \varepsilon_0 \varepsilon_d} H_{0y} e^{-\kappa_s^d z} e^{i(k_x^s x - \omega t)} \tag{9.106}$$

$$E_z^{s,d}(x, z, t) = -\frac{k_x^s}{\omega \varepsilon_0 \varepsilon_d} H_{0y} e^{-\kappa_s^d z} e^{i(k_x^s x - \omega t)} \tag{9.107}$$

Similar expressions obtain on the metallic side with the appropriate change of sign for $i\kappa_s^m$ (Equation 9.104). With the change to pure imaginary arguments, Equation 9.71 for the Bloch vector becomes [15],[16]

$$\cos K\Lambda = \cosh(k_{z1} d_1) \cosh(k_{z2} d_2) + \frac{1}{2} \left(\frac{k_{z2}}{k_{z1}} \frac{\varepsilon_1}{\varepsilon_2} + \frac{k_{z1}}{k_{z2}} \frac{\varepsilon_2}{\varepsilon_1} \right) \sinh(k_{z1} d_1) \sinh(k_{z2} d_2) \tag{9.108}$$

Taking into account the form of k_{z1}, k_{z2} for the evanescent z components of the surface plasmon wave in Equations 9.101 and 9.102, and the fact that $\varepsilon_m = \varepsilon_2 < 0$, the expression for the Bloch vector becomes

$$\cos K\Lambda = \cosh(k_{z1} d_1) \cosh(k_{z2} d_2) - \sinh(k_{z1} d_1) \sinh(k_{z2} d_2) = \cosh(k_{z1} d_1 - k_{z2d} d_2) \tag{9.109}$$

This expression is valid in the limit of weak coupling of the surface plasmon modes between adjacent layers. Equations 9.101 and 9.102 were obtained for an isolated surface plasmon mode in which the metal and dielectric half-spaces extend to infinity. In the stacked layer these modes will couple to produce symmetric and antisymmetric linear combinations that are the eigen modes of the stacked layer system. However, it appears from numerical simulation that use of the weak-coupling expressions are adequate for practical design of a resonant tunnelling stack. The $\cos K\Lambda$ expression in Equation 9.109 must be equal to or less than unity for the cosine argument to be real. When $k_{z1} d_1 = k_{z2} d_2$, $\cos(K\Lambda) = 1$ and $K = 0, 2\pi, 4\pi, \ldots$. Thus, when the phases $k_{z1} d_1, k_{z2} d_2$ are equal, the Bloch vector is real but stationary. The phase velocity of the Bloch wave along z is null although the light fields are linked evanescently from layer to layer. Plane waves incident on the stack emerge from it without change of phase, due to this linked, resonant tunnelling along z. Equations 9.105–9.107 show that the plasmon waves do propagate along the x direction, transverse to z. Figure 9.16 plots $k_{z1} d_1, k_{z2} d_2$, and $\cos K\Lambda$ as a

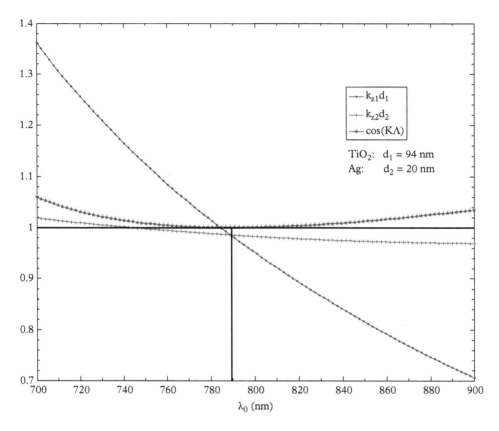

Figure 9.16 *Plots of $k_{z1}d_1, k_{z2}d_2$, and $\cos K\Lambda$ versus λ_0, the wavelength of light incident on a periodic stack of TiO_2 layers ($d_1 = 95\ nm$) and Ag ($d_2 = 20\ nm$). The plots show that for this particular choice of materials and layer thicknesses, $\cos(K\Lambda) = 1$ in a relatively narrow range of wavelengths centred at $\lambda_0 = 788.9\ nm$.*

function of the wavelength λ_0 of incident light for a typical example of alternating layers of a silver–titanium dioxide (Ag-TiO_2) periodic stack. In this particular example the dielectric layer thickness (d_1) is chosen to be 94 nm and metal layer thickness (d_2) is 20 nm. Figures 9.17 and 9.18 show the results of an FDTD simulation of the $\lambda_0 = 785$ nm plane wave incident on a five-layer stack constructed with the materials and geometry used in Figure 9.16. Figure 9.17 shows a plot of the E_x field amplitude in the $X - Z$ plane of the periodic Ag-TiO_2 stack. The plane wave is incident on the top, a 10 nm thick Ag surface at $z = 684$ nm, and at −50.8 degrees from the normal in a slab of silicon (Si) above the stack. This Si slab permits excitation of the surface plasmon modes at Brewster's angle. A corresponding Si slab substrate at the stack exit couples out the propagating plane wave at $z = 0$ nm. For this example of a Ag-TiO_2 system, the zeroth-order wavelength of the surface plasmon, corresponding to an isolated, single interface,

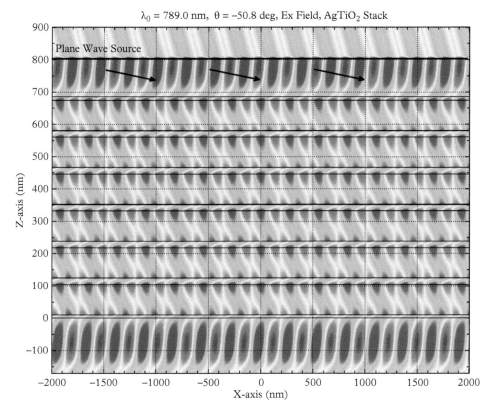

Figure 9.17 *A plane wave source located at z = 800 nm impinges on a periodic stack of alternating Ag layers (20 nm thickness) and TiO₂ layers (94 nm thickness). Top Ag surface, z = 684 nm; bottom Ag surface, z = 0 nm. Top and bottom Ag layers are 10 nm thick. The source and stack are embedded in a layer of Si on top and another on the bottom that ensure efficient coupling into and out of the plasmon modes. The source wavelength λ_0 = 789 nm, with E-field vector in the X – Z plane (TM polarisation) and is incident on the top surface at an angle of –50.8 deg with respect to the stack normal. The plot shows E_x with positive and negative amplitudes indicated by red and blue. Within the stack the fields are linked evanescently along z while propagating as guided waves in the +x direction. Note that absence of phase shift along z indicating a null phase velocity of the Bloch wave. Similar plots obtain for field components E_z and H_y.*

is $\lambda_s pp$ = 278 nm. In the simulated stack of Figure 9.17 the surface plasmon wavelength is 277 nm, very close to the single-interface value. Figure 9.19 shows the amplitude of the SPP mode along x within the TiO_2 layer at z = 670 nm.

A metal-dielectric periodic stack, exhibiting resonant tunnelling along z, also exhibits the properties of an anisotropic 'epsilon-near-zero' (ENZ) metamaterial [17],[18]. According to the *effective medium theory* (EMT) characterising metamaterials, permittivities perpendicular and parallel to the optical axis are given by,

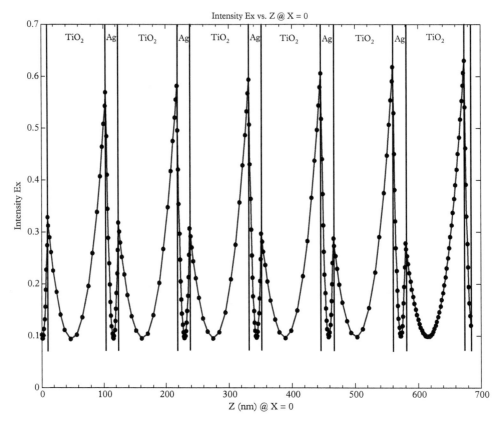

Figure 9.18 *Intensity profile of the illuminated stack shown in Figure 9.17. E-field Intensity along z at x = 0 from the top surface (right) at z = 694 nm to bottom surface (left) at z = 10 nm. The profile shows how the fields are evanescently linked along z from incident (left) to exit (right) surfaces.*

$$\varepsilon_\perp = \frac{d_1\varepsilon_1 + d_2\varepsilon_2}{d_1 + d_2} \tag{9.110}$$

$$\frac{1}{\varepsilon_\parallel} = \frac{1}{d_1 + d_2}\left[\frac{d_1}{\varepsilon_1} + \frac{d_2}{\varepsilon_2}\right] \tag{9.111}$$

Figure 9.20 shows the effective medium permittivities for the Ag-TiO$_2$ stack as a function wavelength around the region where $\varepsilon_\perp \simeq 0$. We see from Equations 9.108, 9.101, and 9.102 and 9.110 that the condition for a real Bloch vector, $k_{z1}d_1 - k_{z2}d_2 = 0$, also implies, at least in the weak-coupling limit, that $\varepsilon_\perp = 0$. Transmission by resonant tunnelling through surface plasmon modes [19] is intimately connected to the 'epsilon-near-zero' (ENZ) metamaterial [20–24] that holds promise for many novel and unexpected material properties. In particular, Engheta *et al.* [23] have pointed out that a point dipole above a surface of ENZ material will be subject to a repulsive force

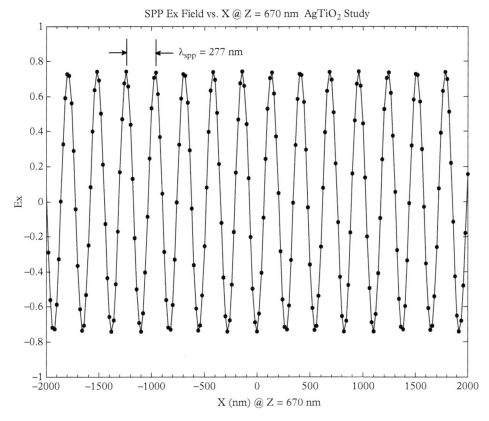

Figure 9.19 *Amplitude of the surface plasmon Ex component along the x direction within a TiO₂ layer at z = 670 nm (see Figure 9.17). The simulated FDTD λ_{spp} = 277 nm, is very close to the wavelength λ_0 = 278 nm of a surface plasmon at a single, isolated interface.*

analogous to the expulsion of a magnetic dipole from a superconductor due to the Meissner effect. Figure 9.21 illustrates the analogy and Figure 9.22 plots an FDTD simulation of the D-field generated by a point dipole close to an ENZ surface. According to [23] the time-averaged vertical component of the repulsive force is given by

$$\langle F_z \rangle = -\sigma \frac{9}{512\pi^4 c} \mathrm{Re} \left[\frac{\varepsilon_{matl} - \varepsilon_0}{\varepsilon_{matl} + \varepsilon_0} \right] P_{rad} \left(\frac{\lambda}{d} \right)^4 \tag{9.112}$$

where $\sigma = 1, 2$ if the alignment of the dipole is horizontal, vertical with respect to the surface plane, and d, λ are the distance above the plane and the wavelength, respectively. The factor P_{rad} is the spatially averaged radiated power of a classical dipole p in free space (see Equation 5.7 in Chapter 5) and is given by

$$P_{rad} = \frac{4\pi^3 c}{3\varepsilon_0 \lambda^4} \cdot |p|^2 \tag{9.113}$$

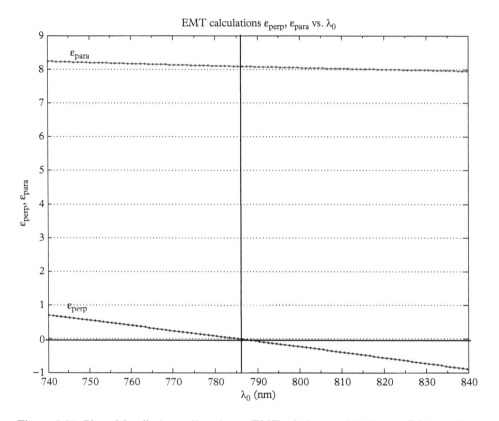

Figure 9.20 *Plots of the effective medium theory (EMT) relative permittivities parallel ($\varepsilon_\parallel$) and perpendicular ($\varepsilon_\perp$) as function of wavelength for the Ag-TiO$_2$ periodic stack characterised in Figures. 9.16–9.19. The terms 'parallel' and 'perpendicular' refer to the optical (z) axis. The perpendicular permittivity is near zero between $\lambda_0 = 785 - 790$ nm .*

When $\varepsilon_{matl} \rightarrow 0$ the effective vertical force is repulsive. In the case of Figure 9.20 we see that $\varepsilon_\perp \rightarrow 0$ around $\lambda = 785$ nm. This wavelength is very close to the resonance transition of a rubidium (Rb) atom, and we can calculate the equilibrium position of an atom subject to this levitation force and the force of gravity. Figure 9.23 shows a plot of the levitation force for the Rb atom as a function of distance from the plane of the anisotropic material. The equilibrium point is where this curve crosses the constant gravitational force.

9.4.8   Anisotropy in stacked layers

Returning to Figure 9.20, we emphasise in this section the anisotropy in the parallel and perpendicular permittivities. In k-space this property can be represented by the geometry of the dispersion equation. For a lossless structure conservation conditions require

(a)

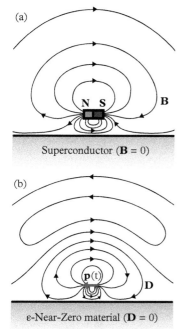

Superconductor (**B** = 0)

(b)

Figure 9.21 *Top diagram (a) shows how the B-field is excluded from a superconductor and results in a strong B-field gradient just above the surface. The magnetic dipole is repelled by a dipole-gradient force. Bottom diagram (b) shows the analogous D-field gradient and repulsive force resulting from an electric dipole above an ENZ material. Figure adapted from [21] and used with permission.*

$$n_x^2 k_x^2 + n_y^2 k_y^2 + n_z^2 k_z^2 = n^2 k_0^2 \qquad (9.114)$$

where we take z to be the optic axis and n_x, n_y, n_z are the refractive indices along x, y, z, and $k_0 = 2\pi/\lambda_0 = \omega/c$ is, as usual, the propagation parameter in free space. We then have

$$\varepsilon_x k_x^2 + \varepsilon_y k_y^2 + \varepsilon_z k_z^2 = \varepsilon_\| \varepsilon_\perp k_0^2 \qquad (9.115)$$

where $\varepsilon_x, \varepsilon_y = \varepsilon_\perp$ and $\varepsilon_z = \varepsilon_\|$ denote the permittivities perpendicular and parallel to the optic axis[1]. Dividing both sides by $\varepsilon_\perp \varepsilon_\|$, we can then write,

$$\frac{k_x^2}{\varepsilon_\|} + \frac{k_y^2}{\varepsilon_\|} + \frac{k_z^2}{\varepsilon_\perp} = k_0^2 \qquad (9.116)$$

Equation 9.116 is conventionally called the dispersion equation for a uniaxial anisotropic material. It defines a 3-D surface–the *isofrequency surface*. Figure 9.24 shows the different classes of surface defined by the dispersion equation as a function of the relative magnitudes and signs of the perpendicular and parallel permittivities. Clearly, our example Ag/TiO$_2$ stacked layer goes from ellipsoidal at wavelengths less than 785 nm to a one-sheet hyperbolic surface at wavelengths greater than 790 nm. There has recently been

---

[1] Some authors take $\varepsilon_\perp, \varepsilon_\|$ perpendicular and parallel to the $x - y$ plane.

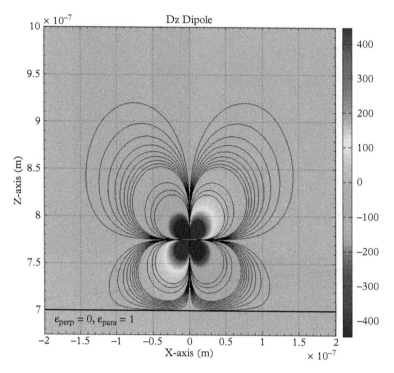

Figure 9.22 *An FDTD numerical simulation of a point dipole placed 650 nm above an anisotropic material with $\varepsilon_{perp} = 0$ and $\varepsilon_{para} = 1$. The contours plot the D_z component of the displacement field. Anisotropic ENZ metamaterials can be realised at a particular wavelength by stacked layers of dielectric and metal of fixed geometry (see Figure 9.20).*

a great deal of interest in hyperbolic metamaterials because a dipole emitting radiation into a hyperbolic material can, in principle, couple to a much higher density of states than when the same dipole couples radiatively to the vacuum. We recall from Chapter 6 that the density of states available for spontaneous emission into the isotropic vacuum increases as k_0^2 and corresponds to the surface of the isofrequency sphere. Suppose now that an emitting dipole is situated just above the surface of an isotropic material. Figure 9.25a shows a schematic representation. At the surface the k-vector components parallel to the $x - y$ plane must be continuous through the surface. Therefore, for the closed-surface isotropic sphere, the upper limit to k_x is k_0 pointing somewhere in the $x - y$ plane. In contrast, if the material isofrequency surface is an open two-sheet hyperboloid, as shown in Figure 9.25b, any k_x vector can satisfy the continuity condition at the boundary, and therefore, in principle, the number of k-states available for emission is greatly enhanced. This condition should give rise to a marked increase in the radiative rate of a quantum emitter. Reference to Figure 9.20 shows that for the Ag-TiO$_2$ stacked layer, it is $\varepsilon_\perp$ that goes negative, and therefore, the appropriate isofrequency surface is

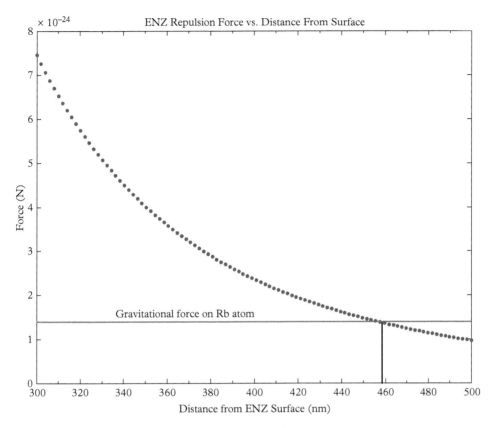

Figure 9.23 *Levitation force $< F_z >$ and gravitational force on a Rb atom as a function of distance from an anisotropic metamaterial with $\varepsilon_\perp \to 0$.*

the one-sheet hyperboloid, Fig. 9.24 (d). In this case the minimum-length k-vector is k_0 pointing in the $x-y$ plane, while for all longer k-vectors the continuity condition can always be satisfied. It is interesting to note that at the wavelength crossover point in Figure 9.20, where $\varepsilon_\perp = 0$, the available k-vectors are also all confined to the $x-y$ plane. Rather than a surface, the k-vectors that satisfy the continuity condition are confined to a ring in the $x-y$ plane of radius k_0. Since the total number of k-states is considerably less than in the isotropic vacuum, one expects a marked *decrease* in the rate of emission.

9.4.9 Validity of effective medium theory

In this section we examine the validity of the effective medium theory used in Section 9.4.7 and in particular the use of Equations 9.110 and 9.111. The effective medium theory (EMT) presupposes that the wavelength of light interacting with a structured material is very long compared to the characteristic length scale of any component of

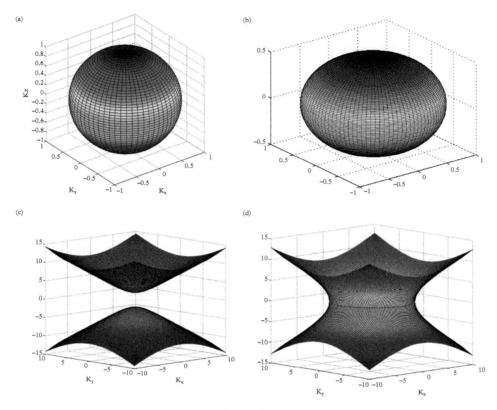

Figure 9.24 *(a) sphere: isotropic, $\varepsilon_\perp = \varepsilon_\parallel$ (b) ellipsoid: $\varepsilon_\perp > \varepsilon_\parallel$ (c) hyperbolic: two-sheet, $\varepsilon_\parallel < 0$ (d) hyperbolic: one-sheet, $\varepsilon_\perp < 0$.*

that composite. This assumption is usually well-founded when the components are all dielectrics whose refractive indices are not greater than, say, 2.5. But in the case of metal-dielectric layers, even when the layer thicknesses are subwavelength, the characteristic plasmon waves at the interfaces may have wavelengths greatly reduced from the incident light. This important wavelength reduction is due to the large imaginary terms of metal refractive indices in the visible and near infra-red spectral regions. The range of legitimate application of the effective medium theory may therefore be greatly reduced. However, we can compare the anisotropic permittivities determined by the simple EMT formulas that *assume* constant field amplitudes within each subwavelength layer, to those extracted by the more accurate and precise transfer matrix theory that numerically matches appropriate field amplitudes at each interface.

Before describing this extraction procedure it is important to note that for some metamaterial design goals, a useful answer can be obtained independently of the EMT assumptions. As Smith *et al.* [25] have pointed out, the Bloch condition for periodic structures, Equation 9.72, is insensitive to the length scale of the individual components.

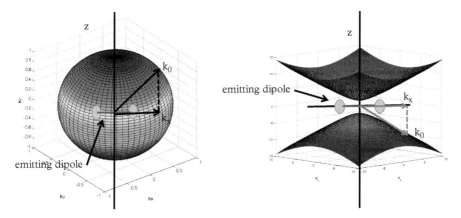

Figure 9.25 *(a) Isofrequency sphere showing the surface to which k_0 must be constrained. The limiting amplitude of any component in the $x - y$ plane is therefore k_0. (b) Isofrequency two-sheet hyperboloid showing the open surface to which, in principle, any k_0 can couple because the continuity condition at the surface can always be satisfied.*

The Bloch vectors associated with these structures should therefore be valid whether or not the EMT formulas apply. In particular, in the case of alternating slabs of TiO$_2$ and Ag, if the goal is to find a condition for ENZ behaviour, Figure 9.16 shows in what wavelength region the Bloch K vector is close to zero. This is also the condition for resonant tunnelling through the stack as illustrated in Figures 9.17 and 9.18. Note that the plot of EMT permittivities in Figure 9.20 *also* shows that $\varepsilon_\perp \simeq 0$ in the same wavelength region as the Bloch vector. For the ENZ property one can say that effective medium theory yields a wavelength region that is at least consistent with a Bloch vector behaviour, itself independent of EMT assumptions. This conclusion, however, does not mean that a stack of alternating TiO$_2$ and Ag slabs with given slab thicknesses is a good approximation to a homogenous anisotropic material with $\varepsilon_\perp$ and $\varepsilon_\parallel$ as indicated in Figure 9.20. In order to answer *that* question the following extraction procedure may prove useful.

The procedure for this extraction [26] begins with the matrix expressions used for the metal layer and the dielectric layer in a periodic stack. For the time being we will consider the individual layers to be isotropic, and the overall transmission matrix T_T corresponds to a uniform material composed of the dielectric and metal slabs. Figure 9.26 shows the schematic layout used for the procedure and the matrices for the individual slabs are

$$T_m = \begin{bmatrix} \cos \varphi_m & \frac{i \cos \theta}{\sqrt{\varepsilon_m}} \sin \varphi_m \\ i \frac{\sqrt{\varepsilon_m}}{\cos \theta} \sin \varphi_m & \cos \varphi_m \end{bmatrix} \tag{9.117}$$

with

$$\varphi_m = \frac{2\pi}{\lambda} l \sqrt{\varepsilon_m} \sqrt{1 - \frac{\varepsilon_a \sin^2 \theta}{\varepsilon_m}} \tag{9.118}$$

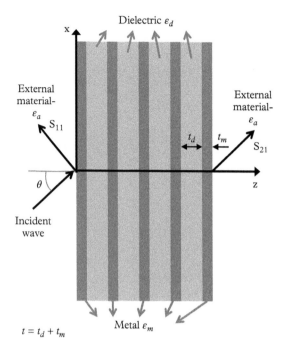

Dielectric ε_d

x

External material- ε_a

S_{11}

External material- ε_a

t_d t_m

S_{21}

θ

z

Incident wave

$t = t_d + t_m$

Metal ε_m

Figure 9.26 *Schematic of the metal-dielectric periodic stack structure. The S-parameters S_{11}, S_{21} correspond to net or measured reflection at the incident port and net or measured transmission at the exit port, respectively, with the angle of incidence θ. The thicknesses for the dielectric and metal are t_d, t_m, respectively. The relative permittivities of the metal, dielectric, and surrounding material (usually air or glass) are $\varepsilon_m, \varepsilon_d$, and ε_a, respectively. Figure adapted from [26] and used with permission.*

where the notation for the various terms are defined in Figure 9.26 and

$$T_d = \begin{bmatrix} \cos\varphi_d & \frac{i\cos\theta}{\sqrt{\varepsilon_d}}\sin\varphi_d \\ i\frac{\sqrt{\varepsilon_d}}{\cos\theta}\sin\varphi_d & \cos\varphi_d \end{bmatrix} \tag{9.119}$$

with

$$\varphi_d = \frac{2\pi}{\lambda}d\sqrt{\varepsilon_d}\sqrt{1 - \frac{\varepsilon_a\sin^2\theta}{\varepsilon_d}} \tag{9.120}$$

The overall transmission through the stack composed of N dielectric slabs and $N + 1$ metal slabs is the matrix product of the matrices for the individual dielectric and metal slabs:

$$T_T = T_m \prod_{i=1}^{N} T_d T_m = \begin{bmatrix} T_{11} & T_{12} \\ T_{21} & T_{22} \end{bmatrix} \tag{9.121}$$

The next step is to write the scattering S-parameters (the measurable quantities) for the 'homogenised' stack of slabs in terms of the transmission matrix elements [25]. For a slab of homogenous material $T_{11} = T_{22} = T_s$ and the S matrix is symmetric:

$$S_{21} = S_{12} = \frac{1}{T_s + \frac{1}{2}\left(ik_0 T_{12} + \frac{T_{21}}{ik_0}\right)} \tag{9.122}$$

$$S_{11} = S_{22} = \frac{\frac{1}{2}\left(\frac{T_{21}}{ik_0} - ik_0 T_{12}\right)}{T_s + \frac{1}{2}\left(ik_0 T_{12} + \frac{T_{21}}{ik_0}\right)} \tag{9.123}$$

With the S_{21}, S_{11} parameters in hand we can obtain the impedance Z and the propagation parameter in the material, k_z', from the following expressions, [25]

$$Z = \sqrt{\frac{(1 + S_{11})^2 - S_{21}^2}{(1 - S_{11})^2 - S_{21}^2}} \tag{9.124}$$

and

$$k_z' = \frac{1}{t}\cos^{-1}\left[\frac{1}{2S_{21}}\left(1 - S_{11}^2 + S_{21}^2\right) + 2\pi n\right] \tag{9.125}$$

where t is the period unit thickness, $t = t_d + t_m$, and n is an integer. But the S-parameters can also be written in terms of the Fresnel reflection coefficient R. We write approximate expressions for the S matrix elements, taking into account multiple reflections and transmission from the component slabs and dropping terms with R^3 or higher powers of R:

$$S_{11} \simeq \frac{R(1 - e^{i2k_z't})}{1 - R^2 e^{2ik_z't}} \tag{9.126}$$

$$S_{21} \simeq \frac{\left(1 - R^2\right)e^{ik_z't}}{1 - R^2 e^{2ik_z't}} \tag{9.127}$$

The reflectivity R is determined from these expressions because S_{11}, S_{21}, k_z' have been previously determined by Equations 9.122, 9.123, and 9.125. Furthermore, we can write the Fresnel reflection and transmission coefficients, R, T of the homogenised material by tangential field component matching (with TM polarisation) at the incident interface, using Equations 7.6–7.8:

$$E_x^i - RE_x^i = TE_x^t \tag{9.128}$$

$$\frac{k_0 \cos\theta \sqrt{\varepsilon_a}}{\varepsilon_0 \varepsilon_a} - R\frac{k_0 \cos\theta \sqrt{\varepsilon_a}}{\varepsilon_0 \varepsilon_a} = T\frac{k_z'}{\varepsilon_0 \varepsilon_x} \tag{9.129}$$

$$T = \frac{k_0 \varepsilon_x}{k_z' \sqrt{\varepsilon_a}}\cos\theta(1 - R) \tag{9.130}$$

$$H_y^i + RH_y^i = TH_y^t \tag{9.131}$$

$$1 + R = T \tag{9.132}$$

Eliminating T from Equations 9.130 and 9.132:

$$R = \frac{\frac{k_0}{k_z'}\frac{\varepsilon_x}{\sqrt{\varepsilon_a}}\cos\theta - 1}{\frac{k_0}{k_z'}\frac{\varepsilon_x}{\sqrt{\varepsilon_a}}\cos\theta + 1} = \frac{1 - \frac{k_z'}{k_0}\frac{\sqrt{\varepsilon_a}}{\varepsilon_x}\left(\frac{1}{\cos\theta}\right)}{1 + \frac{k_z'}{k_0}\frac{\sqrt{\varepsilon_a}}{\varepsilon_x}\left(\frac{1}{\cos\theta}\right)} \tag{9.133}$$

In writing Equation 9.130 we have labelled ε_x explicitly in anticipation of an anisotropy between ε_x and ε_z in a homogenised but *anisotropic* material. The reflectivity can be written in terms of the 'normalised impedance' as

$$R = \frac{1 - \frac{Z}{Z_0}}{1 + \frac{Z}{Z_0}} \tag{9.134}$$

from which we identify that

$$\frac{Z}{Z_0} = \frac{k_z'}{k_0}\frac{\sqrt{\varepsilon_a}}{\varepsilon_x}\left(\frac{1}{\cos\theta}\right) \tag{9.135}$$

The impedance Z_0 corresponds to the impedance of the external material (usually air or glass) with relative permittivity ε_a and Z is the impedance of the homogenised stack. We now suppose that our layered stack exhibits anisotropy in the permittivities with $\varepsilon_x = \varepsilon_y \neq \varepsilon_z$. Figure 9.27 shows how the layered stack of isotropic, alternating material slabs

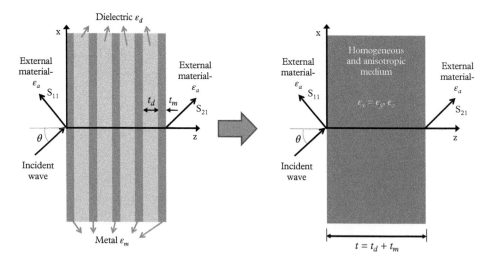

Figure 9.27 *The stacked layers of periodic, alternating dielectric and metal slabs are modelled as a homogenous, anisotropic uniaxial material with optical axis along z. The thickness $t = t_d + t_m$ corresponds to the thickness of a unit cell in the stacked layer. Figure adapted from [26] and used with permission.*

can be modelled as a homogenous *anisotropic* material of thickness t. Finally, we obtain from Equation 9.135 a determination of the effective relative permittivity perpendicular to the optical axis as a function of the unit cell thickness:

$$\varepsilon_x = \frac{Z_0}{Z} \frac{k'_z(t)}{k_0} \left(\frac{1}{\cos \theta} \right)$$
(9.136)

Then, using Equation 9.115, where evidently $\varepsilon_\perp = \varepsilon_x$ and $\varepsilon_\| = \varepsilon_z$,

$$\varepsilon_z(t) = \frac{\varepsilon_x(t) k_x^2}{\varepsilon_x k_0^2 - k_z'^2}$$
(9.137)

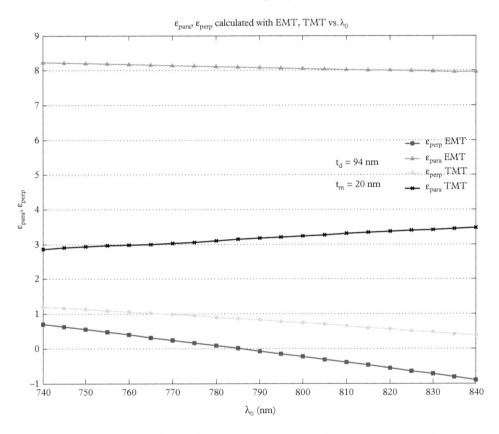

Figure 9.28 *Plots of $\varepsilon_\|, \varepsilon_\perp$ calculated according to TMT extraction using Equations 9.136 and 9.137, and EMT using Equations 9.110 and 9.111 vs λ_0 at normal incidence. The TMT extraction is called the complete parameter retrieval (CPR) approach in [27]. The EMT plots are the same as in Figure 9.20. The 'filling factor' parameter ρ is the metal thickness fraction to the total unit thickness, $t_m/(t_d + t_m)$. The value $\rho = 0.2$ corresponds approximately to the parameters $t_m = 20\,nm$ and $t_d = 94\,nm$ used in the example of Sections 9.4.7 and 9.4.8. Figure data from [27] and used with permission.*

Figures 9.28 and 9.29 show plots of $\varepsilon_\parallel, \varepsilon_\perp$, calculated according to Equations 9.136 and 9.137 as a function of the wavelength λ_0 in the vicinity where the effective medium formulas predict $\varepsilon_\perp \simeq 0$. The plots compare the results of the TMT procedure to the EMT calculations. The results of Figure 9.28 show that the EMT approximation is not reliable with slab thicknesses of 20 nm and 94 nm for Ag and TiO$_2$, respectively; but by reducing both thicknesses by a factor of 3, Figure 9.29 shows that good agreement between EMT and TMT can be obtained. Note that the geometry for 'good agreement' is at the limit of standard metal vapour deposition technology for metal layer thickness. We conclude, therefore, that a stack of alternating slabs of dielectric and metal can indeed mimic a homogenous, anisotropic material but the EMT approximation does not provide a reliable guide for design purposes even when the slab geometries are nominally

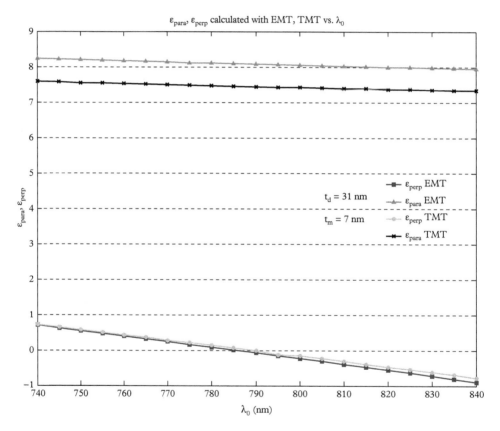

Figure 9.29 *Plots of* $\varepsilon_\parallel, \varepsilon_\perp$, *calculated according to TMT extraction using Equations 9.136 and 9.137, and EMT using Equations 9.110 and 9.111 vs* λ_0 *at normal incidence. The TMT extraction is called the complete parameter retrieval (CPR) approach in [27]. The EMT plots are the same as in Figure 9.20. The filling factor* $\rho = 0.2$ *is the same as in Figure 9.28 but* $t_m = 7$ nm *and* $t_d = 31$ nm. *Figure data from [27] and used with permission.*

well into the 'subwavelength' regime. The TMT procedure should be as reliable as the constituent material properties since it is based only on geometry of the stack, the indices of refraction of the dielectric and metal components, and numerical field matching at the layer interfaces.

9.5 Summary

This chapter begins with a brief summary of the definitions and properties of left-handed materials, then proceeds to a discussion of negative-index waveguides in the subwavelength régime. The focus then turns to reflection and transmission in stacked layers: first of alternating dielectrics and then of periodic dielectric-metal 2-D structures. Bloch waves and bandgaps appear as a consequence of the periodicity. In particular, it is shown that for judicious choices of material and geometry in a periodic dielectric-metal stack, a Bloch vector of $K = 0$ appears over a narrow range of wavelength. This Bloch vector corresponds to resonant evanescent-wave tunnelling through the stack. The result is a near-unity transmission, independent of the number of periods in the stack. It is also shown that in addition to resonant tunnelling, the dielectric-metal periodic stack also exhibits anisotropy with the permittivity parallel to the optical axis different from the permittivity of the plane perpendicular to the optical axis. The resonant tunnelling region also corresponds to an 'epsilon-near-zero' (ENZ) condition for the permittivity perpendicular to the optical axis. The possibility of levitating cold atoms (radiating at the appropriate frequency) above an ENZ material is briefly considered. The chapter ends with a critical examination of the 'effective medium theory' (EMT) for determining effective anisotropies in metal-dielectric stacked layers.

9.6 Bibliography

[1] V. E. Veselago and E. E. Narimanov, *The left hand of brightness: past, present and future of negative index materials*, Nat Mater vol 5, pp. 759–762 (2006).

[2] P. Yeh, *Optical Waves in Layered Media*, Wiley-Interscience, Hoboken, New Jersey (2005).

[3] V. G. Veselago, *The Electrodynamics of Substances with Simultaneously Negative Values of ε and μ*, Sov Phys Uspekhi vol 10, pp. 509–514 (1968).

[4] J. B. Pendry, *Negative Refraction Makes a Perfect Lens*, Phys Rev Lett vol 85, pp. 3966–3969 (2000).

[5] J. A. Dionne, L. A. Sweatlock, H. A. Atwater, and A. Polman, *Planar metal plasmon waveguides: frequency-dependent dispersion, propagation, localization, and loss beyond the free electron model*, Phys Rev B: Condens Matter vol 72, p. 075,405 (2005).

[6] J. A. Dionne, L. A. Sweatlock, H. A. Atwater, and A. Polman, *Plasmon slot waveguides: Towards chip-scale propagation with subwavelength-scale localization*, Phys Rev B: Condens Matter vol 73, p. 035,407 (2006).

[7] H. J. Lezec, J. A. Dionne, and H. A. Atwater, *Negative Refraction at Visible Frequencies, Science* vol 316, pp. 430–432 (2007).

[8] J. A. Dionne, E. Verhagen, A. Polman, and H. A. Atwater, *Are negative index materials achievable with surface plasmon waveguides? A case study of three plasmonic geometries, Opt. Express* vol 16, pp. 19,001–19,017 (2008).

[9] D. M. Pozar, *Microwave Engineering*, 3rd edition John Wiley & Sons, Inc. *The left hand of brightness: past, present and future of negative index materials, Nat Mater* vol 5, pp. Hoboken, New Jersey (2005).

[10] N. W. Ashcroft and N. D. Mermin, *Solid State Physics The Electrodynamics of Substances with Simultaneously Negative Values of ε and μ, Sov Phys Uspekhi* vol 10, pp. Thomson Learning, London (1976).

[11] J. D. Joannopoulos, R. D. Meade, and J. N. Winn, *Photonic Crystals Negative Refraction Makes a Perfect Lens, Phys Rev Lett* vol 85, pp. Princeton University Press, Princeton (1995).

[12] J. Chilwell and I. Hodgkinson, *Thin-films field-transfer matrix theory of planar multilayer waveguides and reflection from prism-loaded waveguides, J Opt Soc Am A* vol 1, pp. 742–753 (1984).

[13] K. H. Schlereth and M. Tacke, *The Complex Propagation Constant of Multilayer Waveguides: An Algorithm for a Personal Computer, IEEE J. Quantum Electron* vol 26, pp. 627–630 (1990).

[14] P. S. Russel, T. A. Birks, and F. D. Lloyd-Lucas, *Photonic Bloch Waves and Photonic Band Gaps*, pp. 585–633. *Planar metal plasmon waveguides: frequency-dependent dispersion, propagation, localization, and loss beyond the free electron model, Phys Rev B: Condens Matter* vol 72, p. Plenum Press, New York (1995).

[15] S. Feng, J. M. Elson, and P. L. Overfelt, *Transparent photonic band in metallodielectric nanostructures, Phys Rev Lett* vol 72, p. 085,117–1–6 (2005).

[16] S. Feng, J. M. Elson, and P. L. Overfelt, *Optical properties of multilayer metaldielectric nanofilms with all-evanescent modes, Opt Express* vol 13, pp. 4113–4124 (2005).

[17] A. Alù and N. Engheta, *Optical nanotransmission lines: synthesis of planar lefthanded metamaterials in the infrared and visible regimes, J Opt Soc Am B: Opt Phys* vol 23, pp.571–583 (2006).

[18] M. Silveirinha and N. Engheta, *Tunneling of Electromagnetic Energy through Subwavelength Channels and Bends using ε -Near -Zero Materials, Phys Rev Lett* vol 97, pp. 157, 403–1–4 (2006).

[19] S. Tomita, T. Yokoyama, H. Yanagi, B. Wood, J. B. Pendry, M. Fujii, and S. Hayashi, *Resonant photon tunneling via surface plasmon polaritons through one-dimensional metal-dielectric metamaterials, Opt Express* vol 16, pp. 9942–9950 (2008).

[20] A. Alù, M. G. Silveirinha, A. Salandrino, and N. Engheta, *Epsilon-near-zero metamaterials and electromagnetic sources: Tailoring the radiation phase pattern, Phys Rev B: Condens Matter* vol 75, pp. 155,410 (2007).

[21] Y. Li and N. Engheta, *Supercoupling of surface waves with ε -near -zero metastructures, Phys Rev B: Condens Matter* vol 90, pp. 201,107 (2014).

[22] R. Maas, J. Parsons, N. Engheta, and A. Polman, *Experimental realization of an epsilon-near-zero metamaterial at visible wavelengths*, Nat Photonics vol 7, pp. 907–912 (2013).

[23] F. J. Rodríguez-Fortuño, A. Vakil, and N. Engheta, *Electric Levitation Using ε - Near - Zero Metamaterials*, Phys Rev Lett vol 112, pp. 033,902 (2014).

[24] F. J. Rodríguez-Fortuño, A. Vakil, and N. Engheta, *Electric Levitation Using ε - Near - Zero Metamaterials: Supplementary Information*, Phys Rev Lett vol 112, pp. 033,902 (2014).

[25] D. R. Smith, D. C. Vier, T. Koschny, and C. M. Soukoulis, *Electromagnetic parameter retrieval from inhomogenous metamaterials*, Phys Rev B: Condens Matter vol 71, pp. 036,617 (2005).

[26] B.-H. Borges and A. F. Mota, *Metal-dielectric layered media and hyperbolic metamaterials* (2015). Presented at the Workshop on Strongly Coupled Field Theories for Condensed Matter and Quantum Information Theory, Natal, Brazil.

[27] A. F. Mota, A. Matins, J. Weiner, F. L. Teixeira, and B.-H. V. Borges, *Constitutive parameter retrieval for uniaxial metamaterials with spatial dispersion*, Phys Rev B 115410 (2016).

10

Momentum in Fields and Matter

The argument has not, it is true, been carried on at high volume, but the list of disputants is very distinguished.[1]

10.1 Introduction

In Chapter 2, Section 2.5.3, we developed the idea that Poynting's vector $\mathbf{S}$ could be interpreted as the flow of electromagnetic energy across a closed surface around some point $\mathbf{r}$. For fields propagating through spaces (vacuum or material) or on surfaces, the Poynting vector describes the field energy flux (S.I. units of Watts m^{-2}) and is identified with the cross product between the electric and magnetic fields,

$$\mathbf{S} = \mathbf{E} \times \mathbf{H} \tag{10.1}$$

Using Maxwell's equations, Equations 2.25–2.28, together with Equation 10.1, it can be shown fairly straightforwardly how energy flows out of the enclosed volume and interacts with free and bound currents that may be present:

$$\nabla \cdot \mathbf{S} + \frac{\partial}{\partial t}\left(\frac{1}{2}\varepsilon_0 \mathbf{E} \cdot \mathbf{E} + \frac{1}{2}\mu_0 \mathbf{H} \cdot \mathbf{H}\right) + \mathbf{E} \cdot \mathbf{J}_{\text{free}} + \mathbf{E} \cdot \frac{\partial \mathbf{P}}{\partial t} + \mathbf{H} \cdot \frac{\partial \mathbf{M}}{\partial t} = 0 \tag{10.2}$$

In the same sense that Equation 2.35 provides a charge-current continuity across a closed surface, Equation 10.2, Poynting's theorem in differential form, states that the energy flux across a closed surface must be equal to the sum of the time rate of energy change residing in the fields, the free current sources, and the bound current sources. Clearly, the two terms in parentheses are the electric and magnetic field energies. A positive flow of energy outwards from a point implies a decrease in the field energies at that point. The $\mathbf{E} \cdot \mathbf{J}_{\text{free}}$ term is interpreted as the work done by an electric field $\mathbf{E}$ on a free current source $\mathbf{J}$ at the point in question. If the Ohm's law constitutive relation holds, $\mathbf{J} = \sigma \mathbf{E}$, ($\sigma$ being the conductivity), then this term can be written as J^2/σ, which indicates a 'Joule

---

[1] Reported in [1] and attributed to E. I. Blount.

Light-Matter Interaction. Second Edition. John Weiner and Frederico Nunes.
© John Weiner and Frederico Nunes 2017. Published 2017 by Oxford University Press.

heating' or irreversible ohmic resistive power dissipation. The last two terms are also field-current interactions, but here they involve the bound currents of polarisation and magnetisation, $\partial \mathbf{P}/\partial t$ and $\partial \mathbf{M}/\partial t$, respectively, rather than the free current $\mathbf{J}_{\text{free}}$.

If Poynting's theorem states how energy can flow between current sources and fields in space and in matter, one might ask if momentum and momentum flow could not also be associated with electromagnetic fields. Newton's second law asserts that a force applied to a body is equal to the rate of change of momentum transferred to or from that body. When we apply Newton's second law of motion, we usually think of a scattering event where one body carrying momentum impinges on another, exerting a force between the two bodies and altering their motion. We normally consider the momentum transfer to be instantaneous upon contact. But suppose the two bodies are electrically charged and are separated by a great distance. Motion of the first body must be transmitted to the second by the Coulomb field, but this transmission cannot be faster than the speed of light. Therefore, the field itself must somehow receive and transmit momentum between the two bodies. From a less classical point of view one might imagine a two-level atom at rest in the upper state that emits a quantum of light. It is well known from cold-atom experiments that the atom suffers a backward recoil motion from the direction of light emission. Clearly, the atom has assumed a mechanical momentum equal to the product of its mass and velocity. After propagating some distance the emitted light encounters a second two-level atom at rest in the lower state. The atom absorbs the light and receives a forward impulsive recoil. The forward recoiling atom assumes a momentum equal to the product of its mass and velocity, which is equal in magnitude but opposite in sign from the initial backward recoil. Conservation of linear momentum requires that the momentum assumed by the second atom must have been transported from the first by the propagating light quantum or 'photon'. Assuming for the moment then, that a quantity of linear momentum might be associated with a temporal light pulse, a 'thought experiment' attributed to Einstein suggests what the relation might be between the exchange of energy and exchange of momentum.

10.2   Einstein Box thought experiment

Consider a pulsed light emitter, for example a pulsed laser, mounted on a frictionless platform. Suppose that on the same platform a light receiver is positioned to receive the light pulses. How is momentum in the light pulse transferred from emitter to receiver in such a way as to be consistent with the principles of momentum conservation and centre-of-mass conservation? The situation of the Einstein Box experiment is shown in Figure 10.1. The principle of centre-of-mass conservation means that if no external forces act on a system, the velocity of the centre-of-mass remains constant no matter how momentum and energy may be rearranged internally. The velocity of the centre-of-mass can of course include $v_{cm} = 0$, in which case the coordinates of the centre-of-mass point remain unchanged. In Figure 10.1 we see that prior to pulse emission,

$$z_{cm} = \frac{m_e z_e + m_r z_r}{m_e + m_r} = \frac{m_r D}{m_e + m_r} \qquad (10.3)$$

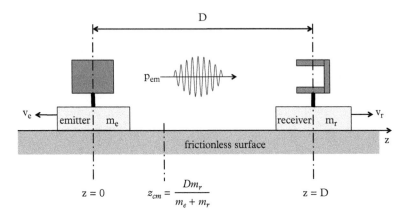

Figure 10.1 *The Einstein Box thought experiment. An emitter of mass m_e sends a light pulse to a receiver of m_r, separated from the emitter by a distance of D along z. Conservation of linear momentum requires that the emitter recoil backwards with momentum $-p_{em}$ and the receiver recoils forwards with momentum p_{em} after the pulse time-of-flight D/c. Since no external forces act on the system, the centre-of-mass must remain unchanged by the internal momentum transfer.*

where m_e, m_r are the masses of the emitter and receiver, z_e, z_r their coordinates along z, and where initially $z_e = 0$ and $z_r = D$. At the initial time $t = 0$, both emitter and receiver are at rest so the momentum of the system is zero. Now the source emits a pulse, sufficiently long that we can consider it essentially monochromatic, but sufficiently short such that pulse length is much less than the distance separating emitter and receiver. If the pulse carries momentum, we expect an initial recoil in the emitter so that as soon as the pulse is in flight, carrying momentum forward, the emitter moves backwards, carrying equal and opposite momentum. When the pulse arrives at the receiver, its momentum is absorbed, and the receiver will recoil forwards. At all times after the transfer, the emitter will move backwards with constant velocity v_e and the receiver will move forwards with constant velocity v_r so that their momenta will sum to zero:

$$- m_e v_e + m_r v_r = 0 \tag{10.4}$$

Furthermore, the centre-of-mass coordinate before and after the transfer must remain constant. After the pulse transfer, for times $t > D/c$, we have

$$\frac{\left[\left(m_e - \frac{\mathscr{E}}{c^2}\right) z_e(t) + \left(m_r + \frac{\mathscr{E}}{c^2}\right) D + \left(m_r + \frac{\mathscr{E}}{c^2}\right) z_r(t)\right]}{m_e + m_r} = \frac{m_r D}{m_e + m_r} \tag{10.5}$$

where the left-hand side of Equation 10.5 expresses the centre-of-mass coordinate after the pulse transfer and the right-hand side is the same coordinate at $t = 0$. We have also, following Einstein, identified a 'mass' with the light pulse energy $\mathcal{E}$ such that $\mathcal{E} = m_{em}c^2$, where m_{em} is the 'electromagnetic' mass and c is the speed of light in vacuum. Now, the momentum acquired by the emitter is $-p_{em}$, the momentum of the light pulse, and the velocity of the emitter after recoil is

$$v_e = \frac{-p_{em}}{(m_e - \mathcal{E}/c^2)} \quad \text{and} \quad z_e(t) = v_e t \tag{10.6}$$

Similarly for the receiver,

$$v_r = \frac{p_{em}}{(m_e + \mathcal{E}/c^2)} \quad \text{and} \quad z_r(t) = v_r(t - D/c) \tag{10.7}$$

Substituting Equations 10.6 and 10.7 into Equation 10.5 and solving for p_{em} results in

$$p_{em} = \frac{\mathcal{E}}{c} \tag{10.8}$$

We see that the momentum associated with the light pulse must be its energy divided by c. But we have already postulated that the *energy flux* of light must be carried by the Poynting vector $\mathbf{S}$. From the S.I. units of energy flux (Watts m^{-2}), it is clear that the energy *density* in the pulse must be $\mathcal{E}/V = S/c$, where V is the pulse volume, and therefore, the electromagnetic momentum density p_{em}/V is

$$\frac{\mathbf{p}_{em}}{V} = \frac{\mathcal{E}/V}{c}\hat{\mathbf{z}} = \frac{\mathbf{S}}{c^2} \tag{10.9}$$

Of course, this thought experiment is just suggestive since the light pulse is collimated (no spatial divergence), travels through free space, undergoes no reflection, and is perfectly absorbed. What would the momentum look like if it travelled through a non-dispersive, lossless material medium with refractive index greater than unity: glass, for example? Another thought experiment due to Balazs can help answer this question.

10.3 Balazs thought experiment

The Balazs thought experiment [2] illustrates how the centre-of-mass principle can be used to determine the momentum carried into an object from a light pulse. Again suppose we have a pulsed light source emitting a quasi-monochromatic pulse, the length of which is short compared to the spatial dimension of its overall transit. And again let us associate a mass m with the pulse, following the Einstein mass-energy equivalence formula, $\mathcal{E} = mc^2$. Figure 10.2 illustrates the two situations. In the first, a light pulse is emitted and travels a distance $D = ct$ through space with velocity c in time t. In the second, the light pulse enters, without reflection, a material slab of mass M with a lossless, real refractive index $n > 1$ and propagates the length of the slab in a time t_s. The light pulse group velocity v_g within the slab is less than in vacuum, and so for the propagation

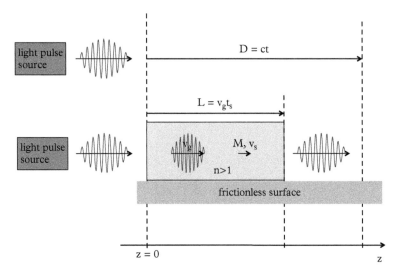

Figure 10.2 *The Balazs though experiment consists of two situations: in the first a quasi-monochromatic light pulse is emitted and travels through space a distance D in time t. In the second situation the pulse travels through a lossless transparent slab of mass M, refractive index n, and length L. The group velocity of the pulse within the slab is $v_g = c/n$. Since no outside forces act on the system, the centre-of-mass coordinate must remain invariant between the two situations. This requirement implies that the slab must move forward some distance to compensate for the slower pulse speed v_g within the slab.*

time t, the distance travelled must necessarily be less. However, whether the pulse travels in free space or whether it travels through the slab, no external forces act on the system, and so the centre-of-mass coordinate at time t must be the same in both cases. Clearly, the centre-of-mass principle requires that the slab displaces forwards some distance to compensate for the lower group velocity of the pulse within the slab. The principal questions are: what is the momentum of the light pulse within the slab, p_{cm_s}, and what is the momentum of the slab, P_s? The centre-of-mass in the first case is simply given by

$$z_{cm} = \frac{m}{m + M}D = \frac{m}{m + M}ct \qquad (10.10)$$

In the second case we have

$$z_{cm} = \left(\frac{m}{m + M}\right)v_g t_s + \left(\frac{M}{m + M}\right)v_s t_s + \left(\frac{m}{m + M}\right)c(t - t_s) \qquad (10.11)$$

Setting the expressions for z_{cm} in Equations 10.10 and 10.11 equal, we find that

$$mc = mv_g + Mv_s \qquad (10.12)$$

and since $P_s = Mv_s = m(c - v_g)$, we have for the momentum of the slab,

$$P_s = \frac{\mathscr{E}}{c}\left(1 - \frac{1}{n}\right) \tag{10.13}$$

The total momentum is just the momentum of the light pulse in the first case, which we found in the Einstein Box thought experiment to be $p_{em} = \mathscr{E}/c$. Therefore, the light pulse momentum in the slab p_{em_s} is

$$p_{em_s} = \frac{\mathscr{E}}{c} - \left[\frac{\mathscr{E}}{c}\left(1 - \frac{1}{n}\right)\right] = \frac{\mathscr{E}}{c}\left(\frac{1}{n}\right) = \frac{p_{em}}{n} \tag{10.14}$$

The momentum of the light pulse in the slab is just the light momentum in free space divided by the slab refractive index. This momentum is called the *Abraham momentum*.

From the momentum of the slab P_s we can deduce its displacement Δz:

$$\Delta z = \frac{P_s}{M} t_s \tag{10.15}$$

and from Figure 10.2 we see that $t_s = L/v_g$. Therefore, we can write

$$\Delta z = \frac{\mathscr{E}/c}{M}\left(1 - \frac{1}{n}\right)\frac{L}{v_g} = \frac{\mathscr{E}/c^2}{M}(n - 1)L \tag{10.16}$$

Even though the Abraham momentum is reduced in the material slab, compared to the same momentum packet in free space, by a factor of $1/n$, the momentum *density* is the same. The pulse volume in space is Act, where A is the pulse cross-sectional area and ct the pulse length along z. In the slab the volume is compressed to Act/n so

$$\frac{p_{em_s}}{V_s} = \frac{p_{em}}{V} = g_A = \frac{\mathbf{E} \times \mathbf{H}}{c^2} = \frac{\mathbf{S}}{c^2} \tag{10.17}$$

where g_A denotes the Abraham momentum density. From these two thought experiments we can tentatively postulate that the momentum density is everywhere given by the Poynting vector divided by the square of the speed of light in vacuum.

10.3.1 Abraham momentum and Minkowski momentum

However, the postulate is tentative because the momentum density can also be plausibly expressed as

$$g_M = \frac{1}{\varepsilon_0 \mu_0}\frac{\mathbf{D} \times \mathbf{B}}{c^2} = \mathbf{D} \times \mathbf{B} \tag{10.18}$$

This formulation is called the Minkowski momentum density, and the original motivation for the Balazs thought experiment was to distinguish the consequences for Abraham momentum transferred to material bodies compared to that of Minkowski. Although the Minkowski and Abraham expressions are equivalent in free space, they are not equivalent as the vector fields pass into the slab. The tangential field components of **E** and **H** are continuous at the vacuum-slab interface. The total Abraham momentum of the pulse inside the slab is therefore

$$\mathcal{P}_{em_{As}} = g_A V_s = g_A \frac{V}{n} \tag{10.19}$$

in agreement with Equation 10.3. The tangential components of the **D** and **B** fields, however, are not continuous at the interface. In general for nonmagnetic materials we have at the interface

$$D_{\|}(free\ space) = \varepsilon_0 E_{\|}, D_{\|}(slab) = \varepsilon E_{\|}\ and\ B_{\|}(free\ space) = \mu_0 H_{\|}, B_{\|}(slab) = \mu H_{\|} \tag{10.20}$$

where the 'parallel' subscripts indicate tangential components and ε, μ are the permittivity and permeability of the slab. In our thought experiment the slab is just a piece of glass, so $\mu_0 = \mu$, but $\varepsilon > \varepsilon_0$. Remembering that $n^2 = \varepsilon$, we find the Minkowski total momentum of the pulse inside the slab:

$$\mathcal{P}_{em_{Ms}} = g_M V_s = n^2 g_A V_s = n g_A V = n \mathcal{P}_{em} \tag{10.21}$$

The Minkowski momentum of the light pulse inside the slab is *greater* than the momentum of the light pulse in free space by a factor of n. In order to conserve momentum and the centre-of-mass coordinate, the slab would have to move *backwards*:

$$\Delta z = \frac{\mathcal{E}/c^2}{M} n(1 - n)L \tag{10.22}$$

Since a collimated quasi-monochromatic light pulse cannot exert a 'tractor-beam', negative light pressure force on a material body (see, however, [3]), the Abraham momentum appears to be the right choice.

But is it? Suppose we conceive another very simple thought experiment. A quasi-monochromatic plane-wave pulse propagating in a lossless medium with refractive index $n = \sqrt{\varepsilon}$ impinges on a perfect mirror embedded in the medium. What is the momentum imparted to the mirror? Figure 10.3 illustrates the situation. The conservation of linear momentum demands that the change in momentum of the light pulse before and after reflection must be equal to the change in momentum of the reflector. Also, by Newton's second law, a change in the momentum of a body is equal to force applied to it. Therefore, a calculation of the Lorentz force at the reflector surface must be equal to the change in momentum. We write the Lorentz force as

$$\mathbf{F} = \sigma \mathbf{E} + \mathbf{J_s} \times \mathbf{B} \tag{10.23}$$

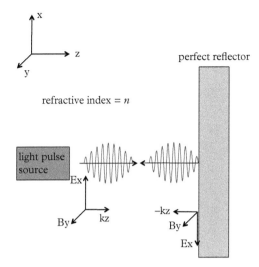

Figure 10.3 *Quasi-monochromatic light pulse impinges on a perfect reflector at normal incidence in a medium with refractive index n. Note that the reflected E-field is π out of phase with the incident E-field and that the incident and reflected B-fields are in phase. The net E-field at the surface is therefore null and the B-field amplitude at the surface is twice that of the incident B-field.*

where σ is the surface charge density, $\mathbf{J_s}$ the surface current density in the $x-y$ plane, and $\mathbf{E, B}$ the electric field and magnetic induction field at the surface. The reflector is a perfect conductor so the amplitude of the reflected E-field changes sign at the surface. The sum of the incident and reflected E-fields is therefore null at the surface, and no field penetrates into the bulk of the reflector . The B-field, however, does not change sign, and the net B-field at the surface is twice the amplitude of $\mathbf{B_i}$, the incident B-field. From the Maxwell–Ampère law, integrating along z through the interface, we find $J_{s_x} = H_y$. Then the Lorentz force normal to the surface, averaged over an optical cycle is

$$F_z = \frac{1}{2}\left[J_{s_x} \cdot B_y^*\right] = \frac{1}{2}\mu_0\left[H_y \cdot 2H_y\right] = \varepsilon\varepsilon_0 E_x^2 \tag{10.24}$$

Now we compare this Lorentz-force result to the change in the Minkowski momentum of the light pulse upon reflection. Suppose the incident pulse carries Minkowski momentum density $\mathbf{g_M} = \mathbf{D} \times \mathbf{B}$. Taking the absolute value and averaging over an optical cycle we have

$$g_M = \frac{1}{2}\left[D_x \cdot B_y^*\right] = \frac{1}{2}\varepsilon\varepsilon_0\mu_0 E_x H_y = \frac{1}{2}\varepsilon\varepsilon_0 \frac{E_x^2}{v} \tag{10.25}$$

where in the right-most term $v = c/n$. At the surface the momentum reverses direction but maintains the same magnitude. Therefore, the *change* in momentum with respect to z evaluated at the surface is

$$\frac{d\mathbf{g_M}}{dz} = 2\left[\frac{1}{2}\varepsilon\varepsilon_0 \frac{E_x^2}{v}\right] \tag{10.26}$$

and the time rate of change of momentum is

$$\frac{d\mathbf{g_M}}{dt} = \frac{d\mathbf{g_M}}{dz} \cdot \frac{dz}{dt} = 2\left[\frac{1}{2}\varepsilon\varepsilon_0\frac{E_x^2}{v}\right] \cdot v = \varepsilon\varepsilon_0 E_x^2 \tag{10.27}$$

The expression for the Abraham momentum density, averaged over an optical cycle, is

$$\mathbf{g_A} = \frac{1}{2}\left[\frac{E_x \cdot H_y^*}{c^2}\right] \tag{10.28}$$

and following through the same reasoning as for the Minkowski expression, we find the Abraham force density applied to the reflector is

$$\frac{d\mathbf{g_A}}{dt} = \varepsilon_0 E_0^2 \tag{10.29}$$

We see that the force calculated from the Lorentz expression, Equation 10.24, is equal to the time rate of change of the Minkowski momentum, *not* the Abraham momentum.

Both the Balazs and the reflection thought experiment seem to exhibit impeccable logic but each leads to a different expression for the momentum transfer. So what is the correct answer? A number of real experiments [4–6] have attempted to decide the issue, but it turns out that it is extremely difficult to design and execute an experiment with results that can be interpreted in only one way. In fact, to this day there is no wide-spread consensus, despite a number of proposals, to resolve the Abraham–Minkowski controversy [7–9]. To make matters worse there is a simmering, related controversy concerning the interpretation of 'hidden momentum' in magnetic materials (materials with magnetisation field **M**) in the presence of charged currents. It turns out that invoking the Minkowski expression leads to an extra term in the momentum content that must be reconciled with momentum conservation. We will discuss two of the key experiments and their varying interpretations in Section 10.4.11.

10.4   Field equations and force laws

In order to frame these issues as clearly as possible, we digress to a general, but summary, discussion of the relation between Maxwell's field equations, energy-momentum flux, and force laws. Although this discussion will not settle these controversies, the hope is that it will lead to a deeper understanding of how the macroscopic field equations inform our interpretation of electromagnetic interaction within ponderable media.

10.4.1   Maxwell's fields in matter

In this section we follow an approach due to Mansuripur [10, 11]. As a point of depart-ure, we write down again Maxwell's equations appropriate to electromagnetic fields in matter:

$$\nabla \cdot D = \rho_{\text{free}} \tag{10.30}$$

$$\nabla \cdot B = 0 \tag{10.31}$$

$$\nabla \times E = -\frac{\partial B}{\partial t} \tag{10.32}$$

$$\nabla \times H = J_{\text{free}} + \frac{\partial D}{dt} \tag{10.33}$$

where, in polarisable, magnetic matter, D and B are related to E and H by

$$D = \varepsilon_0 E + P \tag{10.34}$$

$$B = \mu_0 H + M \tag{10.35}$$

and ρ_{free} and J_{free} are the free charge density and free current density, respectively.[2]

10.4.2   Electric dipole or 'Lorentz' Model

As written, Equations 10.30–10.33, express the field relations using all four D,B,E,H; but, by invoking Equations 10.34 and 10.35, we can rewrite Maxwell's equations in terms of E,B and the two 'matter' fields, P, M. The result is

$$\varepsilon_0 \nabla \cdot E = \rho_{\text{free}} - \nabla \cdot P \tag{10.36}$$

$$\nabla \times B = \mu_0 \left(J_{\text{free}} + \frac{\partial P}{\partial t} + \frac{1}{\mu_0} \nabla \times M \right) + \varepsilon_0 \mu_0 \frac{\partial E}{\partial t} \tag{10.37}$$

$$\nabla \times E = -\frac{\partial B}{\partial t} \tag{10.38}$$

$$\nabla \cdot B = 0 \tag{10.39}$$

From Equation 10.36 we interpret the term $-\nabla \cdot P$ as the 'bound charge',

$$\rho_{\text{bound}} = -\nabla \cdot P \tag{10.40}$$

and from Equation 10.37 we interpret the terms $\partial P/\partial t + 1/\mu_0 (\nabla \times M)$ as the 'bound current':

$$J_{\text{bound}} = \frac{\partial P}{\partial t} + \frac{1}{\mu_0} \nabla \times M \tag{10.41}$$

The bound charge density is the negative divergence of the polarisation field and leads us to a material model in which the medium consists of microscopic dipoles, small with respect to the wavelength of light, but greater than atomic dimensions. The polarisation field itself has units corresponding to dipole density. If the dipoles are oriented within

---

[2] Note that the definition in Equation 10.35 is not universal. Many authors use the definition $B = \mu_0(H+M)$ that has the advantage of resulting in a bound current term $J_{\text{bound}} = \nabla \times M$ without the annoying $1/\mu_0$ factor, but has the disadvantage of an annoying lack of symmetry with Equation 10.34.

a given volume of the material then by Gauss's law the integral of the divergence of the polarisation field is equal to the integral of the polarisation itself over the surface of the volume:

$$-\int \nabla \cdot P \, dV = -\int P \cdot d\sigma = \int \rho_{\text{bound}} \, dV = q_{\text{bound}} \qquad (10.42)$$

where $d\sigma$ is the surface element. If the dipoles are randomly oriented, this surface integral will evaluate to zero. Figure 10.4 shows how the polarisation field can give rise to bound charge.

Figure 10.4 *Material consisting of microscopic dipoles, the length of which is much shorter than optical wavelengths but much longer than the atomic scale. The dipoles are arranged so that the net divergence is non-zero. Note that within the volume the positive and negative ends of the dipoles cancel the charged ends of their neighbours, but outward from the surface there are no neighbours and the negative integral of the charge density over the surface is equal to the total bound charge at the volume centre.*

(a)

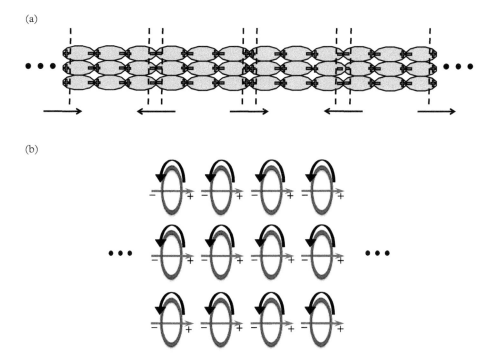

(b)

Figure 10.5 *(a) Bound current in a polarisation wave built up from microscopic electric dipole elements. Regions of positive charge density travel in the opposite direction to regions of negative charge density to assure the continuity conditions of current and charge. (b) Bound current in microscopic Amperian current loops that give rise to individual magnetic moments (horizontal arrows). The density and orientation of the Amperian current loops produce the net, macroscopic magnetisation* **M**.

The bound current consists of two terms: the first term, dP/dt, can be thought of as an electric polarisation current resulting from a polarisation wave travelling through the material. We imagine the second term, proportional to the curl of the magnetisation, to consist of microscopic loops of current producing magnetic dipole moments. The units of **M** in the SI system is the same as **B**, the magnetic induction field, $W\,m^{-2}$ or $N/(m{\cdot}A)$. These current loops are often called 'Amperian current loops'. Figure 10.5 shows how the microscopic-dipoles model of **P** and **M** give rise to bound currents.

This picture of **P** and **M** arising from microscopic electric dipoles and Amperian current loops can be termed the 'Lorentz model' [11] because a generalised Lorentz force F_L can be written with a total charge density term and a total current density term:

$$F_L = (\rho_{\text{free}} + \rho_{\text{bound}})\,E + (J_{\text{free}} + J_{\text{bound}}) \times B \qquad (10.43)$$

10.4.3   Magnetic dipole model

Another way to rearrange Equations 10.30–10.33 is to use Equation 10.34 and Equation 10.35 to eliminate D and B. The result is

$$\varepsilon_0 \nabla \cdot E = \rho_{\text{free}} - \nabla \cdot P \tag{10.44}$$

$$\nabla \times H = \left(J_{\text{free}} + \frac{\partial P}{\partial t}\right) + \varepsilon_0 \frac{\partial E}{\partial t} \tag{10.45}$$

$$\nabla \times E = -\frac{\partial M}{\partial t} - \mu_0 \frac{\partial H}{\partial t} \tag{10.46}$$

$$\mu_0 \nabla \cdot H = -\nabla \cdot M \tag{10.47}$$

We see from Equation 10.44 and 10.45 that a bound electric charge and current can still be interpreted in terms of microscopic electric dipoles, but Equations 10.46 and 10.47 introduce a bound magnetic current, $\partial M/\partial t$, and a bound magnetic charge, $-\nabla \cdot M$, *as if* magnetic matter was composed of microscopic permanent magnetic dipoles. Note the similarity between Equation 10.44 and Equation 10.47. Free magnetic monopoles are absent from Equation 10.47, but the negative divergence of the magnetisation field suggests a bound magnetic charge, $\rho_{\text{mag}} = -\nabla \cdot M$, arising from fixed magnetic dipoles analogous to the picture in Figure 10.4. The Amperian current loops are no longer in evidence and have been replaced by these microscopic permanent magnetic dipoles. Both the Lorentz model and the magnetic dipole model are simply two different ways to write Maxwell's equations.

These ball-and-stick models of material composition are not to be taken literally. The components of matter are assembled at the atomic and molecular level and must be described by quantum mechanics. The models simply give us a way to visualise microscopic origins for the macroscopic polarisation and magnetisation fields. From the Lorentz and magnetic-dipole formal rearrangements of Maxwell's equations we can write expressions for the energy and momentum flux between fields and matter.

10.4.4   Energy conservation—Lorentz model

The well-known expression for charge conservation

$$\nabla \cdot J + \frac{\partial \rho}{\partial t} = 0 \tag{10.48}$$

states that the positive divergence of the current density from a point in space is equal to the time rate of charge density decrease from that point. We seek something similar for energy flowing between fields and matter:

$$\nabla \cdot S + \frac{\partial \mathcal{E}}{\partial t} = 0 \tag{10.49}$$

where S is the energy flux $(\mathrm{J\,s^{-1}\,m^{-2}})$ and $\nabla \cdot S$ is the time rate of energy passing through a unit of area normal to the flux. The second term in Equation 10.49 is the time rate of change of energy density corresponding to the spatial flux. The conservation of energy expression states that the spatial energy flux out of an enclosing volume must be equal to the rate of decrease in energy density within that volume. A field divergence expression can be obtained from Equations 10.37 and 10.38 by operating on the first with $E\cdot$ and the second with $B\cdot$ and then subtracting Equation 10.37 from Equation 10.38. The result is

$$B \cdot (\nabla \times E) - E \cdot (\nabla \times B) = B \cdot \left[-\frac{\partial B}{\partial t} \right] - E \cdot \left[\mu_0 \left(J_{\text{free}} + \frac{\partial P}{\partial t} + \frac{1}{\mu_0} \nabla \times M \right) + \varepsilon_0 \mu_0 \frac{\partial E}{\partial t} \right]$$

or

$$\nabla \cdot \left[\frac{1}{\mu_0}(E \times B) \right] + \frac{1}{2} \left(\frac{1}{\mu_0} \frac{\partial B \cdot B}{\partial t} + \varepsilon_0 \frac{\partial E \cdot E}{\partial t} \right) + E \cdot \left(J_{\text{free}} + \frac{\partial P}{\partial t} + \frac{1}{\mu_0} \nabla \times M \right) = 0$$

$$(10.50)$$

The first term on the left we interpret as the divergence of the energy flux in the Lorentz model:

$$S_L = \frac{1}{\mu_0}(E \times B) \tag{10.51}$$

The middle term is the rate of change of energy density stored in the fields B and E. Absent free charges or ponderable matter, these two terms state that the energy flow away from a point in space is equal to the time rate of field energy density decrease at that point. When matter is present, the third term on the left adds the time rate of work due to the E-field interacting with the current density (both free and bound currents). We usually think of this third term as some kind of 'Joule heating' where incoming light flux excites currents within the material and leads to a dissipative energy sink. But this term can also be a source of energy flux as well as a sink. If the motion of the currents is oscillatory, for example, they can be a source of radiation.

10.4.5 Energy conservation—magnetic dipole model

An analogous field divergence expression can be obtained from Equations 10.45 and 10.46 resulting in

$$\nabla \cdot (E \times H) + \frac{1}{2} \left(\mu_0 \frac{\partial H \cdot H}{\partial t} + \varepsilon_0 \frac{\partial E \cdot E}{\partial t} \right) + \left[E \cdot \left(J_{\text{free}} + \frac{\partial P}{\partial t} \right) + H \cdot \frac{\partial M}{\partial t} \right] = 0 \tag{10.52}$$

Again we identify the first term on the left with the divergence of the energy flux in the magnetic dipole model and recognise the usual expression for the Poynting vector:

$$S_{MD} = E \times H \tag{10.53}$$

The middle term is again the rate of change of field energy density, and the third term represents the time rate of work done, on or by, the E-field acting on electric current density (free and bound) plus time rate of work done, on or by, the H-field acting on the magnetic current density. In free space with no sources or ponderable matter, the third terms in Equations 10.50 and 10.52 vanish and the expressions become identical. When magnetic matter ($\mu \neq \mu_0, M \neq 0$) is present, the two expressions for the energy flux, Equations 10.51 and 10.53, differ by a term involving the permeability and the magnetisation,

$$S_L = S_{MD} + \frac{1}{\mu_0}(E \times M) \tag{10.54}$$

The extra term in the Lorentz version of the energy flux is sometimes called 'hidden energy' [12] and arises at the interface between two materials with different permeabilities. We will discuss hidden energy and hidden momentum at greater length in Section 10.4.10.

10.4.6   Force law and momentum conservation in the Lorentz and magnetic dipole models

It was mentioned above in Equation 10.43 that the generalised Lorentz force law could be written down using $\rho_{\text{total}} = \rho_{\text{free}} + \rho_{\text{bound}}$ and $J_{\text{total}} = J_{\text{free}} + J_{\text{bound}}$. Carrying out the relevant substitutions and making use of vector field identities, we find that

$$F_L = (\varepsilon_0 \nabla \cdot E) E + \left(\frac{1}{\mu_0} \nabla \times B - \varepsilon_0 \frac{\partial E}{\partial t} \right) \times B \tag{10.55}$$

The cross-product term on the right can be expanded so that Equation 10.55 can be written as

$$F_L = (\varepsilon_0 \nabla \cdot E) E + \left(\frac{1}{\mu_0} \nabla \times B - \varepsilon_0 \frac{\partial E}{\partial t} \right) \times B$$

$$= (\varepsilon_0 \nabla \cdot E) E + \frac{1}{\mu_0} (\nabla \times B) \times B - \frac{\partial (\varepsilon_0 E \times B)}{\partial t} + \varepsilon_0 E \times \frac{\partial B}{\partial t} \tag{10.56}$$

where we have used

$$\frac{\partial \varepsilon_0 E \times B}{\partial t} = \varepsilon_0 E \times \frac{\partial B}{\partial t} + \frac{\partial \varepsilon_0 E}{\partial t} \times B$$

Substituting Faraday's law in the last term on the right in Equation 10.56 we can now write it as

$$F_L = (\varepsilon_0 \nabla \cdot E) E + \frac{1}{\mu_0} (\nabla \times B) \times B - \frac{\partial (\varepsilon_0 E \times B)}{\partial t} + \varepsilon_0 (\nabla \times E) \times E \tag{10.57}$$

and using vector-field identity, Equation C.59, in the second and fourth terms of Equation 10.57, we have

$$F_L = (\varepsilon_0 \nabla \cdot E) E - \frac{1}{\mu_0} \left[\frac{1}{2} \nabla (B \cdot B) - (B \cdot \nabla) B \right] - \frac{\partial (\varepsilon_0 E \times B)}{\partial t} - \varepsilon_0 \left[\frac{1}{2} \nabla (E \cdot E) - (E \cdot \nabla) E \right]$$

(10.58)

Regrouping terms we have

$$F_L = \varepsilon_0 [(\nabla \cdot E) E + (E \cdot \nabla) E] - \left[\varepsilon_0 \frac{1}{2} \nabla (E \cdot E) \right] - \left[\frac{1}{\mu_0} \frac{1}{2} \nabla (B \cdot B) \right] + \frac{1}{\mu_0} (B \cdot \nabla) B - \frac{\partial (\varepsilon_0 E \times B)}{\partial t}$$

(10.59)

The first two terms on the right can be written as the divergence of the second rank tensor resulting from the direct product of E with itself:

$$\nabla \cdot \overleftrightarrow{EE} = (\nabla \cdot E) E + (E \cdot \nabla) E$$

(10.60)

It is possible to get a similar divergence relation for the B field by adding $1/\mu_0 (\nabla \cdot B) B$ to Equation 10.59. The addition is permissible because $\nabla \cdot B = 0$. We then can write Equation 10.59 as

$$F_L = \varepsilon_0 \nabla \cdot \overleftrightarrow{EE} + \frac{1}{\mu_0} \nabla \cdot \overleftrightarrow{BB} - \frac{1}{2} \left[\varepsilon_0 \nabla (E \cdot E) + \frac{1}{\mu_0} \nabla (B \cdot B) \right] - \frac{\partial (\varepsilon_0 E \times B)}{\partial t}$$

(10.61)

Finally, the third square-bracketed term on the right can be included in the tensor divergence by converting $E \cdot E$ and $B \cdot B$ to tensor form by use of the identity tensor $\overleftrightarrow{I}$. The generalised Lorentz force density expression is then

$$F_L = \nabla \cdot \left[\overleftrightarrow{EE} + \overleftrightarrow{BB} - \frac{1}{2} (E \cdot E + B \cdot B) \overleftrightarrow{I} \right] - \frac{\partial (\varepsilon_0 E \times B)}{\partial t}$$

(10.62)

The last term on the right of Equation 10.62 is the time rate of change of the field momentum density $G_L = \varepsilon_0 E \times B$. Note that $G_L = S_L / c^2$. The term in brackets is called the Maxwell stress tensor denoted by $\overleftrightarrow{T}$. Note that the first term on the right is the divergence of a *tensor*, not a vector. Finally, we can write the conservation of momentum condition in analogy with the conservation expressions for charge, Equation 10.48, and for energy, Equation 10.49:

$$\nabla \cdot \overleftrightarrow{T} - \frac{\partial (\varepsilon_0 E \times B)}{\partial t} - F_L = 0$$

(10.63)

The divergence of the stress tensor can be considered the time rate of momentum passing through a unit of area normal to the momentum flux. Notice that the signs of the spatial and temporal derivatives are opposite (compare Equation 10.49) so that the *negative* of the momentum flux flowing *out of* a closed volume ($\nabla \cdot -\overleftrightarrow{T}$) results in an *decrease* in the field momentum, $-\partial(\varepsilon_0 E \times B)/\partial t$, within the volume. We will return to discuss the physical significance of this generalised Lorentz force density expression after an interlude on tensor analysis.

10.4.7   Digression on tensor calculus

This section sketches some essential properties of tensors, their application to electrodynamics, and the meaning of the notation first introduced in Equation 10.60. For readers already familiar with tensors and their physical significance, this section can be skipped and they can proceed directly to Section 10.4.8. Others may find the section worthwhile as a brief summary or review of tensor properties.

In 3-D space a second rank tensor is nothing more than a 3×3 array consisting of nine terms that transform under rotation according to certain rules. It can be generated by the direct or outer product of two vectors. For example, the direct product of a column vector and a row vector results in a 3×3 matrix:

$$\begin{pmatrix} a_1 \\ a_2 \\ a_3 \end{pmatrix} \times \begin{pmatrix} b_1 & b_2 & b_3 \end{pmatrix} = \begin{pmatrix} ab_{11} & ab_{12} & ab_{13} \\ ab_{21} & ab_{22} & ab_{23} \\ ab_{31} & ab_{32} & ab_{33} \end{pmatrix} \tag{10.64}$$

where the first index indicates the row and the second index the column of matrix term.

Not every 3×3 matrix, however, is a second rank tensor. Suppose that we have two field vectors E, D in a 3-D space that are related by a linear transformation in the following way:

$$\begin{aligned} D_1 &= T_{11}E_1 + T_{12}E_2 + T_{13}E_3 \\ D_2 &= T_{21}E_1 + T_{22}E_2 + T_{23}E_3 \\ D_3 &= T_{31}E_1 + T_{32}E_2 + T_{33}E_3 \end{aligned} \tag{10.65}$$

A second rank tensor is a linear transformation of the components of vector E into the components of vector D *that is invariant to coordinate rotation*[3]. In matrix form Equation 10.65 is written as

$$\begin{pmatrix} D_1 \\ D_2 \\ D_3 \end{pmatrix} = \begin{pmatrix} T_{11} & T_{12} & T_{13} \\ T_{21} & T_{22} & T_{23} \\ T_{31} & T_{32} & T_{33} \end{pmatrix} \cdot \begin{pmatrix} E_1 \\ E_2 \\ E_3 \end{pmatrix} \tag{10.66}$$

or more compactly,

$$D_j = \sum_k T_{jk}E_k \tag{10.67}$$

---

[3] Here, vector E and vector D are just generic vectors, *not* the electric and displacements fields of Maxwell's equations.

Under coordinate rotation, the coefficients of the vector-field components change as the direction cosines with the condition that the absolute value (length) of the vector remain invariant. In 3-D space a position vector r, specifying a point P with respect to an origin O, can be written as

$$r = x_1 i_1 + x_2 i_2 + x_3 i_3 \tag{10.68}$$

After a coordinate rotation the same vector can be written as

$$r = x_1' i_1' + x_2' i_2' + x_3' i_3' \tag{10.69}$$

The coordinates of P after rotation are

$$x_j' = r \cdot i_j' = x_1 i_1 \cdot i_j' + x_2 i_2 \cdot i_j' + x_3 i_3 \cdot i_j' \tag{10.70}$$

In order to preserve the vector length we must have

$$\sum_{j=1}^{3} x_j^2 = \sum_{j=1}^{3} x_j'^{2} \tag{10.71}$$

Now the coordinate rotation can be written in the compact notation as

$$x_j' = \sum_{k=1}^{3} a_{jk} x_k \qquad (j = 1, 2, 3) \tag{10.72}$$

where a_{jk} are the terms of the 3×3 rotation matrix. So from Equation 10.70 we see that

$$a_{jk} = i_j' \cdot i_k \tag{10.73}$$

are the direction cosines of the rotation.

Now we can determine the constraints on the coordinate rotation matrix, using Equations 10.71 and 10.72:

$$\sum_{j=1}^{3} x_j'^{2} = \sum_{j=1}^{3} \left(\sum_{i=1}^{3} a_{ji} x_i \right) \left(\sum_{k=1}^{3} a_{jk} x_k \right) = \sum_{i=1}^{3} \sum_{k=1}^{3} x_i x_k \left(\sum_{j=1}^{3} a_{ji} a_{jk} \right) \tag{10.74}$$

But Equation 10.74 can only be true if

$$\sum_{j=1}^{3} a_{ji} a_{jk} = \delta_{ik} \qquad \delta_{ik} = 1, i = k \qquad \delta_{ik} = 0, i \neq k \tag{10.75}$$

Linear coordinate transformations Equation 10.72 subject to Equation 10.75 are called *orthogonal* transformations and the matrices are called orthogonal as well. Written out in matrix notation Equation 10.75 states that if the matrix A is orthogonal,

$$A = \begin{pmatrix} a_{11} & a_{12} & a_{13} \\ a_{21} & a_{22} & a_{23} \\ a_{31} & a_{32} & a_{33} \end{pmatrix} \tag{10.76}$$

then A has the following properties:

$$\begin{aligned} a_{11}^2 + a_{21}^2 + a_{31}^2 &= 1 \\ a_{12}^2 + a_{22}^2 + a_{32}^2 &= 1 \\ a_{13}^2 + a_{23}^2 + a_{33}^2 &= 1 \end{aligned} \tag{10.77}$$

but

$$\begin{aligned} a_{11}a_{12} + a_{21}a_{22} + a_{31}a_{32} &= 0 \\ a_{11}a_{13} + a_{21}a_{23} + a_{31}a_{33} &= 0 \\ a_{12}a_{13} + a_{22}a_{23} + a_{32}a_{33} &= 0 \end{aligned} \tag{10.78}$$

The orthogonality property can be easily remembered by considering the matrix as composed of three column vectors. The 'dot product' of a column with itself or with a neighbour reproduces the orthonormal rule Equation 10.75. Strictly speaking, Equation 10.75 applies to the product of A with itself (by multiplying columns and summing rows as indicated by Equation 10.74), and it is evident from Equations 10.77 and 10.78 that the determinant of the product matrix is unity:

$$\det (A \cdot A) = 1 \tag{10.79}$$

but

$$\det (A \cdot A) = \det A \cdot \det A = 1 \tag{10.80}$$

and therefore

$$\det A = \pm 1 \tag{10.81}$$

When $\det A = 1$, the orthogonal matrix corresponds to a true geometric rotation. When $\det A = -1$ the matrix operation corresponds to an inversion followed by a rotation.

The transpose of A is given by

$$A' = \begin{pmatrix} a_{11} & a_{21} & a_{31} \\ a_{12} & a_{22} & a_{32} \\ a_{13} & a_{23} & a_{33} \end{pmatrix} \tag{10.82}$$

and using ordinary matrix multiplication we find that

$$A \cdot A' = I \tag{10.83}$$

where I is the identity matrix. Multiplying both sides of Equation 10.83 by A^{-1} shows that the transpose of A is equal to its inverse:

$$A' = A^{-1} \tag{10.84}$$

Therefore, the reverse rotation from Equation 10.72 is

$$x_k = \sum_{j=1}^{3} a_{kj} x'_j \tag{10.85}$$

the transpose of the original rotation matrix.

Clearly, the orthogonal transformation must apply to vectors as well as point coordinates. From Equation 10.73 and taking into account that $i'_j \cdot i_k = i_k \cdot i'_j$:

$$E'_j = \mathbf{E} \cdot i'_j = \sum_{k=1}^{3} E_k i_k \cdot i'_j = \sum_{k=1}^{3} a_{jk} E_k \tag{10.86}$$

In order for second rank tensors to correspond to physical entities they must also be invariant to coordinate rotations. Start with our linear transformation Equation 10.67:

$$D_j = \sum_{k=1}^{3} T_{jk} E_k \qquad (j = 1, 2, 3) \tag{10.87}$$

We seek a second rank tensor so that, after a coordinate rotation, the following relation applies:

$$D'_i = \sum_{l=1}^{3} T'_{il} E'_l \qquad (i = 1, 2, 3) \tag{10.88}$$

where the vector components D_j, D'_i and E_k, E'_l are already related by an orthogonal transformation. Thus, we need to find the linear transformation,

$$T_{jk} \rightarrow T'_{il} \tag{10.89}$$

where T'_{il} has the property of Equation 10.88. We can relate Equation 10.87 to Equation 10.88 by multiplying Equation 10.87 by a_{ij} and summing over the index j:

$$\sum_{j=1}^{3} a_{ij} D_j = \sum_{j=1}^{3} \sum_{k=1}^{3} a_{ij} T_{jk} E_k \tag{10.90}$$

But since D'_i, D_j and E'_l, E_k are related simply by rotations, we can also write

$$D'_i = \sum_{j=1}^{3} a_{ij} D_j \quad \text{and} \quad E_k = \sum_{l=1}^{3} a_{lk} E'_l \tag{10.91}$$

Note the sum on the row and not the column in the reverse operation on the right. Now substitute the expression for E_k into Equation 10.90:

$$\sum_{j=1}^{3} a_{ij} D_j = \sum_{j=1}^{3} \sum_{k=1}^{3} a_{ij} T_{jk} \sum_{l=1}^{3} a_{lk} E'_l \tag{10.92}$$

$$D'_i = \sum_{l=1}^{3} \sum_{j=1}^{3} \sum_{k=1}^{3} a_{ij} a_{lk} T_{jk} E'_l \tag{10.93}$$

and we identify the 'rotated' second rank tensor, T'_{il}, as

$$T'_{il} = \sum_{j=1}^{3} \sum_{k=1}^{3} a_{ij} a_{lk} T_{jk} \qquad (i, l = 1, 2, 3) \tag{10.94}$$

The second rank tensor T_{jk} that transforms as Equation 10.94 is invariant to coordinate rotation and is admissible to represent physical quantities.

We can easily show that the scalar product of two vectors is invariant to coordinate transformation as well by invoking Equations 10.75 and 10.85:

$$\mathbf{D} \cdot \mathbf{E} = \sum_{k=1}^{3} D_k E_k = \sum_{k=1}^{3} \left(\sum_{j=1}^{3} a_{jk} D'_j \right) \left(\sum_{i=1}^{3} a_{ik} E'_i \right) = \sum_{k=1}^{3} \left(\sum_{i,j=1}^{3} a_{jk} a_{ik} D'_j E'_i \right) = \sum_{k=1}^{3} D'_k E'_k \tag{10.95}$$

The gradient of a scalar function φ is invariant to orthogonal transformations:

$$\nabla \varphi = \sum_{k=1}^{3} \frac{\partial \varphi}{\partial x_k} i_k \tag{10.96}$$

Consider one of the terms in the sum and set

$$D_i = \frac{\partial \varphi}{\partial x_i} \tag{10.97}$$

and take the derivative of x_k with respect to x_i' in Equation 10.85:

$$\frac{\partial x_k}{\partial x_i'} = a_{ik} \tag{10.98}$$

Then,

$$D_i' = \frac{\partial \varphi}{\partial x_i'} = \sum_{k=1}^{3} \frac{\partial \varphi}{\partial x_k} \frac{\partial x_k}{\partial x_i'} = \sum_{k=1}^{3} a_{ik} D_k \tag{10.99}$$

and

$$\sum_{i=1}^{3} D_i' i_i' = \nabla' \varphi = \sum_{j=1}^{3} \sum_{k=1}^{3} a_{ik} D_i i_j = \sum_{k=1}^{3} a_{ik} \nabla \varphi \tag{10.100}$$

This last expression shows that the gradient of scalar function transforms as a vector.

In addition to the scalar product and gradient operations we can show that the divergence operation also is preserved under orthogonal transformations. Consider two vectors **D** and **E** and suppose that

$$E_i = \frac{\partial D_i}{\partial x_i} \tag{10.101}$$

Then

$$E_i' = \frac{\partial D_i'}{\partial x_i'} = \sum_{k=1}^{3} a_{ik} \frac{\partial D_i'}{\partial x_k} \tag{10.102}$$

where we have used from Equation 10.98:

$$\frac{\partial}{\partial x_i'} = \sum_{k=1}^{3} \frac{\partial}{\partial x_k} \cdot \frac{\partial x_k}{\partial x_i'} = \sum_{k=1}^{3} a_{ik} \frac{\partial}{\partial x_k} \tag{10.103}$$

Now making a substitution from Equation 10.86 into the right-hand side of Equation 10.102 we have

$$E_i' = \frac{\partial D_i'}{\partial x_i'} = \sum_{k=1}^{3} a_{ik} \frac{\partial D_i'}{\partial x_k} = \sum_{k=1}^{3} a_{ik} \sum_{j=1}^{3} a_{ij} \frac{\partial D_j}{\partial x_k} = \sum_{k=1}^{3} \sum_{j=1}^{3} a_{ik} a_{ij} \frac{\partial D_j}{\partial x_k} \tag{10.104}$$

Finally, summing over the index i in this last expression and using the Kronicker δ property of Equation 10.75 we have:

$$\nabla' \cdot D' = \sum_{i=1}^{3} \frac{\partial D'_i}{\partial x_i} = \sum_{j=1}^{3} \frac{\partial D_j}{\partial x_j} = \nabla \cdot D \tag{10.105}$$

So we see that the divergence of a vector is invariant to orthogonal transformations. The divergence of a *tensor* is defined as

$$\left(\nabla \cdot \overset{\leftrightarrow}{T}\right)_j = \sum_{k=1}^{3} \frac{\partial T_{jk}}{\partial x_k} = D_j \qquad (j = 1, 2, 3) \tag{10.106}$$

We can show that the divergence operation on a second rank tensor results in a vector (first rank tensor), and that therefore, since we have shown above that a vector is invariant to orthogonal coordinate transformation, the same can be said of the tensor divergence. Equation 10.106 shows that the divergence operation on the tensor is similar to the operation on a vector except the derivative is taken on each row and summed on the columns. The result is three terms that constitute the vector components D_j. In order to show the invariance of the divergence with respect to coordinate rotations we follow the same steps as for the vector divergence, applying the procedure to each row:

$$D'_i = \sum_{l=1}^{3} \frac{\partial T'_{il}}{\partial x'_l} = \sum_{k=1}^{3}\sum_{j=1}^{3} a_{lk} a_{ij} a_{lk} \frac{\partial T_{jk}}{\partial x_l} = \sum_{k=1}^{3}\sum_{j=1}^{3} a_{ij} a_{lk} a_{lk} \frac{\partial T_{jk}}{\partial x_k}$$

$$= \sum_{j=1}^{3} a_{ij} \left(\sum_{k=1}^{3} \frac{\partial T_{jk}}{\partial x_k}\right) = \sum_{j=1}^{3} a_{ij} D_j \tag{10.107}$$

In the middle term above we have again taken advantage of the Kronicker δ property. The expression Equation 10.107 shows that each component of the vector resulting from the divergence operation on the tensor transforms as a coordinate rotation. Therefore,

$$\nabla' \cdot \overset{\leftrightarrow}{T'} = \nabla \cdot \overset{\leftrightarrow}{T} \tag{10.108}$$

Finally, the origin of the double-arrow notation arises from considering the second rank tensor a *dyadic*. A dyadic is similar to a vector but is flanked by unit vectors on the right and the left. Thus, writing the tensor as a dyadic,

$$\overset{\leftrightarrow}{T} = \sum_{i=1}^{3}\sum_{j=1}^{3} \varepsilon_i T_{ij} \varepsilon_j \tag{10.109}$$

where $\varepsilon_i, \varepsilon_j$ are unit vectors along the 3-D orthogonal directions. An individual element of the tensor matrix is found by performing the scalar product of the unit vectors on the left and right:

$$T_{ij} = \varepsilon_i \cdot \overleftrightarrow{T} \cdot \varepsilon_j \tag{10.110}$$

It is easily seen that taking the divergence of a dyadic tensor results in a vector:

$$\nabla \cdot \overleftrightarrow{T} = \sum_{i=1}^{3} \frac{\partial}{\partial x_i} \varepsilon_i \cdot \sum_{i=1}^{3} \varepsilon_i T_{ij} \sum_{j=1}^{3} \varepsilon_j = \sum_{j=1}^{3} \frac{\partial T_{ij}}{\partial x_i} \varepsilon_j \qquad (i = 1, 2, 3) \tag{10.111}$$

The dyadic notation is reminiscent of Dirac bra, ket notation in quantum mechanics where the basis space is represented by kets, the 'dual' space represented by bras, and the operators are flanked on the left and right by kets and bras, respectively.

10.4.8   Return to Lorentz force law

Now armed with a better understanding of second rank tensors and dyadic notation we can reconsider the physical significance of Equation 10.62, reproduced here for convenience:

$$F_L = \nabla \cdot \left[\overleftrightarrow{EE} + \overleftrightarrow{BB} - \frac{1}{2} (E \cdot E + B \cdot B) \overleftrightarrow{I} \right] - \frac{\partial (\varepsilon_0 E \times B)}{\partial t}$$

We imagine a body in space subject to external forces and enclosed by a volume. We obtain the mechanical total force on the body by integrating Equation 10.62 over the volume:

$$F_{L\text{total}} = \int F_L \, dV = \int \overleftrightarrow{T} \cdot da - \int \frac{\partial (\varepsilon_0 E \times B)}{\partial t} \, dV \tag{10.112}$$

where da is the surface differential of the enclosing volume, and we have used the tensor version of the divergence theorem:

$$\int \nabla \cdot \overleftrightarrow{T} \, dV = \int \overleftrightarrow{T} \cdot da$$

Note that the Lorentz energy flux Equation 10.51 and the Lorentz field momentum $\varepsilon_0 \, E \times B$ are related by

$$\frac{1}{c^2} \frac{1}{\mu_0} (E \times B) = \varepsilon_0 \, E \times B \tag{10.113}$$

Equation 10.112 states that the total mechanical force within the volume is equal to the difference between the force applied to the enclosing surface and the time rate of change of field momentum within the volume. Or, alternatively, by bringing the last term on the right over to the left, we can write an expression for the conservation of momentum stating that the sum of the rate of change of mechanical momentum ($F_{L\text{total}}$) and the rate of change of electromagnetic field momentum is equal to the total stresses (compressive, tensile, and shear) at the enclosing surface:

$$F_{L\text{total}} + \int \frac{\partial (\varepsilon_0 E \times B)}{\partial t}\, dV = \int \overset{\leftrightarrow}{T} \cdot da \tag{10.114}$$

In general, the enclosed volume does not have to be commensurate with the body, but it makes the physical interpretation easier if we assume that the enclosing volume is just an infinitesimal distance outside the body. Then the term on the right of Equation 10.114 describes the integral of the stress tensor over the positive *outward* normal body surface element *da*. The forces on the body are of two kinds: the force components aligned along the surface element components normal to the coordinate axes (diagonal tensor terms) and force components tangent to these surface elements (off-diagonal tensor terms). The former constitute components of compression or tension (depending on the sign) while the latter are shear forces. In static equilibrium the stress tensor must be symmetric, $T_{ij} = T_{ji}$. However, we do not really have to imagine a ponderable body inside the enclosing volume on which tensile and shear forces bear. Even if the volume only encloses the vacuum, Equation 10.114 is valid. In any case, we interpret a positive rate of *increase* in the field momentum (second term on the left-hand side) as due to the flow of momentum across the surface in the *negative* direction to the surface normal. A Maxwell stress tensor element $-T_{ij}$ can be considered the momentum per unit time in the *i* direction crossing the surface element *da* whose normal is oriented along the *j*-axis.

10.4.9   Einstein–Laub force law

The Lorentz force law is almost universally applied to connect Newtonian mechanics to Maxwellian electromagnetics. However, it is not the only force law extant. Several other stress tensors have been proposed by such notables as Minkowski, Abraham, Peierls, and L. J. Chu. A recent review [13] tabulates five principal proposals. Among these, in the first decade of the twentieth century, A. Einstein and J. Laub [14] suggested a force law that, from the perspective of physical interpretation, may have certain advantages over the conventional Lorentz law [12]:

$$F_{EL} = \rho_{\text{free}}\, E + J_{\text{free}} \times \mu_0 H + (P \cdot \nabla) E + \left(\frac{\partial P}{\partial t} \times \mu_0 H \right) + (M \cdot \nabla) H - \left(\frac{\partial M}{\partial t} \times \varepsilon_0 E \right) \tag{10.115}$$

Following the procedure analogous to the development of the generalised Lorentz force expression (Equations 10.55–10.62) we can cast Equation 10.115 into a similar form involving only vector fields and stress tensors:

$$F_{EL} = \nabla \cdot \left[\overleftrightarrow{DE} - \overleftrightarrow{BH} - \frac{1}{2} \left(\varepsilon_0 E \cdot E + \mu_0 H \cdot H \right) \overleftrightarrow{I} \right] - \frac{\partial \left(E \times H/c^2 \right)}{\partial t} \tag{10.116}$$

where the expression in square brackets is the Einstein–Laub stress tensor:

$$\overleftrightarrow{T}_{\mathcal{EL}} = \nabla \cdot \left[\overleftrightarrow{DE} - \overleftrightarrow{BH} - \frac{1}{2} \left(\varepsilon_0 E \cdot E + \mu_0 H \cdot H \right) \overleftrightarrow{I} \right] \tag{10.117}$$

Integrating over the volume on both sides of Equation 10.116, applying the divergence theorem and rearranging terms, results in a conservation of momentum expression similar to Equation 10.114, and we obtain

$$F_{EL\,total} + \int \frac{\partial \left(E \times H/c^2 \right)}{\partial t} \, dV = \int \overleftrightarrow{T}_{EL} \cdot da \tag{10.118}$$

Note that the rate of change of field momentum is now expressed in terms of the cross product of E and H rather than E and B. The time rate of change of field momentum G_{EL} is related to the Abraham energy flux or the familiar Poynting vector S:

$$G_{EL} = \frac{E \times H}{c^2} = \frac{S}{c^2} \tag{10.119}$$

which we have labelled S_{MD} in Equation 10.53 to indicate that it arises naturally from the magnetic dipole model for ponderable matter with magnetisation (Section 10.4.3). In fact, the Einstein–Laub energy flux and force expressions are 'natural' to the Maxwell equations, Equations 10.44–10.47, since they all use E and H. In contrast, the Lorentz energy flux and force expressions are appropriate to the Maxwell equations, Equations 10.36–10.39, since they all use E and B.

It is important to bear in mind that although the Lorentz and Einstein–Laub laws are formulated from the same elements (charges, currents, and fields) as the Maxwell equations, they are not deduced or derived from them. An equation relating electromagnetic fields to mechanical forces is in fact an additional postulate to the four equations of Maxwell. Forces entail the time derivative of momenta, and we have already seen that the Abraham and Minkowski versions of field momentum still have their ardent proponents over a century after their introduction. In addition to the Lorentz and Einstein–Laub force laws, therefore, we can expect to, and do, encounter others. They have been discussed in recent reviews, [13] [15], [16]. It turns out that the global principals of conservation of energy and momentum enforce the same result for the total, integrated energy, momentum, and force expressions over a given volume, but the force *densities* (and therefore the momentum density distributions) within a given volume are *not* identical. Experiments of light forces acting on deformable bodies, such as liquids or soft solids, should in principle, enable comparison of the different predicted force distributions, but so far designing the definitive experiment has proved elusive.

10.4.10   Hidden energy and hidden momentum

The terms 'hidden energy' and 'hidden momentum' [17] do not refer to a single, well-defined phenomenon, but have been invoked from time to time to account for apparently missing or unobservable quantities needed to preserve the energy and momentum conservation laws. Hidden energy arises from the choice of expression for energy flux (Poynting vector) in electromagnetic fields, and hidden momentum is generally associated with a body whose centre-of-mass (or centre-of-energy) is at rest but has an internal structure that somehow possesses momentum. Hidden quantities are related to the long-running dispute over Abraham and Minkowski momentum. The most refractory controversies arise from the different, legitimate ways to parse and interpret certain terms in Maxwell's equations, as well as the consequent choice of energy flux expression and force law. A thorough discussion, analysis, and evaluation of the various points of view is beyond the scope of this book and has been extensively examined elsewhere [9], [11]–[13], [15]–[21]. We will content ourselves, here, with an illustrative example (closely analysed and discussed at greater length by [12]) of the difficulties encountered in the presence of magnetic media ($\mu \neq \mu_0, M \neq 0$).

Imagine again a quasi-monochromatic light pulse of time length T, propagating in free space and then impinging on a material slab with permittivity ε and permeability μ so that $\mu = \varepsilon$ but both are greater than ε_0, μ_0. The vacuum and the material are impedance-matched so there will be no reflection at the interface. Figure 10.6 depicts the situation. The E, H field components parallel to the slab surface are continuous through the interface but the B-field is not. Therefore, if we use the Lorentz form of the Poynting vector, $S_L = \mu_0^{-1}(E \times B)$, the power flux will also be discontinuous at the interface. In fact, there will be a sudden increase in power flux as the light pulse propagates from free space into the material slab due to the $\mu_0^{-1}(E \times M)$ term of Equation 10.54. In order to maintain energy conservation there must be an amount of 'hidden energy' somewhere to account for the discrepancy. In contrast, the Poynting vector associated with the magnetic dipole

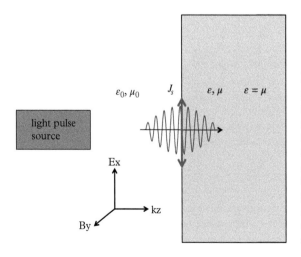

Figure 10.6 *A quasi-monochromatic light pulse of duration T propagates from vacuum into a material slab with permittivity and permeability ε and μ, respectively, and with ε = μ. The media are impedance-matched, and there are no reflections at the surface. A bound surface current J_s appears at the interface while the pulse enters the slab. The pulse edges should be considered abrupt with Heaviside-like changes in the field amplitudes and with near-constant amplitude during the pulse length T.*

model of magnetisation (Section 10.4.3) or the Einstein–Laub force law, $S = E \times H$, is continuous across the boundary and maintains energy conservation without recourse to hidden sources. In the case of the Lorentz form for the Poynting vector the 'missing energy' is supplied by a bound current induced at the slab surface as the light pulse passes into it. The term responsible is $E \cdot \left(\mu_0^{-1} \nabla \times M\right)$ in Equation 10.50. As the light pulse moves into the material, the scalar product of the E-field and the surface current supply the energy needed to maintain the energy balance at the interface. When the pulse exits the slab, this energy is returned to the material via the same surface current interaction. In the magnetic dipole model, the magnetisation contributes to the energy flux through the $H \cdot \partial M / \partial t$ term of Equation 10.52, but since the magnetisation is time-independent, the contribution is null, and the induced surface current does not couple to the pulse energy. Thus, we see that the conventional $S = E \times H$ provides a much simpler route to energy conservation. The origin of material magnetisation in the Lorentz model is the amperian current loop. One imagines a tiny circulating disc with charge current i and fixed diameter r giving rise to a magnetic moment $m = \mu_0 iA$, where A is the area of the circulating disk plane. The macroscopic magnetisation $M = \mathcal{N} m$, where $\mathcal{N}$ is the number density of the individual magnetic moments. But even if this model is admitted, it is difficult to understand how to interpret the flow of energy into or out of such an amperian current loop. Adopting $S = E \times H$ avoids hidden energy and the current loop model, but the price to be paid is a force law that is *not* a straightforward generalisation of the microscopic Lorentz force law.

Further close analysis [12] reveals that, under the Lorentz model, the slab is subject to an impulsive force of equal amplitude but opposite sign at the leading and trailing edges of the light pulse. In addition, as the pulse enters the slab, the vector product of the surface bound current and the B-field produces a force that exactly cancels the impulsive force of the leading edge. Altogether no net force is applied to the slab as the pulse passes through it and therefore no momentum is transferred to the slab. The initial momentum of the light pulse prior to incidence on the slab is $1/2\varepsilon_0 E_0^2$. But according to the Balazs thought experiment (Section 10.3) the momentum carried by the pulse inside the slab must be reduced by a factor of $\sqrt{\mu\varepsilon}$. Therefore, the Lorentz bound-current model results in a momentum deficit of $1/2\varepsilon_0 E_0^2 (1 - 1/\sqrt{\mu\varepsilon})$, the 'missing momentum'. Using the alternative force law of Einstein–Laub, the bound surface currents J_s do not couple to the light, and therefore, the cancelling force on the slab due to $J_s \times B$ is not present. The impulsive force due to the pulse leading edge is, however, still operative, and so during the time of the pulse entry, the slab does receive a forward kick. The momentum acquired by the slab is $1/4\varepsilon_0 [(\varepsilon + \mu - 2)/\mu] E_0^2 T \hat{z}$, where T is the temporal pulse length. Remembering that we have set $\mu = \varepsilon$, the mechanical 'kick' momentum acquired is therefore $1/2\varepsilon_0 (1 - 1/\sqrt{\varepsilon\mu}) E_0^2 T \hat{z}$. The sum of the slab mechanical and light pulse momentum is therefore:

$$\frac{1}{2}\varepsilon_0 \left(1 - \frac{1}{\sqrt{\varepsilon\mu}}\right) E_0^2 T \hat{z} + \frac{1}{2}\varepsilon_0 \frac{1}{\sqrt{\varepsilon\mu}} E_0^2 T \hat{z} = \frac{1}{2}\varepsilon_0 E_0^2 T \hat{z} \qquad (10.120)$$

the momentum of the light pulse prior to impinging on the slab. Therefore, the Einstein–Laub force law results in a full accounting of momentum without the need to resort to 'hidden' terms. Shockley and James [22] first drew attention to missing momentum in magnetic media and identified it with amperian current loops responsible for the magnetisation in the Lorentz model.

10.4.11 Key experiments

In this section we will describe two experiments that have attempted to provide decisive evidence to resolve the Abraham–Minkowski controversy (Section 10.3.1).

10.4.11.1 *Radiation pressure on a mirror immersed in a dielectric fluid*

The first experiment involves measuring the radiation pressure on a mirror immersed in a variety of dielectric fluids with different indices of refraction. The experiment was first carried out by Jones and Richards in 1954 [23] and later repeated with much improved technology in 1978 [5]. The idea is to measure the radiation pressure on the mirror as a function of the index of refraction so as to determine if the force, and therefore the rate of change of momentum of the light reflecting from the mirror, varies directly with the refractive index (Minkowski) or inversely (Abraham). Figure 10.7 shows in some detail the torsion balance setup to measure the force on the immersed mirror. A laser beam impinges alternately on the left and right sides of the radiation pressure mirror, which itself is surrounded by a dielectric organic liquid of accurately known refractive index at the wavelength of the helium-neon laser. The ratio of the radiation

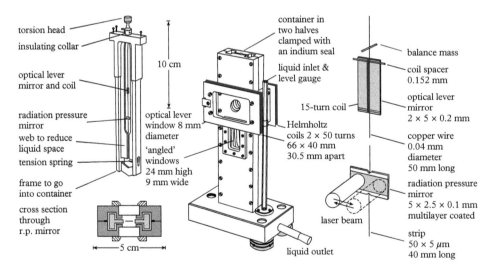

Figure 10.7 *Detail of the torsion balance: (left) torsion balance suspension frame, (centre) container into which the frame is introduced, (right) detail of the radiation pressure mirror and suspension. Figure from ref. [5] and used with permission.*

pressure on the deflection mirror in the liquid to that in air was measured for seven different organic liquids, and the ratios were compared to the phase refractive indices. The ratios agreed to within a 'mean discrepancy, averaged over the seven liquids, of $-(0.003 \pm 0.053)\%$' [5], which would appear to confirm that the light propagates in the dielectric with the Minkowski momentum. Or does it? Analysis and interpretation of these results has led to widely varying conclusions. Mansuripur [24] has pointed out that while the experimental results are certainly convincing, the force on the submerged mirror is a sensitive function of the phase shift between the incident and reflected light. A mirror with a Fresnel reflection coefficient near unity in which the E-field changes phase by π at the surface (the usual case) will indeed result in a Minkowski-like momentum; but the complex reflection coefficient is expressed by $r = |r|e^{i\varphi}$, and if $\varphi = 0$, then the mirror would experience a force consistent with the Abraham momentum. In fact, the measured radiation pressure would vary continuously between the two limiting cases as a function of the phase angle of the mirror. The conclusion is that the experiment does not measure the intrinsic momentum of the light wave in the dielectric fluid independently of the phase of the Fresnel reflection coefficient. The expression for force density on the mirror is given by [24]:

$$\langle F_z \rangle = \left[1 + \left(n_0^2 - 1 \right) \sin^2 \left(\frac{\varphi}{2} \right) \right] \varepsilon_0 E_0^2 \tag{10.121}$$

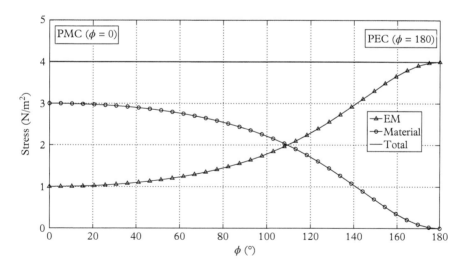

Figure 10.8 *Stress (Nm^{-2}) vs. mirror phase angle ϕ. Triangles are the electromagnetic stress, open circles the material stress of the dielectric liquid, and the solid line is the sum = $2n\varepsilon_0 E_0^2$, twice the Minkowski momentum. PMC=perfect magnetic conductor, PEC=perfect electric conductor. The index of refraction $n = 2$. Figure taken from [13] and used with permission.*

where F_z is the force density on the mirror averaged over an optical cycle and integrated over the mirror penetration depth, n_0 the refractive index of the dielectric fluid, φ the Fresnel phase, and E_0 the light wave E-field amplitude.

But that interpretation and conclusion have not gone unchallenged. Kemp and Grzegorczyk [25], and later Kemp [13], agree that the Mansuripur calculation for the *electromagnetic* force is correct but that it is not the only force bearing on the mirror. Calculating the forces from the Minkowski stress tensor, they claim the presence of an additional force at the mirror-dielectric interface, due to the dielectric itself, that effectively restores the Minkowski result for all mirror phases. Their expression [25] for the force density on the mirror is

$$F_{\text{total}} = F_l + F_m = 2n_0\varepsilon_0 E_0^2 \tag{10.122}$$

where F_l, F_m are the forces due to the dielectric liquid and the EM field, respectively. Figure 10.8 shows the calculation of the two forces as a function of mirror phase and their sum, seen to be independent of phase. Mansuripur [26] disputes the validity of the material force term which is dependent on the choice of stress tensor. The two calculations do not use the same approach to obtain the net force. Mansuripur calculates the force on the mirror directly from the Lorentz force law while Kemp calculates it from the Minkowski stress tensor. In principle both approaches are valid. It appears that the disagreement could be finally resolved by an experiment measuring the radiation pressure in a series of mirrors with varying phases and little loss. The technology for fabricating such mirror coatings is readily available.

10.4.11.2   *Atom recoil in a Bose–Einstein condensate*

The immersed-mirror experiments involve light propagating through condensed, continuous media. At the other extreme of matter density, light of wavelength λ and frequency ν in the optical range scatters from individual atoms. Since the light interacts with quantised atomic internal states, one usually considers the light as photons with energy $\hbar\omega$ and momentum $\hbar k$, where ω is the angular light frequency, $\omega = 2\pi/\nu$ and $k = 2\pi/\lambda$, even though the light field itself is classical. If a contained ensemble of boson atoms is cold enough and reaches a critical density in the containing volume, it will condense into a collective quantum state called a Bose–Einstein condensate (BEC). Although the BEC can be characterised by a wave function with a unique phase, it is still a highly dilute gas. Typical densities are $\simeq 10^{14}$ atoms cm^3 implying an average distance between the atoms of the condensate $\simeq 200$ nm. Considering that the atomic Bohr diameter $\simeq 10^{-10}$ m, we see that the condensate consists mostly of empty space.

Here we consider an experiment in which two separate light pulses interacting with a BEC probe the atom recoil momentum by atom interferometry [6]. The schematic of the experiment is shown in Figure 10.9. The experiment consists of a two-pulse atom interferometer. The first pulse sets up a standing wave in the ^{87}Rb BEC. The laser is tuned near (above and below but not directly on) the $5^2S_{1/2}, F = 1 \leftrightarrow 5^2S_{3/2}, F = 1$ transition at $\lambda = 785$ nm. The pulse duration is long enough such that the laser linewidth

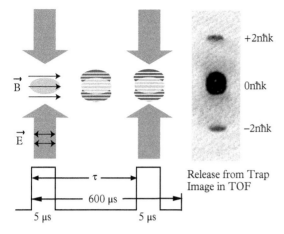

+2nℏk

0nℏk

−2nℏk

$\vec{B}$

$\vec{E}$

τ

600 μs

5 μs

5 μs

Release from Trap
Image in TOF

Figure 10.9 *Schematic of the two-pulse, Ramsey-type interferometer (Kapitza-Dirac interferometer). The first laser pulse on the left sets up a standing wave and scatters a small fraction of the ^{87}Rb BEC atoms into $|\pm 2n\hbar k >$ momentum states. The phases of the ground- and excited-state wave functions evolve at different rates. After the variable interval τ a second pulse projects the excited states back onto the ground state. The laser propagates perpendicular to the plane of the figure with the E-field polarised along the long axis of the BEC. A static magnetic field B, also aligned along the same axis, provides an axis of quantisation. The probability of finding the atoms in the $|\pm 2n\hbar nk >$ momentum states is measured by a time-of-flight (TOF) imaging technique shown on the right. Figure reproduced from [6] with permission.*

is much smaller than the spontaneous emission linewidth of the atom, and the laser frequency offset from the atom resonance centre frequency is always much greater than the natural linewidth. In the presence of the standing wave, a small fraction of the atoms in the $|0n\hbar k >$ ground momentum state scatter into $|\pm 2n\hbar k >$ excited momentum states, experiencing a momentum 'kick' and gaining a recoil energy. In free space an isolated atom absorbing a photon gains a recoil energy of $E_R = \hbar\omega_R = \hbar^2 k^2/2m$, where $k = 2\pi/\lambda$ and m is the atomic mass. The BEC, although a very dilute gas, nevertheless presents a non-negligible index of refraction n, and the essential idea of the experiment is to measure how the recoil momentum depends on n. The standing-wave light field produces a momentum grating that scatters the atoms of the BEC into a mixed momentum-state wave function,

$$|\psi(\tau) >= J_0|0n\hbar k > +J_1|\pm 2n\hbar k > e^{-i\hbar\omega_R\tau} \qquad (10.123)$$

where J_0^2 is the population of atoms in the $|0\hbar k >$ ground state and J_1^2 is the population of atoms in the $|\pm 2n\hbar k >$ momentum states. The coefficients J_0, J_1 are zero- and first-order Bessel functions. In fact, there are higher order terms corresponding to higher momentum states but their probabilities are negligible. The phase of the $|\pm 2n\hbar k >$ evolves in time with the recoil energy as indicated in the second term on the right of Equation 10.123. The second light pulse arriving at time τ projects the mixed momentum state back onto the ground state with a probability modulated by this phase evolution. The expression for the modulation is

$$P_0 = J_0^4 + 4\left[J_0^2 J_1^2 + J_1^4\right]\cos(4n^2\omega_R\tau) \qquad (10.124)$$

assuming that the n dependence of the recoil momentum is $2n\hbar k$, and therefore, $E_R = 4n^2\hbar\omega_R$. The measured points together with a fitted analytic curve of the form of

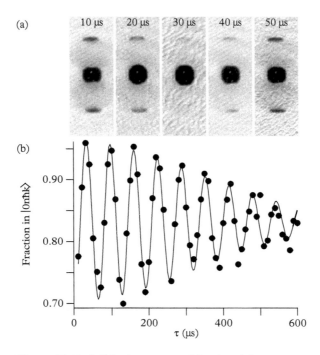

Figure 10.10 *Solid points: measured fraction of the atom population in the |0nℏk > state as a function of τ. Solid line: Functional form of Equation 10.124 with damping fitted to take into account time dependence of the wave function overlap. Figure reproduced from [6] with permission.*

Equation 10.124 for a fixed laser detuning is shown in Figure 10.10. The fit is seen to be very good for the assumed form of the n dependence of the recoil momentum, $|2n\hbar k >$. Furthermore, because the BEC 'material' exhibits dispersion around the atomic transition resonance line, the refractive index will have a pronounced frequency dependence in this spectral region. A series of measurements with the laser frequency set to several 'detunings', $\Delta = \omega - \omega_0$ around the resonance frequency ω_0, confirm that the recoil momentum is directly proportional to n (Minkowski form).

10.4.12 Interpretation of the key experiments

But does this result mean that the 'photons' in the BEC carry momentum $n\hbar k$? With the case of the immersed-mirror experiment in mind, Mansuripur and Zakharian [27] calculated the energy flux and Lorentz force carried into a generic material slab with refractive index $n_2 + i\kappa_2$ from a dielectric medium of index n_1 by a plane wave propagating along $\hat{z}$. They obtain an expression for the momentum transferred,

$$g_2 = \frac{2n_1\left(1 + n_2^2 + \kappa_2^2\right)}{(n_1 + n_2)^2 + \kappa_2^2}\hbar k_z \hat{z} \qquad (10.125)$$

For the highly reflective, lossless mirror case, κ_2 is pure imaginary and its absolute value is much greater than n_1 or n_2. Application of Equation 10.125 to this case yields $g_2 \simeq 2n_1 \hbar k_z \hat{z}$, in agreement with the previous discussion of the Fresnel reflection coefficient phase [24]. According to Reference [27] in the Bibliography in Section 10.7 [27], the *apparent* Minkowski result is a consequence of the high impedance mismatch between the two media, the liquid dielectric n_1 and the mirror $n_2 + i\kappa_2$, rather than the intrinsic momentum carried by the light in the dielectric fluid. In the case of the BEC, $n_1 = 1$, $n_2 \approx 1.2$ (see Figure 3 and Equation 4 of Reference [6] in the Bibliography in Section 10.7), and $\kappa_2 \approx 0$. Taking the refractive index of the BEC as $n_2 \simeq n_1 + \alpha$ (with $\alpha \ll n_1$), the *apparent* light momentum in the BEC, according to Equation 10.125, appears as

$$g_2 = \frac{2n_1\left[1 + (n_1 + \alpha)^2\right]}{(2n_1 + \alpha)^2}\hbar k_z \hat{z} \simeq \frac{1}{2}\left[\frac{1 + n_1^2}{n_1} + 2\alpha\right]\hbar k_z \hat{z} \qquad (10.126)$$

then

$$\lim_{n_1 \to 1} g_2 \to (1 + \alpha)\hbar k_z \hat{z} \simeq n_2 \hbar k_z \hat{z} \qquad (10.127)$$

Although both experiments yield the Minkowski form in the measurements, there is an important difference between the immersed-mirror experiment and the BEC experiment. The quantum-gas measurement senses the recoil *energy* (via $n^2 \omega_R$ in Equation 10.124), not the force, and therefore, is directly sensitive to the wavelength of light in the BEC. The data fit $E_R = 4n_2^2 \hbar \omega_R$, not $E_R = 4n_1^2 \hbar \omega_R$, and provides a fundamentally different measure of the light momentum in the BEC than the force measurement on the mirror immersed in the liquid dielectric. Although it might appear that Minkowski wins the day, at least in this interferometric measurement, even here appearances can be deceiving. In addition to the impedance arguments of Mansuripur, the early study by Gordon [1] shows that even in a highly dilute, weakly polarisable classical gas, the true electromagnetic momentum is still the Abraham term *even though in force* calculations it is the Minkowski form that appears. Loudon [28] reached similar conclusions from a general quantum treatment of light forces on dielectrics (including lossy homogenous materials). Mansuripur [29], from a purely classical treatment, also found that the momentum within dielectrics consists of a mechanical momentum term with the Minkowski form and an electromagnetic term of the Abraham form. All three authors calculate the momentum density and total momentum starting from the Lorentz force expression, avoiding the stress tensor formalism altogether. There is general agreement among these authors that experiments measuring force do not directly probe the *electromagnetic* momentum content contained within a ponderable medium and transferred to it from an initial light field. Therefore, it is not permissible to infer directly the form of the total momentum content (electromagnetic plus mechanical) from radiation pressure experiments. Although the Minkowski form persistently appears in the results of very careful measurements, there is no gainsaying the validity of the Balazs thought experiment or the consistent conclusions from classical and quantum analysis, all of which point to the Abraham form for field momentum in dielectrics. The nagging question still remains: is there any resolution to this seemingly intractable conundrum of which momentum, Minkowski or Abraham, is 'correct'?

10.4.13   Kinetic and canonical momentum

Barnett and Loudon have proposed [19] that in fact there are two distinct momenta associated with light: the kinetic momentum and the canonical momentum. They identify the kinetic momentum as the Newtonian momentum equal to the product of the mass and velocity of a point particle. The canonical momentum is associated with the Lagrange formulation of classical mechanics in which the position and momentum are said to be 'canonical' variables. They are related by

$$p_i = \frac{\partial L}{\partial \dot{q}_j} \qquad i,j\text{th degrees of freedom} \tag{10.128}$$

where $L = T - V$ is the Lagrangian of a system, p_i is the momentum, and q_i the generalised coordinate. The kinetic energy T and potential energy V constitute the two terms of L. At the microscopic level of a point particle with mass m and charge q the Lagrangian can be written more explicitly as

$$L = \frac{1}{2}mv^2 + qv \cdot A - qV \tag{10.129}$$

where v is the particle velocity and A is the vector potential. From Equation 10.128, the canonical momentum is then

$$p = mv + qA \tag{10.130}$$

The canonical variables can be used to form a function called the 'Lagrange bracket' which possesses special invariance attributes in classical mechanics. The *fundamental Lagrange brackets* [30] are given by

$$\{q_i, q_j\} = 0$$
$$\{p_i, p_j\} = 0$$
$$\{q_i, p_j\} = \sum_k \left(\frac{\partial q_k}{\partial q_i} \frac{\partial p_k}{\partial p_j} - \frac{\partial q_k}{\partial p_j} \frac{\partial p_k}{\partial q_i} \right) \tag{10.131}$$

The second term on the right vanishes because p, q are independent coordinates. Therefore,

$$\{q_i, p_j\} = \sum_k \delta_{jk}\delta_{ki} = \delta_{ij} \tag{10.132}$$

The Lagrange bracket is related to the more familiar Poisson bracket by

$$\sum_i^n \{p_l, q_i\}[p_l, p_j] + \sum_i^n \{q_l, q_i\}[q_l, p_j] = 0 \tag{10.133}$$

From which it can be shown that the canonical variables p, q obey the Poisson bracket expression

$$[q_i, p_j] = \delta_{ij} \tag{10.134}$$

which in quantum mechanics goes over into the familiar commutation relation

$$[x, p_x] = i\hbar \tag{10.135}$$

Of course in quantum mechanics p_x is a component of the momentum operator $\hat{p} = i\hbar\nabla$, and this form of the momentum operator was intended to correspond to the de Broglie hypothesis on matter waves, $p \rightarrow h/\lambda$. Therefore, in quantum mechanics one may say that the canonical momentum operator is associated with the wavelength of the matter wave. Reasoning by analogy, Barnett and Loudon [19] suggest that since the Minkowski momentum appears in experiments involving diffraction and interference (the BEC experiment of Section 10.4.11, for example), where the wavelength of light is the determining property, it might be fruitful to associate the Minkowski momentum to the canonical momentum. In contrast, as illustrated by the Balazs thought experiment, where the light simply passes from one refractive index to another 'kinematically' (without reflection or absorption), the Abraham momentum might be the appropriate choice.

10.4.14   Illustrative example—light force on a point dipole

For solid, polarisable dielectric objects with no free charges or currents, the light force expressions arising from the Lorentz force law can be written in two equivalent ways [31, 32],

$$F_c = -(\nabla \cdot P)\,E + \left(\frac{\partial P}{\partial t}\right) \times B \tag{10.136}$$

$$F_d = (P \cdot \nabla)\,E + \left(\frac{\partial P}{\partial t}\right) \times B \tag{10.137}$$

where F_c indicates the force derived from bound charges and currents, and F_d assumes the material is composed of point dipoles[4][1]. For most practical calculations of experimental interest, the dipole form, Equation 10.137, is the one of choice. In Chapter 11 we will discuss in some detail, from a semiclassical point of view, light forces bearing on a two-level atom (see especially Section 11.4), in which the two levels are coupled by a transition dipole matrix element. As a precursor to that chapter we summarise here the forces and momenta acquired by a point dipole as a pulse of light passes through it. Considered optically, the electric point dipole stands in for the most primitive of atoms. The discussion is based on an important article by Hinds and Barnett [33] that illustrates how the Abraham and Minkowski momenta may be associated with qualitatively different aspects of the light–dipole interaction.

To summarise the preceding discussion (Sections 10.4.1–10.4.6, and 10.4.8), the Abraham and Minkowski expressions for the momentum density of light are,

$$G_A = \frac{1}{c^2} E \times H \quad \text{and} \quad G_M = D \times B \tag{10.138}$$

In free space these expressions are equivalent, but when light travels in a medium of refractive index n, the momentum of the light takes on two different expressions. Integrating over a defined volume and normalising the light energy to the photon, $\hbar\omega$, the Abraham and Minkowski momenta become

$$p_A = \frac{p_0}{n} \quad \text{and} \quad p_M = p_0 n \tag{10.139}$$

where p_0 is the free-space momentum of the light. The question we wish to answer is: what is the relation between p_A and p_M and the kinetic and canonical momenta in the case of light interacting with a point dipole? We can begin to answer this question by citing an important relation between the kinetic and canonical momenta [19] when an optical field interacts with a dipole d,

$$p_{\text{kin}} = p_{\text{can}} + d \times B \tag{10.140}$$

and integrating over the volume of a medium consisting of dipoles d,

$$p_{\text{kin}} - p_{\text{can}} = \int (g_M - g_A) \, dV = d \times B \tag{10.141}$$

or

$$p_{\text{can}} + \int g_M \, dV = p_{\text{kin}} + \int g_A \, dV \tag{10.142}$$

---

[4] The form of the first term in Equation 10.137 is always used in atomic physics and arises from first considering the dipole-E-field interaction energy, $\mathscr{E} = -d \cdot E$, followed by taking the negative of the spatial derivative to get the force while taking into account that the spatial extent of the optical wave is always much greater than that of the atomic dipole. See Reference 7 in Further reading, Section 10.6.

The canonical momentum as introduced in Section 10.4.13, is essentially a quantum operator, not a classical entity, the form of which depends on the gauge of the corresponding vector field. Equations 10.140–10.142 are valid for the Coulomb gauge. According to Barnett [34] the canonical momentum can also be cast as a classical entity, and the choice of Coulomb gauge rather than some other arbitrary gauge does not materially effect the results. The choices here are for convenience and clarity.

Starting from Equation 10.137, the force component F_i acting on a point dipole d by a quasi-monochromatic optical pulse $E = E_0(z)\hat{e}\cos(\omega t - kz)$, propagating along z and with unit vector $\hat{e}$ in the $x - y$ plane, can be expressed in the form of a Lorentz force as,

$$F_i = (d \cdot \nabla) E_i + (\dot{d} \times B)_i \qquad (10.143)$$

or, invoking the identity,

$$\frac{\partial}{\partial t}(d \times B)_i = (\dot{d} \times B)_i + (d \times \dot{B})_i \qquad (10.144)$$

and Faraday's law,

$$\dot{B} = -\nabla \times E \qquad (10.145)$$

we can, after using a vector triple product identity, write

$$F_i = d \cdot \frac{\partial}{\partial x_i} E + \frac{\partial}{\partial t}(d \times B)_i \qquad (10.146)$$

Now substitute the E-field of the pulse, $E = E_0(z)\hat{e}\cos(\omega t - kz)$ into Equation 10.146. The result is

$$F_z = F_{1z} + F_{2z} = d \cdot \hat{e}\left[\frac{\partial E_0(z)}{\partial z}\cos(\omega t - kz) + E_0(z)k\sin(\omega t - kz)\right] + \frac{\partial}{\partial t}(d \times B)_i \qquad (10.147)$$

We will see in Chapter 11 that the light force on an atom appears in a form similar to the first term in square brackets in Equation 10.147. The second term does not appear in the conventional force expressions for atom optical manipulation, but we can call it the Lorentz term, because $\dot{d}$ is analogous to a bound polarisation current and $d \times B$ is a Lorentz force term. Up to this point we have been able to consider the fields and the dipole as classical quantities. The development of Hinds and Bartlett [33] now proceeds semiclassically by quantising the dipole but keeping the fields classical. We consider the point dipole a quantum object with two internal states: one lower (1) and one upper (2). The force F_z must now be calculated by taking matrix elements according to conventional quantum mechanics as described in detail in Chapter 11, Section 11.4. Here, we simply write the result,

$$\langle F_{1z}\rangle = \langle d \cdot \hat{e}\rangle \left[\frac{\partial E_0(z)}{\partial z}\cos(\omega t - kz) + E_0(z)k\sin(\omega t - kz)\right] \qquad (10.148)$$

where $\langle d \cdot \hat{e}\rangle = d_{12}$ is the transition matrix element of the dipole between states (1) and (2). Around the dipole transition frequency ω_0, Section 11.5.3 shows how the E-field of the incoming pulse drives the dipole as a forced oscillator according to

$$\langle d \cdot \hat{e}\rangle = 2d_{12}\left\{\left[\frac{\delta(\omega)\frac{1}{2}\Omega}{\delta^2 + \gamma^2 + \frac{1}{2}\Omega^2}\right]\cos(\omega t - kz) - \left[\frac{\gamma\frac{1}{2}\Omega}{\delta^2 + \gamma^2 + \frac{1}{2}\Omega^2}\right]\sin(\omega t - kz)\right\}$$
$$(10.149)$$

where $\delta = \omega - \omega_0$ is the detuning, γ is the generic loss rate from the upper state[5], and $\hbar\Omega = -d_{12}E_0$ defines the Rabi frequency[6] Ω. The first term on the right is the 'dispersive' term that goes to zero at zero detuning and is antisymmetric on either side of ω_0, while the second term is the 'absorptive' term that peaks on resonance. The dispersive term is in phase with the driving field while the absorptive term is in quadrature. The susceptibilities χ' and χ'' show the same behaviour and are plotted as a function of frequency in Figures 11.2 and 11.3. Substitution of Equation 10.149 into Equation 10.148 and averaging over an optical cycle results in a two-term expression for $\bar{F}_{1z}$,

$$\bar{F}_{1z} = d_{12}\left[\frac{\delta(\omega)\frac{1}{2}\Omega}{\delta^2 + \gamma^2 + \frac{1}{2}\Omega^2}\right]\frac{\partial E_0(z)}{\partial z} - d_{12}\left[\frac{\gamma\frac{1}{2}\Omega}{\delta^2 + \gamma^2 + \frac{1}{2}\Omega^2}\right]kE_0(z) \qquad (10.150)$$

The dispersive term, proportional to the E-field amplitude gradient along z, is called the 'dipole-gradient force', and the absorptive term is called the 'radiation-pressure force'. Now we concentrate on the dipole-gradient force term and integrate it over the time required for the light pulse to fully overlap the dipole. The result will be the momentum delivered to the dipole by the dipole-gradient force, p_{dg}. We set the detuning to the 'red' $(-\delta)$ so that the dipole-gradient force will point in the direction of high light intensity:

$$p_{dg} = \int_0^t d_{12}\left[\frac{\delta(\omega)\frac{1}{2}\Omega}{\delta^2 + \gamma^2 + \frac{1}{2}\Omega^2}\right]\frac{\partial E_0(z)}{\partial z}\,dt \qquad (10.151)$$

Taking into account that Ω is a function of $E_0(t)$, varying from $\Omega = 0$ before the pulse encounters the dipole to $\Omega = \Omega_0$ at full overlap, and after substituting $dt = dz/c$, we obtain

$$p_{dg} = -\frac{\hbar\delta(\omega)}{2c}\ln\left[1 + \frac{\frac{1}{2}\Omega_0^2}{\delta^2 + \gamma^2}\right] \simeq -\frac{1}{2}\frac{d_{12}E_0}{c}\left[\frac{\delta(\omega)\frac{1}{2}\Omega_0}{\delta^2 + \gamma^2}\right] \simeq -\frac{1}{2}\frac{d_{12}E_0}{c}\left[\frac{\delta(\omega)\frac{1}{2}\Omega_0}{\delta^2 + \gamma^2 + \frac{1}{2}\Omega_0^2}\right]$$
$$(10.152)$$

---

[5] The spontaneous emission loss rate of population from the upper state, $\Gamma = 2\gamma$.
[6] In Chapter 11 the definition of the on-resonant Rabi frequency is $\hbar\Omega = d_{12}E_0$.

We have added $(1/2)\Omega_0^2$ to the denominator in the last term on the right so that the factor in square brackets is the same as in Equations 10.151 and 10.150, and under the assumption that $\Omega_0^2 \ll \delta^2 + \gamma^2$.

Now consider F_{2z}, the second term on the right of Equation 10.147. The time integration can be carried out directly. Using Equation 10.149 and $B = (E_0/c)\cos(\omega t - kz)$, and averaging over an optical cycle, we find the $\overline{F}_{2z}$ term contribution to the momentum,

$$p_2 = \frac{d_{12}E_0}{c}\left[\frac{\delta(\omega)\frac{1}{2}\Omega_0}{\delta^2 + \gamma^2 + \frac{1}{2}\Omega_0^2}\right] \tag{10.153}$$

We see that p_2 is twice as great as p_{dg} in absolute value and opposite in sign. As the leading edge of the pulse passes through the dipole, it receives a positive net kick equivalent to

$$p_{net} = p_{dg} + p_2 \simeq \frac{1}{2}\frac{d_{12}E_0}{c}\left[\frac{\delta(\omega)\frac{1}{2}\Omega_0}{\delta^2 + \gamma^2 + \frac{1}{2}\Omega_0^2}\right] \tag{10.154}$$

The same result can be obtained from using Equation 10.143. Here the first term, $(d \cdot \nabla)E_i$, does not contribute because there is no z component of the E-field in the incident pulse. The second term can be regarded as a Lorentz force with d as a bound current. Using Equation 10.149, this Lorentz force can be written as

$$F_L = 2d_{12}\left[\dot{u}\cos(\omega t - kz) - u\omega\sin(\omega t - kz) - \dot{v}\sin(\omega t - kz) - v\omega\cos(\omega t - kz)\right]\frac{E_0}{c}\cos(\omega t - kz) \tag{10.155}$$

where we have set

$$u = \left[\frac{\delta(\omega)\frac{1}{2}\Omega}{\delta^2 + \gamma^2 + \frac{1}{2}\Omega^2}\right] \quad \text{and} \quad v = \left[\frac{\gamma\frac{1}{2}\Omega}{\delta^2 + \gamma^2 + \frac{1}{2}\Omega^2}\right] \tag{10.156}$$

Averaging over an optical cycle and integrating over the leading edge of the pulse[7] yields

$$p_L = d_{12}\left(u\frac{E_0}{2c} - v\omega\frac{E_0}{c}t\right) \tag{10.157}$$

The first term on the right is the same as Equation 10.153 while the second term expresses the momentum due to atom recoil absorptive scattering.

Now the energy in the pulse is $(1/2)\varepsilon_0 E_0^2 V$, where V is the pulse spatial volume. The number of photons in the pulse is $(1/2)\varepsilon_0 E_0^2 V/\hbar\omega$, and the Lorentz momentum p_L per photon transferred to the dipole is

$$\frac{p_L}{\hbar\omega} = \frac{d_{12}u}{\varepsilon_0 E_0 V}p_0 \tag{10.158}$$

---

[7] The factor of 1/2 appearing in the first term of Equation 10.157 arises from integrating over the discontinuity of B at the dipole position.

where $p_0 = \hbar\omega/c$ is the total momentum originating in the light pulse. The momentum remaining in the light pulse after the transfer of p_L to the dipole is

$$p_L = p_0 \left(1 - \frac{d_{12}u}{\varepsilon_0 E_0 V} \right) \simeq \frac{p_0}{1 + \frac{d_{12}u}{\varepsilon_0 E_0 V}} \qquad (10.159)$$

We can identify the susceptibility of the in-phase driving term for a single dipole from Equation 11.87 in Chapter 11 as

$$\chi' = \frac{2d_{12}u}{\varepsilon_0 E_0 V} \qquad (10.160)$$

and therefore,

$$p_L \simeq \frac{p_0}{1 + \frac{1}{2}\chi'} \qquad (10.161)$$

Finally, we recognise that the real part of the refractive index $n = \sqrt{\varepsilon} = \sqrt{1 + \chi'} \simeq 1 + 1/2\chi'$. So that

$$p_L \simeq \frac{p_0}{n} \qquad (10.162)$$

which we associate with the Abraham momentum.

The field momentum *gained* by the transfer p_{dg} is

$$p_{dg} = p_0 \left(1 + \frac{d_{12}u}{\varepsilon_0 E_0 V} \right) = p_0 \left(1 + \frac{\chi'}{2} \right) \simeq p_0 n \qquad (10.163)$$

which we associate with the Minkowski momentum.

We now have shown that the Lorentz force of Equation 10.143 results in a transfer of Abraham momentum to the dipole, and we can identify the dipole gain in momentum as p_{kin}, the kinetic momentum. What about the dipole gradient momentum? With red detuning the dipole is pulled by the gradient to a region of higher field intensity, and we have seen that the momentum transferred to the field is the Minkowski momentum. According to Equation 10.142, the Minkowski momentum gained by the field must be equal to the loss of canonical momentum by the dipole. Therefore, we must associate the dipole gradient momentum to the canonical momentum. This association can be rationalised by remembering that the canonical momentum operator is given by $i\hbar\nabla$, and the dipole-gradient force results from the field gradient (Equation 10.150).

10.4.15   Conclusions

The Einstein thought experiment establishes the necessity of momentum in the electromagnetic field, and there can be no doubt that the Abraham momentum and Minkowski momentum are valid expressions. There is almost universal agreement that the Poynting vector, $S = E \times H$, describes the energy flux of the electromagnetic field both in free space and in materials. The momentum in free space is therefore given by $G = (1/c^2)S$. The question then is: what is the momentum of the electromagnetic field propagating within ponderable media? The Balazs thought experiment says it must be the Abraham momentum, although another thought experiment due to Padgett [35] concludes equally convincingly that in the case of diffraction it should be the Minkowski momentum. Appeal to experiment is not as conclusive as one might think because close analysis of the submerged-mirror project reveals that the force measured on the mirror is not *just* the change of the electromagnetic momentum upon reflection in the medium, but contains a force term due to the material itself acting on the mirror; or may depend on the phase of the reflection. At this writing (early 2016), experiments required to decide this issue have not yet been carried out. The suggestion that both forms of the electromagnetic momentum may be valid under different circumstances is appealing, but rigorous criteria specifying the use of the Abraham or Minkowski form have yet to be worked out. Furthermore, specifying the form of the canonical momentum in the Coulomb gauge means that the Minkowski-canonical association is not gauge invariant. The Abraham form seems to be definitely associated with Newtonian kinetic momentum, but the conditions appropriate to associating the Minkowski form with the canonical momentum are more difficult to pin down (interference, diffraction, intensity gradients?). The BEC experiment uses a standing-wave two-field Ramsey interferometer with the individual atoms scattering off the gradients of the periodic light potential. According to the Minkowski-canonical association it is not surprising, therefore, that the Minkowski momentum appears. At this time, theory appears to favour the Abraham form of electromagnetic momentum in ponderable media when light forces due to field gradients are not present, and the Minkowski form when they are. A way forward in the submerged-mirror experiments is to repeat them with a series of low-loss reflectors of varying phase.

10.5 Summary

This chapter treats the question of how momentum is transported from a light source to and through material objects by propagating light fields. The idea of what form momentum might take in fields is first established by the Einstein thought experiment. The Balazs thought experiment then concludes that momentum must take the Abraham form as light propagates through dielectrics with refractive index greater than vacuum. A second thought experiment based on reflection, however, seems to indicate that the momentum, the rate of change of which gives rise to the radiative force on a mirror, must take the Minkowski form. The question of radiative force and force laws is then opened and examined. Maxwell's equations can be rearranged to fit various physical interpretations. The Lorentz model expresses Maxwell's equations entirely in terms of the fields E and B, and suggests that the source of material magnetisation consists of a density of Amperian current loops. The magnetic dipole model expresses the same Maxwell equations in terms of E and H, but the interpretation of bound magnetic charge and bound magnetic current invites a picture of magnetic dipole density analogous to the electric dipole density of the polarisation field. These two models of the constitution of magnetic matter give rise to two different expressions for the energy flux density. The two are linked by Equation 10.54. In contrast to energy conservation, momentum conservation must be expressed in terms of the divergence of a tensor, not a vector field. For readers not yet initiated into the mysteries of second-rank tensors, we allow a digression into the exploration of their properties before returning to Lorentz and Einstein–Laub force laws. The next topic is a somewhat cursory discussion of hidden

298 Momentum in Fields and Matter

energy and hidden momentum, the motivation being the clarification of issues without being drawn into a debilitating controversy. After the discussion of hidden energy and hidden momentum, the focus passes to key experiments that have attempted to answer the question of the correct form of momentum in light fields within ponderable media. Light forces applied to mirrors submerged in a series of liquids of refractive index above unity indicates that the momentum varies as the refractive index of the material. The experimental finding supports the Minkowski form, but according to Mansuripur and Kemp, the issue is far from settled because either a phase difference at the reflection plane or an extra force term at the dielectric-mirror interface itself may contribute to the apparent light force on the mirror. A second experiment involving a Bose–Einstein condensate (BEC) and a two-field Ramsey interferometer setup is not really any more successful at finding a definitive, once-and-for-all answer. Although again, the measurement supports the Minkowski form, this result may be a consequence of the close-to-unity refractive index of the BEC. The chapter ends with the proposition of Barnett that the 'correct' form for the momentum depends on the specific circumstances of the measurement, and relies on drawing a fundamental distinction between kinetic and canonical forms of the momentum. An example of resonant light interacting with a two-level atom coupled by a dipole interaction illustrates how the kinetic momentum can be associated with the Abraham form and the canonical momentum can be associated with the Minkowski form.

10.6   Further reading

1. D. J. Griffiths, *Introduction to Electrodynamics*, 3rd edition, Chapters 4–9, Pearson-Addison-Wesley (1981).
2. L. D. Landau, E. M. Lifshitz, and L. P. Pitaevskii, *Electrodynamics of Continuous Media*, Chapter IX, 2nd edition, Elsevier (1982).
3. J. A. Stratton, *Electromagnetic Theory*, Chapters I and II, McGraw-Hill (1941).
4. J. D. Jackson, *Classical Electrodynamics*, Chapter 6, John Wiley & Sons (1962).
5. M. Mansuripur, *Field, Force, Energy, and Momentum in Classical Electrodynamics*, Chapters 2 and 10, Bentham Books (2011).
6. I. S. Sokolnikoff and R. M. Redheffer, *Mathematics of Physics and Modern Engineering*, g 4, McGraw-Hill (1958).
7. C. Cohen-Tannoudji, J. Dupont-Roc, and G. Grynberg, *Processus d'interaction entre photons et atomes*, Chapter 5, Editions du CNRS (1988).

10.7   Bibliography

[1] J. P. Gordon, *Radiation Forces and Momenta in Dielectric Media. Phys Rev A: At Mol Opt Phys* vol 8, pp. 14–21 (1973).
[2] N. Balazs, *The Energy-Momentum Tensor of the Electromagnetic Field inside Matter. Phys Rev* vol 91, pp. 408–411 (1953).
[3] M. Mansuripur, *Momentum exchange effect*, Nat Photonics vol 7, pp. 765–766 (2013).
[4] A. Ashkin and J. M. Dziedzic, *Radiation Pressure on a Free Liquid Surface. Phys Rev Lett* vol 30, pp. 139–142 (1973).

[5] R. V. Jones and B. Leslie, *The measurement of optical radiation pressure in dispersive media. Proc R Soc London Ser A* vol 360, pp. 347–363 (1978).

[6] G. K. Campbell, A. E. Leanhardt, J. Mun, M. Boyd, E. W. Streed, W. Ketterle, and D. E. Pritchard, *Photon Recoil Momentum in Dispersive Media. Phys Rev Lett* vol 94, pp. 170,403 (2005).

[7] S. M. Barnett, *Resolution of the Abraham-Minkowski Dilemma. Phys Rev Lett* vol 104, pp. 070,401 (2010).

[8] M. Mansuripur, *Resolution of the Abraham-Minkowski controversy. Opt Commun* vol 283, pp. 1997–2005 (2010).

[9] B. A. Kemp, *Resolution of the Abraham-Minkowski debate: Implications for the electromagnetic wave theory of light in matter. J Appl Phys* vol 109, pp. 111,101 (2011).

[10] M. Mansuripur, *On the Foundational Equations of the Classical Theory of Electrodynamics. Resonance* vol 18, pp. 130–155 (2013).

[11] M. Mansuripur, *The Force Law of Classical Electrodynamics: Lorentz versus Einstein and Laub. Proc SPIE* vol 8810, pp. 88,100K (2013).

[12] M. Mansuripur, *Nature of electric and magnetic dipoles gleaned from the Poynting theorem and the Lorentz force law of classical electrodynamics. Opt Commun* vol 284, pp. 594–602 (2011).

[13] B. A. Kemp, *Macroscopic Theory of Optical Momentum. Prog Optics* vol 60, pp. 437–488 (2015).

[14] A. Einstein and J. Laub, *Über die im elektromagnetischen Felde auf ruhende Körper aus geübten ponderomotorischen Kräte. Ann Phys* vol 331, pp. 541–550 (1908).

[15] D. F. Nelson, *Momentum, pseudomomentum, and wave momentum: Toward resolving the Minkowski-Abraham controversy. Phys Rev* vol 44, pp. 3985–3996 (1991).

[16] P. W. Milonni and R. W. Boyd, *Momentum of Light in a Dielectric Medium. Adv Opt Photonics* vol 2, pp. 519–553 (2010).

[17] D. J. Griffiths, *Resource Letter EM-1: Electromagnetic Momentum. Am J Phys* vol 80, p. 7 (2012).

[18] C. Baxter and R. Loudon, *Radiation pressure and the photon momentum in dielectrics. J Mod Opt* vol 57, pp. 830–842 (2010).

[19] S. M. Barnett and R. Loudon, *The enigma of optical momentum in a medium. Philos Trans R Soc London Ser A* vol 368, pp. 927–939 (2010).

[20] P. L. Saldanha, *Division of the energy and of the momentum of electromagnetic wave in linear media into electromagnetic and material parts. Opt Commun* vol 284, pp. 2653–2657 (2011).

[21] M. Mansuripur, A. R. Zakharian, and E. M. Wright, *Electromagnetic-force distribution inside matter. Phys Rev A: At Mol Opt Phys* vol 88, pp. 023,826 (2013).

[22] W. Shockley and R. P. James, *'Try Simplest Cases' Discovery of 'Hidden Momentum' Forces on 'Magnetic Currents'. Phys Rev Lett* vol 18, pp. 876–879 (1967).

[23] R. V. Jones and J. C. S. Richards, *The pressure of radiation in a refracting medium. Proc R Soc London. Ser A* vol 221, pp. 480 (1954).

[24] M. Mansuripur, *Radiation Pressure on Submerged Mirrors: Implications for the Momentum of Light in Dielectric Media. Opt Express* vol 15, pp. 2677–2682 (2007).

[25] B. A. Kemp and T. M. Grzegorczyk, *The observable pressure of light in dielectric fluids. Opt Lett* vol 36, pp. 493–495 (2011).

[26] M. Mansuripur. Personal communication (2015).

[27] M. Mansuripur and A. R. Zakharian, *Whence the Minkowski momentum. Opt Commun* vol 283, pp. 3557–3563 (2010).

[28] R. Loudon, *Theory of the radiation pressure on dielectric surfaces. J Mod Opt* vol 49, pp. 821 (2002).

[29] M. Mansuripur, *Radiation pressure and the linear momentum of the electromagnetic field. Opt Express* vol 12, pp. 5375–5401 (2004).

[30] H. Goldstein, *Classical Mechanics*, Addison-Wesley, Reading, Massachusetts (1950).

[31] S. M. Barnett and R. Loudon, *On the electromagnetic force on a dielectric medium. J Phys B: At Mol Opt Phys* vol 39, pp. S671–S684 (2006).

[32] A. R. Zakharian, P. Polynkin, M. Mansuripur, and J. V. Moloney, *Single-beam trapping of micro-beads in polaized light: Numerical simulations. Opt Express* vol 14, pp. 3660–3676 (2006).

[33] E. A. Hinds and S. M. Barnett, *Momentum Exchange between Light and a Single Atom: Abraham or Minkowski? Phys Rev Lett* vol 102, pp. 050,403 (2009).

[34] S. M. Barnett. Personal communication (2015).

[35] M. J. Padgett, *On diffraction within a dielectric medium as a example of the Minkowski formulation of optical momentum. Opt Express* vol 16, pp. 20,864–20,868 (2008).

11

Atom-Light Forces

11.1 Introduction

As we have seen in Chapter 10, a light beam carries momentum, and light scattering transfers some or all of that momentum to a ponderable object. The time rate of change of momentum produces a force on the object. This property of light was first demonstrated through the observation of a very small transverse deflection (3×10^{-5} rad) in a beam of sodium atoms exposed to light from a resonance lamp. A single-mode laser source tuned to the atomic resonance transitions greatly facilitates the deflection measurement due to spectral purity, intensity, and directionality of the laser beam. Although these results kindled interest in using light forces to control the motion of neutral atoms, the basic groundwork for the understanding of light forces acting on atoms was not laid out before the end of the 1970s. Application of light forces to atom cooling and trapping was not accomplished before the mid-1980s. In this chapter we discuss some fundamental aspects of light forces and schemes employed to cool and trap neutral atoms.

The light force exerted on an atom can be of two types: a dissipative, *spontaneous force* and conservative, *dipole force*[1]. The spontaneous force arises from the impulse experienced by an atom when it absorbs or emits a quantum of photon momentum. When an atom scatters light, the resonant scattering cross section can be written as

$$\sigma_{0a} = \frac{g_1}{g_2} \frac{\pi \lambda_0^2}{2} \tag{11.1}$$

where λ_0 is the resonant wavelength. In the optical region of the spectrum wavelengths are on the order of a few hundred nanometres, so resonant scattering cross sections become relatively large, $\sim 10^{-9}$ cm^2. Each photon absorbed transfers a quantum of momentum $\hbar k$ to the atom in the direction of the light beam propagation. Spontaneous emission following the absorption occurs in random directions, and over many absorption-emission cycles, it averages to zero. As a result, the *net* spontaneous force

---

[1] In this chapter we drop the second term in Equation 10.146 of Section 10.4.13 since in the usual realisations of light forces on atoms this term is negligible compared to the field-gradient term.

Light-Matter Interaction. Second Edition. John Weiner and Frederico Nunes.

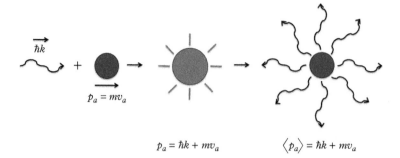

$$p_a = \hbar k + m v_a \qquad \langle p_a \rangle = \hbar k + m v_a$$

Figure 11.1 *Left: atom moves to the right with mass m, velocity v_a, and absorbs a photon propagating to the left with momentum ℏk. Centre: excited atom experiences a change in momentum $p_a = m v_a + \hbar k$. Right: photon isotropic re-emission results in an average momentum change for the atom, after multiple absorptions and emissions, of $\langle p_a \rangle = m v_a + \hbar k$.*

acts on the atom in the direction of the light propagation, as shown schematically in Figure 11.1. The saturated rate of photon scattering by spontaneous emission (the reciprocal of the excited-state lifetime) fixes the upper limit to the force magnitude. The force is often called the *radiation pressure force*.

The dipole-gradient force can be readily understood by considering the light as a classical wave. It is simply the time-averaged force arising from the interaction of the transition dipole, induced by the oscillating electric field of the light, with the spatial gradient of the electric field amplitude (see, however, Section 10.4.13 of Chapter 10). Focusing the light beam controls the magnitude of this gradient, and detuning the optical frequency below or above the atomic transition controls the sign of the force acting on the atom. Tuning the light below resonance attracts the atom to the region of maximum intensity (usually along the light beam longitudinal axis), while tuning above resonance repels the atom from this region. The dipole force is a stimulated process in which no net exchange of *energy* between the field and the atom takes place. Photons are absorbed from one mode and reappear by stimulated emission in another. Momentum conservation requires that the change of photon propagation direction from initial to final mode imparts a net recoil to the atom. Unlike the spontaneous force, there is, in principle, no upper limit to the magnitude of the dipole force since it is a function only of the field gradient and detuning. This statement is strictly true only for two-level atoms. In real atoms or molecules, excited upper states or the ionisation or dissociation continuum will eventually couple into the optical interaction.

11.2   The atom as a damped harmonic oscillator

In order to bring these introductory comments into focus, we first discuss atomic dipole emission as a classical damped harmonic oscillator. In Section 2.8 we considered a classical electric dipole aligned along the z-axis and wrote the dipole as

$$\boldsymbol{p}(t) = q_0 a \cos \omega t \hat{\boldsymbol{z}}$$

If we consider an atom as an electron bound to a massive proton, subject to a radial restoring force centred at the proton position, then we can write down a similar dipole expression,

$$\boldsymbol{p}(t) = -e r(t) \qquad p_0 = -e r_0 \qquad (11.2)$$

where $-e$ is the electron charge, r is the radial distance from the atom centre, and r_0 is the maximum radial distance. The electron is pictured as vibrating radially through the proton position and in this simple model exhibits no angular momentum from orbital motion about the charge centre. Using the result for power emitted by a classical dipole, Equation 2.157, we can write the optical-cycle-averaged power emitted by the atom as

$$\langle W \rangle = \frac{1}{4\pi\varepsilon_0} \frac{\omega_0^4 p_0^2}{3c^3} = \frac{e^2}{4\pi\varepsilon_0} \frac{\omega_0^4 r_0^2}{3c^3} \qquad (11.3)$$

For harmonic oscillation the maximum acceleration is $a_0 = -\omega_0^2 r_0$ and

$$\langle W \rangle = \frac{e^2}{4\pi\varepsilon_0} \frac{a_0^2}{3c^3} \qquad (11.4)$$

Writing the cycle-averaged acceleration as $\langle \boldsymbol{a} \cdot \boldsymbol{a} \rangle = \langle a^2 \rangle = (1/2)a_0^2$, we have

$$\langle W \rangle = \frac{e^2}{4\pi\varepsilon_0} \frac{2\langle a^2 \rangle}{3c^3} \qquad (11.5)$$

and we see that the average power emitted by the oscillating electron is proportional to the average square of the electron acceleration. A charged particle must be accelerating to emit radiation; an electron moving at constant velocity, not subject to external forces, does not emit.

The energy (cycle-averaged) emitted by the oscillator must come from mechanical (kinetic plus potential) energy, $\mathcal{E}_{\text{mech}}$:

$$\mathcal{E}_{\text{mech}} = \frac{1}{2} m_e \omega_0^2 r_0^2 \qquad (11.6)$$

where m_e is the electron mass. The average power emitted by the oscillator can be equated to the diminution of mechanical energy with time:

$$\langle W \rangle = -\frac{\mathcal{E}_{\text{mech}}}{\tau} \qquad (11.7)$$

From Equations 11.3, 11.6, and 11.7 we see that

$$\tau = \left(\frac{e^2}{4\pi\varepsilon_0}\right)^{-1}\frac{3m_ec^3}{2\omega_0^2} \tag{11.8}$$

The 'lifetime' of the oscillator τ can be written a little more simply by introducing the *classical radius* of the electron,

$$r_c = \frac{e^2}{4\pi\varepsilon_0}\frac{1}{m_ec^2} \simeq 2.8 \times 10^{-15} \text{ m} \tag{11.9}$$

and

$$\tau = \frac{3c}{2r_c\omega_0^2} \tag{11.10}$$

In differential form, Equation 11.7 is written as

$$\langle W \rangle = -\frac{d\mathcal{E}_{\text{mech}}}{dt}$$

so

$$\frac{d\mathcal{E}_{\text{mech}}}{\mathcal{E}_{\text{mech}}} = -\frac{dt}{\tau} \tag{11.11}$$

and

$$\mathcal{E}_{\text{mech}}(t) = \mathcal{E}_{\text{mech}}(0)e^{-t/\tau} \tag{11.12}$$

We see that the internal energy of the atom, modelled as a classical electron oscillating around a positive charge centre, decreases exponentially with time as the energy is radiated away.

A specific example provides some feeling for the relevant time and energy scales. Consider a sodium atom, structured with 10 electrons in a 'closed shell' and one electron outside. This exposed or 'active' electron is subject to an effective positive charge at the nucleus roughly equal to that of one proton. The potential and electric field experienced by the active electron is radial and the simple dipole oscillator model might therefore be appropriate. Emission from the first excited state to the ground state is concentrated in a narrow band of wavelengths centred at 590 nm, the well-known 'D line' of atomic spectroscopy. Considering this emission as emanating from a dipole oscillator with resonant frequency ω_0,

$$\omega_0 = \frac{2\pi c}{\lambda} = 3.2 \times 10^{15} \text{ s}^{-1}$$

and

$$\tau = \frac{3c}{2r_c\omega_0^2} \simeq 16 \times 10^{-9} \text{ s}$$

This result is actually quite close to the measured lifetime of the first excited state of the sodium atom and lends credence to the harmonic oscillator model. In reality, however, the atom must be treated quantum mechanically because there is no lower limit to the classical radiative emission, and the classical electron would fall into the nucleus as its mechanical energy diminishes to zero. Furthermore, the first excited state with a quantum-mechanically 'allowed' transition possesses one quantum of orbital angular momentum, and this angular momentum does not appear in the simple oscillator model.

11.3   Radiative damping and electron scattering

Nevertheless, the damped harmonic oscillator model provides some insight into the physics of atomic emission, absorption of radiation, and atomic scattering. We can write down the equation of motion of the classical, radiatively damped, externally driven harmonic oscillator, moving along the z-axis, as

$$\frac{d^2z}{dt^2} + \frac{1}{\tau}\frac{dz}{dt} + \omega_0^2 z = -\frac{eE_z}{m_e}e^{-i\omega t} \tag{11.13}$$

where as before ω_0 is the characteristic frequency of the oscillator, e and m_e are the electron charge and mass, respectively, τ the classical oscillator lifetime, and E_z is the amplitude of the external driving field oscillating at frequency ω. The solution to this equation of motion is

$$z = \frac{-eE_z e^{-i\omega t}}{m_e\left[(\omega_0^2 - \omega^2) - i\omega/\tau\right]} \tag{11.14}$$

From this result we can write down the frequency-dependent dipole moment,

$$p = -ez\hat{z} = \frac{-eE_z e^{-i\omega t}}{m_e\left[(\omega_0^2 - \omega^2) - i\omega/\tau\right]}\hat{z} \tag{11.15}$$

From Equation 11.3 we have, for the average dissipated power,

$$\langle W \rangle = \frac{e^4 E_z^2 \omega^4}{12\pi\varepsilon_0 m_e^2\left[(\omega_0^2 - \omega^2)^2 + (\omega/\tau)^2\right]} \tag{11.16}$$

Using the classical electron radius, Equation 11.9, we can write this expression for the dissipated power as the product of an energy flux and a scattering cross section,

$$\langle W \rangle = \underbrace{\left(\frac{\varepsilon_0 E_z^2}{2} \right) c}_{\text{energy flux}} \cdot \underbrace{\left\{ \left[\frac{8\pi r_c^2}{3} \right] \left[\frac{\omega^4}{\left(\omega_0^2 - \omega^2 \right)^2 + (\omega/\tau)^2} \right] \right\}}_{\text{cross section}} \tag{11.17}$$

where the scattering cross section is defined as

$$\sigma(\omega) = \left[\frac{8\pi r_c^2}{3} \right] \left[\frac{\omega^4}{\left(\omega_0^2 - \omega^2 \right)^2 + (\omega/\tau)^2} \right] \tag{11.18}$$

11.4   The semiclassical two-level atom

We have seen in Chapter 6 that we can infer a purely phenomenological relation between stimulated and spontaneous emission by invoking the Einstein rate expression, Equation 6.9. We found that, assuming thermal equilibrium between states 1, 2, the rate constants of *stimulated* absorption and emission, B_{12}, B_{21} are equal, and the rate of *spontaneous* emission is related to B_{21} by Equation 6.15:

$$\frac{A_{21}}{B_{21}} = \frac{\hbar \omega_0^3}{\pi^2 c^3}$$

We will now develop a quantum semiclassical expression for the stimulated rate of absorption and relate it back to B_{21} and A_{21}. Then we will interpret this semiclassical result in terms of a classical radiating dipole.

11.4.1   Coupled equations of a two-state system

We start with the time-dependent Schrödinger equation,

$$\hat{H} \Psi(\mathbf{r}, t) = i\hbar \frac{\Psi}{dt} \tag{11.19}$$

and write the stationary-state solution of level n as

$$\Psi_n(\mathbf{r}, t) = \psi_n(\mathbf{r}) e^{-iE_{nt}/\hbar} = \psi_n(\mathbf{r}) e^{-i\omega_n t} \tag{11.20}$$

The time-independent Schrödinger equation then becomes

$$\hat{H}_A \psi_n(\mathbf{r}) = E_n \psi_n(\mathbf{r}) \tag{11.21}$$

where the subscript A indicates 'atom'. Then for the two-level system we have

$$\hat{H}_A\psi_1 = E_1\psi_1 = \hbar\omega_1\psi_1$$
$$\hat{H}_A\psi_2 = E_2\psi_2 = \hbar\omega_2\psi_2$$

and write

$$\hbar\omega_0 \equiv \hbar(\omega_2 - \omega_1) = E_2 - E_1 \tag{11.22}$$

Now we add a time-dependent term to the Hamiltonian that will turn out to be proportional to the oscillating classical field with frequency not far from ω_0:

$$\hat{H} = \hat{H}_A + \hat{V}(t) \tag{11.23}$$

With the field turned on, the state of the system becomes a time-dependent linear combination of the two stationary states

$$\Psi(\mathbf{r}, t) = C_1\psi_1 e^{-i\omega_1 t} + C_2(t)\psi_2 e^{-i\omega_2 t} \tag{11.24}$$

which we require to be normalised,

$$\int |\Psi(\mathbf{r}, t)|^2 \, d\tau = |C_1(t)|^2 + |C_2(t)|^2 = 1 \tag{11.25}$$

Now, if we substitute the time-dependent wave function (Equation 11.24) back into the time-dependent Schrödinger equation (Equation 11.19), multiply on the left with $\psi_1^* e^{i\omega_1 t}$, and integrate over all space, we get

$$C_1 \int \psi_1^* \hat{V}\psi_1 \, d\mathbf{r} + C_2 e^{-i\omega_0 t} \int \psi_1^* \hat{V}\psi_2 \, d\mathbf{r} = i\hbar\frac{dC_1}{dt} \tag{11.26}$$

From now on we will denote 'matrix elements' $\int \psi_1^* \hat{V}\psi_1 \, d\mathbf{r}$ and $\int \psi_1^* \hat{V}\psi_2 \, d\mathbf{r}$ as V_{11} and V_{12}, so we have

$$C_1 V_{11} + C_2 e^{-i\omega_0 t} V_{12} = i\hbar\frac{dC_1}{dt} \tag{11.27}$$

and similarly for C_2, we obtain

$$C_1 e^{i\omega_0 t} V_{21} + C_2 V_{22} = i\hbar\frac{dC_2}{dt} \tag{11.28}$$

These two coupled equations define the quantum-mechanical expression, and their solutions, C_1 and C_2, define the time evolution of the state wave function, Equation 11.24. Of course, any measurable quantity is related to $|\Psi(\mathbf{r}, t)|^2$; consequently we are really more interested in $|C_1|^2$ and $|C_2|^2$ than the coefficients themselves.

11.4.2   Field coupling operator

A single-mode radiation source, such as a laser, aligned along the z-axis, will produce an electromagnetic wave with amplitude E_0, polarisation $\hat{\mathbf{e}}$, and frequency ω:

$$\mathbf{E} = \hat{\mathbf{e}}E_0 \cos(\omega t - kz) \tag{11.29}$$

with the magnitude of the wave vector

$$k = \frac{2\pi}{\lambda} \quad \text{and} \quad \omega = 2\pi\nu = 2\pi\frac{c}{\lambda} = kc \tag{11.30}$$

An optical wavelength in the visible region of the spectrum, say, $\lambda = 600\,\text{nm} \simeq 1.1 \times 10^4\,a_0$, is evidently several orders of magnitude greater than the characteristic atomic scale ($\simeq a_0$). Therefore, over the spatial extent of the atom-field interaction, the kz term in Equation 11.29 is negligible, and we can consider the field amplitude to be constant over the scale length of the atom. We can make, therefore, the *dipole approximation* in which the leading interaction term between the atom and the optical field is the scalar product of the atom dipole $\mathbf{d}$, defined as

$$\hat{\mathbf{d}} = -e\hat{\mathbf{r}} = -e\sum_j \hat{\mathbf{r}}_j \tag{11.31}$$

and the field coupling operator is

$$\hat{V} = -\hat{\mathbf{d}} \cdot \mathbf{E} \tag{11.32}$$

The operator $\hat{V}$ has odd parity with respect to the electron coordinate $\mathbf{r}$ so that the matrix elements V_{11} and V_{22} vanish, and only atomic states of opposite parity can be coupled by the dipole interaction. The explicit expression for V_{12} is

$$V_{12} = eE_0 r_{12} \cos \omega t \tag{11.33}$$

with

$$r_{12} = \int \psi_1^* \left(\sum_j \hat{\mathbf{r}}_j \cdot \hat{\mathbf{e}} \right) \psi_2 \, d\tau \tag{11.34}$$

Equation 11.34 describes the matrix element of the electronic coordinate vector operator summed over all electrons (assuming a two-state, multi-electron atom) and projected onto the E-field direction of the optical wave. The transition dipole moment matrix element is defined as

$$p_{12} \equiv er_{12} \tag{11.35}$$

It is convenient to collect all these scalar quantities into one term Ω_0 with units of frequency, often called the 'on-resonance Rabi frequency':

$$\Omega_0 \equiv \frac{p_{12}E_0}{\hbar} = \frac{eE_0 r_{12}}{\hbar} \tag{11.36}$$

So finally we have

$$V_{12} = \hbar\Omega_0 \cos \omega t \tag{11.37}$$

11.4.3 Calculation of B_{12}

Now we rewrite Equations 11.27 and 11.28 in terms of the Rabi frequency:

$$\Omega_0 \cos \omega t e^{-i\omega_0 t} C_2 = i\frac{dC_1}{dt} \tag{11.38}$$

$$\Omega_0^* \cos \omega t e^{i\omega_0 t} C_1 = i\frac{dC_2}{dt} \tag{11.39}$$

We take the initial conditions to be $C_1(t = 0) = 1$ and $C_2(t = 0) = 0$ and recall that $|C_2(t)|^2$ expresses the probability of finding the population in the excited state at time t. The time rate of probability increase, of finding the system in state 2, is given by

$$\frac{|C_2(t)|^2}{t} \tag{11.40}$$

This expression can be set equal to the stimulated absorption term in the Einstein relation Equation 6.9:

$$\frac{|C_2(t)|^2}{t} = B_{12}\rho(\omega)d\omega \tag{11.41}$$

Now we seek $C_2(t)$ from Equation 11.39 and the initial conditions. In the weak-field regime where only terms linear in Ω_0 are important, we have

$$C_2(t) = \frac{\Omega_0^*}{2}\left[\frac{1 - e^{i(\omega_0 + \omega)t}}{\omega_0 + \omega} + \frac{1 - e^{i(\omega_0 - \omega)t}}{\omega_0 - \omega}\right] \tag{11.42}$$

If the frequency ω of the driving wave approaches the transition resonant frequency ω_0, the exponential in the first term in brackets will oscillate at about twice the atomic resonant frequency ω_0 ($\sim10^{15}$ s^{-1}), very fast compared to the characteristic rate of weak-field optical coupling ($\sim10^8$ s^{-1}). Therefore, over the time of the transition, the first term in Equation 11.42 will be negligible compared to the second. To a quite good approximation we can write,

$$C_2(t) \simeq \frac{\Omega_0^*}{2} \left[\frac{1 - e^{i(\omega_0 - \omega)t}}{\omega_0 - \omega} \right]$$

(11.43)

This expression is sometimes called the *rotating wave approximation* (RWA) because it corresponds to the solution for C_2 in a coordinate frame rotating at frequency ω_0. From Equation 11.43 we have

$$|C_2(t)|^2 = |\Omega_0|^2 \frac{\sin^2 \left[(\omega_0 - \omega)\frac{t}{2}\right]}{(\omega_0 - \omega)^2}$$

(11.44)

In order to arrive at a useful expression relating $|C_2(t)|^2$ to the Einstein B coefficients, we have to take into account the fact that there is always a finite width in the spectral distribution of the excitation source. The source might be, for example, an incoherent broadband arc lamp, the output from a monochromator coupled to a synchrotron, or a narrowband mono-mode laser whose spectral width would probably be narrower than the natural width of the atomic transition. So if we write the field energy as an integral over the spectral energy density of the excitation source in the neighbourhood of the transition frequency,

$$\frac{1}{2} \varepsilon_0 E_0^2 = \int_{\omega_0 - \frac{1}{2}\Delta\omega}^{\omega_0 + \frac{1}{2}\Delta\omega} \rho_\omega \, d\omega$$

(11.45)

where the limits of integration, $\omega_0 \pm \frac{1}{2}\Delta\omega$, refer to the spectral width of the excitation source, and recognise from Equation 11.43 that

$$|C_2(t)|^2 = \left(\frac{eE_0 r_{12}^2}{\varepsilon_0 \hbar} \right)^2 \frac{\sin^2 \left[(\omega_0 - \omega)\frac{t}{2}\right]}{(\omega_0 - \omega)^2}$$

(11.46)

we can then substitute Equation 11.45 into Equation 11.46 to find that

$$|C_2(t)|^2 = \frac{e^2 2 r_{12}^2}{\varepsilon_0 \hbar^2} \int_{\omega_0 - \frac{1}{2}\Delta\omega}^{\omega_0 + \frac{1}{2}\Delta\omega} \rho_\omega \frac{\sin^2 \left[(\omega_0 - \omega)\frac{t}{2}\right]}{(\omega_0 - \omega)^2} \, d\omega$$

(11.47)

For conventional broadband excitation sources we can safely assume that the spectral density is constant over the line width of the atomic transition and take ρ_ω outside the integral operation and set it equal to $\rho(\omega_0)$. Note that this approximation is *not* valid for narrow band mono-mode lasers. Let us assume a fairly broadband continuous excitation so that $t(\omega_0 - \omega) \gg 1$. In this case

$$\int_{\omega_0 - \frac{1}{2}\Delta\omega}^{\omega_0 + \frac{1}{2}\Delta\omega} \rho_\omega \frac{\sin^2 \left[(\omega_0 - \omega)\frac{t}{2}\right]}{(\omega_0 - \omega)^2} \, d\omega = \frac{\pi t}{2}$$

(11.48)

and the expression for probability of finding the atom in the excited state becomes

$$|C_2(t)|^2 = \frac{e^2 \pi r_{12}^2}{\varepsilon_0 \hbar^2} \rho(\omega_0) t \qquad (11.49)$$

Remembering that Equation 11.41 provides the bridge between the quantum and classical expressions for the excitation rate, we can now write the Einstein B coefficient in terms of the transition moment as

$$B_{12}\rho(\omega_0) = \frac{|C_2(t)|^2}{t} = \frac{e^2 \pi r_{12}^2}{\varepsilon_0 \hbar^2} \rho(\omega_0) \qquad (11.50)$$

or

$$B_{12} = \frac{e^2 \pi r_{12}^2}{\varepsilon_0 \hbar^2} \qquad (11.51)$$

The square of the transition moment matrix element r_{12}^2 averaged over all spatial orientations is

$$\langle |r_{12}|^2 \rangle = r_{12}^2 \langle \cos^2 \theta \rangle = \frac{1}{3} r_{12}^2 \qquad (11.52)$$

where θ is the angle between the transition moment and the E-field of the plane wave excitation. We have, finally,

$$B_{12} = \frac{e^2 \pi r_{12}^2}{3\varepsilon_0 \hbar^2} \qquad (11.53)$$

and from the definition of the dipole moment p_{12}, Equation 11.35:

$$B_{12} = \frac{\pi p_{12}^2}{3\varepsilon_0 \hbar^2} \qquad (11.54)$$

Assuming that states 1 and 2 are not degenerate, $B_{12} = B_{21}$, and we have from Equation 6.15 an expression for the spontaneous emission rate in terms of the dipole transition moment calculated from our two-level model:

$$A_{21} = \frac{\omega_0^3 p_{12}^2}{3\pi \varepsilon_0 \hbar c^3} \qquad (11.55)$$

11.4.4 Spontaneous emission as loss

Everything we have developed in Section 11.4 up to this point involves the coupling of one optical field mode to a two-level atom. The Schrödinger equation is adequate to describe the time evolution of this system because it can always be expressed by a pure-state wave function. Spontaneous emission, however, cannot be characterised by the time evolution to a final-state wave function, but only as a *probability distribution* of final-state wave functions. The time evolution of a probability distribution requires a *density matrix* description, and the governing equation of motion is the *Liouville equation*. Nevertheless, within the two-level picture we can treat spontaneous emission as a phenomenological, dissipative loss term, and thereby avoid a density matrix treatment that would carry us beyond the scope of this section.

We take account of this dissipative loss by modifying Equation 11.39 to include a radiative loss rate constant γ:

$$\Omega_0^* \cos \omega t e^{i\omega_0 t} C_1 - i\gamma C_2 = i\frac{dC_2}{dt} \tag{11.56}$$

If the driving field ($\Omega^* = 0$) is turned off at $t = t_0$,

$$-i\gamma C_2 = i\frac{dC_2}{dt} \tag{11.57}$$

and

$$C_2(t) = C_2(t_0)e^{-\gamma(t-t_0)} \tag{11.58}$$

So the probability of finding the atom in the excited state is

$$|C_2(t)|^2 = |C_2(t_0)|^2 e^{-2\gamma t} \tag{11.59}$$

In an ensemble of N atoms in a volume, the number N_2 in the excited state 2 is

$$N_2 = N|C_2(t)|^2 = N_2(t_0)e^{-2\gamma t} \tag{11.60}$$

If we compare this behaviour to the result obtained from the Einstein rate equation, we see that

$$A_{21} = 2\gamma \equiv \Gamma \tag{11.61}$$

Now the steady-state solution for our new, improved $C_2(t)$ coefficient (Equation 11.42) is

$$C_2(t) = -\frac{1}{2}\Omega_0^* \left[\frac{e^{i(\omega_0+\omega)t}}{\omega_0 + \omega - i\gamma} + \frac{e^{i(\omega_0-\omega)t}}{\omega_0 - \omega - i\gamma} \right] \tag{11.62}$$

Using Equations 11.24 and 11.34 we can write the net transition dipole $\langle \mathbf{p}_{12} \rangle$ by summing over the electronic coordinate vectors of our multi-electron, two-level atom:

$$\langle \mathbf{p}_{12} \rangle = -e \int \Psi^* \sum_j \mathbf{r}_j \Psi d\tau = -e \langle \mathbf{r}_{12} \rangle \tag{11.63}$$

and the time dependence of the atomic transition dipole moment in terms of the time dependence of $C_2(t)$:

$$\langle \mathbf{p}_{12} \rangle = -e \left[C_1^* C_2(t) \langle \mathbf{r}_{12} \rangle e^{-i\omega_0 t} + C_2(t)^* C_1 \langle \mathbf{r}_{21} \rangle e^{i\omega_0 t} \right] \tag{11.64}$$

Substituting $C_2(t)$ from Equation 11.62 and assuming a weak-field excitation ($C_1 \simeq 1$) we can then write

$$\langle \mathbf{p}_{12}(t) \rangle = \frac{e^2 \, |\mathbf{r}_{12}|^2 \, \mathbf{E}_0}{2\hbar} \left[\frac{e^{i\omega t}}{\omega_0 + \omega - i\gamma} + \frac{e^{-i\omega t}}{\omega_0 - \omega - i\gamma} + \frac{e^{-i\omega t}}{\omega_0 + \omega + i\gamma} + \frac{e^{i\omega t}}{\omega_0 - \omega + i\gamma} \right] \tag{11.65}$$

11.4.5 Polarisation and transition dipole moment

The driving field $\mathbf{E}(t)$ can also be expressed in terms of positive and negative frequency terms:

$$\mathbf{E}(t) = \mathbf{E}_0 \cos \omega t = \frac{1}{2} \mathbf{E}_0 \left[e^{i\omega t} + e^{-i\omega t} \right] \tag{11.66}$$

and from Equation 2.17 the polarisation field in extended matter can be written as

$$\mathbf{P}(t) = \frac{1}{2} \varepsilon_0 \mathbf{E}_0 \left[\chi(\omega) e^{-i\omega t} + \chi(-\omega) e^{i\omega t} \right] \tag{11.67}$$

with $\chi(\pm\omega)$, the susceptibility. Now, as we have seen in Chapter 4, Sections 4.2 and 4.3, the polarisation vector field can be considered a dipole density:

$$\mathbf{P}(t) = \frac{N}{V} \langle \mathbf{p}_{12}(t) \rangle = n \langle \mathbf{p}_{12}(t) \rangle \tag{11.68}$$

where N are the number of dipoles in volume V and n is the number density. After replacing $|\mathbf{r}_{12}|^2$ with its orientation averaged value, $\frac{1}{3}|\mathbf{r}_{12}|^2$, we have for the polarisation vector,

$$\mathbf{P}(t) = \frac{n \, |\langle \mathbf{p}_{12} \rangle|^2}{6\hbar} \mathbf{E}_0 \left[\left(\frac{1}{\omega_0 - \omega - i\gamma} + \frac{1}{\omega_0 + \omega + i\gamma} \right) e^{-i\omega t} + \left(\frac{1}{\omega_0 + \omega - i\gamma} + \frac{1}{\omega_0 - \omega + i\gamma} \right) e^{i\omega t} \right] \tag{11.69}$$

Comparison of Equations 11.67 and 11.69 allows us to write the susceptibility in terms of the transition dipoles:

$$\chi(\omega) = \frac{n \, |\langle \mathbf{p}_{12} \rangle|^2}{3\varepsilon_0 \hbar} \left(\frac{1}{\omega_0 - \omega - i\gamma} + \frac{1}{\omega_0 + \omega + i\gamma} \right) \tag{11.70}$$

Identify the real χ' and imaginary parts χ'' of the susceptibility, $\chi = \chi'(\omega) + i\chi''(\omega)$:

$$\chi(\omega) = \frac{n \, |\langle \mathbf{p}_{12} \rangle|^2}{3\varepsilon_0 \hbar} \left[\left(\frac{\omega_0 - \omega}{(\omega_0 - \omega)^2 + \gamma^2} + \frac{\omega_0 + \omega}{(\omega_0 + \omega)^2 + \gamma^2} \right) \right.$$
$$\left. + i\gamma \left(\frac{1}{(\omega_0 - \omega)^2 + \gamma^2} - \frac{1}{(\omega_0 - \omega)^2 + \gamma^2} \right) \right] \tag{11.71}$$

and apply the RWA:

$$\chi'(\omega) = \frac{n \, |\langle \mathbf{p}_{12} \rangle|^2}{3\varepsilon_0 \hbar} \left(\frac{\omega_0 - \omega}{(\omega_0 - \omega)^2 + \gamma^2} \right) \tag{11.72}$$

$$\chi'' = \frac{n \, |\langle \mathbf{p}_{12} \rangle|^2}{3\varepsilon_0 \hbar} \left(\frac{\gamma}{(\omega_0 - \omega)^2 + \gamma^2} \right) \tag{11.73}$$

We are usually interested in frequencies ω tuned not too far from ω_0, and the term 'detuning' is often used to denote the frequency difference,

$$\Delta\omega = \omega - \omega_0 \tag{11.74}$$

This choice for $\Delta\omega$ means that 'blue detuning' ($\omega > \omega_0$) is positive and 'red detuning' ($\omega < \omega_0$) is negative. With this convention in mind, and remembering that $\Gamma = 2\gamma$ (Equation 11.61), we write the real and imaginary parts of the susceptibility as

$$\chi'(\omega) = -\frac{n \, |\langle \mathbf{p}_{12} \rangle|^2}{3\varepsilon_0 \hbar} \left(\frac{\Delta\omega}{\Delta\omega^2 + \left(\frac{\Gamma}{2}\right)^2} \right) \tag{11.75}$$

$$\chi''(\omega) = \frac{n \, |\langle \mathbf{p}_{12} \rangle|^2}{3\varepsilon_0 \hbar} \left(\frac{\Gamma/2}{\Delta\omega^2 + \left(\frac{\Gamma}{2}\right)^2} \right) \tag{11.76}$$

Around the resonance frequency ω_0, the real part of the susceptibility χ' exhibits a 'dispersive' profile. Right on resonance χ' is zero. It is positive to the red of the resonance and negative to the blue, falling off with the square of the detuning. In contrast, χ'' exhibits an 'absorptive' profile. It is positive and peaks at resonance, falling off with the square of the detuning on each side. Figures 11.2 and 11.3 trace the two susceptibility terms around the detuning resonance.

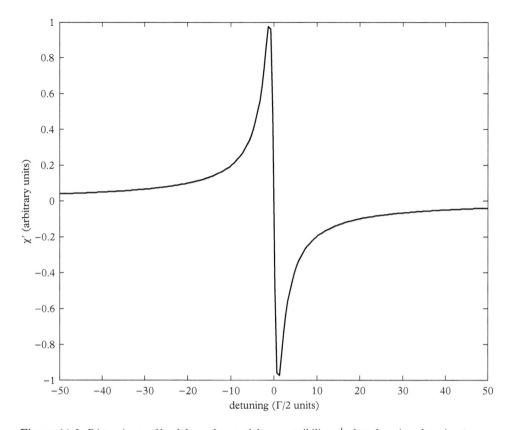

Figure 11.2 *Dispersive profile of the real part of the susceptibility χ' plotted against detuning $\Delta\omega$.*

From Equation 11.71 we see that χ' is symmetric with respect to the interchange of ω and $-\omega$ while χ'' is antisymmetric:

$$\chi'(\omega) = \chi'(-\omega) \qquad (11.77)$$
$$\chi''(\omega) = -\chi''(-\omega) \qquad (11.78)$$

Substituting these symmetry relations into Equation 11.67 allows us to separate the time dependence of real and imaginary parts of the susceptibility:

$$\mathbf{P}(t) = \varepsilon_0 \mathbf{E}_0 \left(\chi' \cos \omega t + \chi'' \sin \omega t \right) \qquad (11.79)$$

The time dependence of the polarisation field shows that the dispersive term is driven by the 'in-phase' real part of the E-field $\cos \omega t$ while the absorptive term is driven by the 'in-quadrature' $\sin \omega t$.

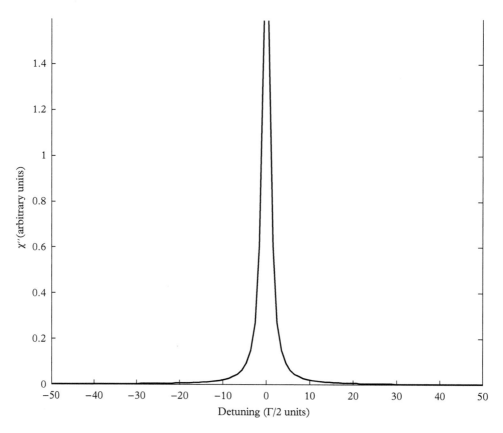

Figure 11.3 *Absorptive profile of the imaginary part of the susceptibility* χ'' *plotted against detuning* $\Delta\omega$.

Using the real form of the driving field $\mathbf{E} = \mathbf{E}_0 \cos\omega t$, the interaction energy between the driving E-field and the polarisation field is

$$\mathscr{E} = -\mathbf{P}(t) \cdot \mathbf{E}(t) = -\varepsilon_0 \mathbf{E}_0 \left[\chi' \cos\omega t + \chi'' \sin\omega t\right] \cdot \mathbf{E}_0 \left[\cos\omega t\right] \tag{11.80}$$

and averaging over a optical cycle we have

$$\langle\mathscr{E}\rangle = -\frac{1}{2}\varepsilon_0 E_0^2 \chi' \tag{11.81}$$

The dipole-field interaction energy separates into a real, dispersive part associated with χ' and an imaginary, absorptive part associated with χ''.

11.5   The dipole-gradient and radiation pressure forces

Light interacting with the two-level atom also gives rise to forces. A light beam carries momentum, and scattering from an atom produces a transfer of momentum resulting in a force on the atom, $\mathbf{F} = d\mathbf{p}/dt$. The light force exerted on an atom can be of two types: a conservative *dipole-gradient force* and a dissipative *spontaneous force*.

11.5.1   Dipole-gradient force

The dipole-gradient force can be readily understood by considering the light as a classical wave. It is simply the optical-cycle-averaged force arising from the interaction of the transition dipole, induced by the oscillating electro-magnetic field, with the spatial gradient of the electric field amplitude. The dipole-gradient force does not apply to propagating plane waves because the amplitudes are constant. The light must be focused or spatially modulated in some way. Focusing the light beam controls the gradient of the electric field amplitude; detuning the frequency below or above the atomic transition controls the sign of the force. Tuning the light to the red of the resonance frequency results in a net attractive force on the atom towards the intensity maximum. Blue-detuned light pushes the atom to the intensity minimum. The dipole-gradient force is a stimulated process in which no net exchange of energy between the field and the atom takes place. There is, in principle, no upper limit to the magnitude of the dipole-gradient force since it is a function only of the field gradient and detuning. The general expression for light forces is given by the scalar product of the polarisation field and the gradient operating on the applied E-field:

$$\mathbf{F} = (\mathbf{P} \cdot \nabla)\mathbf{E} \tag{11.82}$$

$$\mathbf{F} = \mathbf{P} \cdot \nabla \left[\mathbf{E}_0(\mathbf{r}) \cos(\mathbf{k} \cdot \mathbf{r} - \omega t) \right] \tag{11.83}$$

$$\mathbf{F} = \varepsilon_0 \chi \mathbf{E} \cdot \nabla \left[\mathbf{E}_0(\mathbf{r}) \cos(\mathbf{k} \cdot \mathbf{r} - \omega t) \right] \tag{11.84}$$

$$\mathbf{F} = \varepsilon_0 \mathbf{E}_0 \left[\chi'(\omega) \cos \omega t + \chi''(\omega) \sin \omega t \right] \cdot \nabla \left[\mathbf{E}_0(\mathbf{r}) \cos(\mathbf{k} \cdot \mathbf{r} - \omega t) \right] \tag{11.85}$$

The gradient operator acts on the product of the field amplitude and phase, resulting in two terms:

$$\nabla \left[\mathbf{E}_0(\mathbf{r}) \cos(\mathbf{k} \cdot \mathbf{r} - \omega t) \right] = \nabla \mathbf{E}_0(\mathbf{r}) \cos(\mathbf{k} \cdot \mathbf{r} - \omega t) + \mathbf{E}_0(\mathbf{r}) \nabla \cos(\mathbf{k} \cdot \mathbf{r} - \omega t) \tag{11.86}$$

The first of these terms gives rise to the dipole-gradient force and the second to the radiation pressure force. Substituting the first term on the right of Equation 11.86 into Equation 11.85, and taking the optical-cycle average, results in

$$\mathbf{F}_{\text{dg}} = \frac{1}{2} \varepsilon_0 \mathbf{E}_0 \nabla \mathbf{E}_0(\mathbf{r}) \chi' \tag{11.87}$$

$$\mathbf{F}_{\text{dg}} = \frac{1}{4} \varepsilon_0 \nabla E_0^{\,2}(\mathbf{r}) \chi' \tag{11.88}$$

In writing Equations 11.87 and 11.88 we have made use of the fact that $\mathbf{k} \cdot \mathbf{r}$ is negligible over the spatial extent of the atom ($r \sim 2a_0$), where a_0 is the Bohr radius, the atomic unit of length. Now substituting Equation 11.75 for χ' into 11.88:

$$\mathbf{F}_{dg} = \frac{1}{4}\varepsilon_0 \nabla E_0^{\,2}(\mathbf{r}) \left[-\frac{|\langle \mathbf{p}_{12}\rangle|^2}{3\varepsilon_0 \hbar} \left(\frac{\Delta\omega}{\Delta\omega^2 + \left(\frac{\Gamma}{2}\right)^2} \right) \right] \tag{11.89}$$

We see that in the expression for $\mathbf{F}_{dg}$ the χ' factor includes a negative sign, but due to our convention for $\Delta\omega$, Equation 11.74, the detuning to the red is also negative. The dipole gradient force with red detuning points therefore to a higher intensity gradient, while blue detuning results in a 'repulsive' force, away from the intensity maximum. Figure 11.4 shows how a focused Gaussian light beam can produce the dipole gradient force, confining a collection of cold atoms near the beam axis. If the atoms have kinetic energy less than the potential arising from the dipole gradient force (Equation 11.101) it acts as an atom light trap.

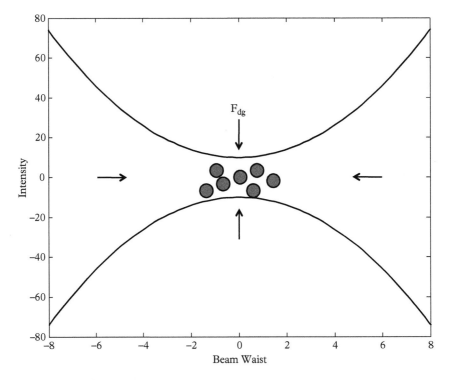

Figure 11.4 *Gaussian light beam focused to a 'beam waist' radius. The intensity gradient transverse to and along the beam axis acts as a confining force for two-level atoms when the light beam is detuned far $[(\Delta\omega)^2 \gg (\Gamma/2)^2]$. Note that the dipole gradient force transverse to the beam is significantly greater than the force confining the atoms along the longitudinal axis.*

11.5.2   Radiation pressure force

Substituting the second term on the right of Equation 11.86 into Equation 11.85 results in an expression for the radiation pressure force:

$$\mathbf{F_r} = \frac{1}{2}\varepsilon_0 E_0^2 \mathbf{k}\chi'' \tag{11.90}$$

and inserting Equation 11.76:

$$\mathbf{F_r} = \frac{1}{2}\varepsilon_0 E_0^2 \left[\mathbf{k}\frac{|\langle\mathbf{p}_{12}\rangle|^2}{3\varepsilon_0\hbar}\left(\frac{\Gamma/2}{\Delta\omega^2 + \left(\frac{\Gamma}{2}\right)^2} \right) \right] \tag{11.91}$$

The radiation pressure force peaks at zero detuning and is always in the direction of light propagation $\mathbf{k}$. The radiation force can be regarded as an impulse experienced by an atom when it absorbs or emits a quantum of light momentum $\hbar\mathbf{k}$, although classically it arises from the Lorentz force, $\mathbf{F} = \mathbf{v} \times \mathbf{B}$, where $\mathbf{B}$ is the magnetic induction field of the light wave.

11.5.3   Atom forces from the optical Bloch equations

So far we have handled spontaneous emission as a phenomenological loss term in Equation 11.56. In fact, spontaneous emission results in a statistical distribution of final states that should be properly expressed in a set of coupled equations of the *density matrix*. Applied to atomic transitions, this density matrix results in the *Optical Bloch Equations*. A digression into the density matrix formalism would take us too far afield for the purposes of this chapter, so we will just state the pertinent results. Instead of an expression for the probability of finding the two-level system in state 2 as the square of the absolute value of the coefficient, $|C_2|^2$, we define the *population* of state 2 as

$$\rho_{22} = \frac{\frac{1}{4}|\Omega_0|^2}{\Delta\omega^2 + \left(\frac{\Gamma}{2}\right)^2 + \frac{1}{2}|\Omega_0|^2} \tag{11.92}$$

where γ is the same loss term we have been using and Ω_0 is the same on-resonance Rabi frequency previously defined (Equation 11.36). Using ρ_{22} instead of $|C_2|^2$ leads to a modification in the denominators of our expressions for the susceptibilities, χ', χ'', and consequently for the dipole gradient and radiation pressure forces. The modified expressions are

$$\mathbf{F_{dg}} = \frac{1}{4}\varepsilon_0 \nabla E_0^2(\mathbf{r}) \left[-\frac{|\langle\mathbf{p}_{12}\rangle|^2}{3\varepsilon_0\hbar} \cdot \frac{\Delta\omega}{\Delta\omega^2 + \left(\left(\frac{\Gamma}{2}\right)^2 + \frac{1}{2}|\Omega_0|^2\right)} \right] \tag{11.93}$$

and

$$F_r = \frac{1}{2}\varepsilon_0 E_0^2 \mathbf{k} \left[\frac{|\langle \mathbf{p}_{12}\rangle|^2}{3\varepsilon_0 \hbar} \cdot \frac{\Gamma/2}{\Delta\omega^2 + \left(\left(\frac{\Gamma}{2}\right)^2 + \frac{1}{2}|\Omega_0|^2\right)} \right] \tag{11.94}$$

The force expressions retain their basic form: dispersive for the dipole gradient force and absorptive for the radiation pressure force; but the denominators exhibit an extra term that produces a broader effective width for each profile:

$$\frac{1}{2}|\Omega_0|^2 = \frac{1}{2}\left(\frac{p_{12}E_0}{\hbar}\right)^2 \tag{11.95}$$

This term, one half the square of the Rabi frequency, is proportional to the electric field intensity E_0^2 and is often called 'power broadening' because the effective line profile broadens as the intensity of the applied light field increases. In fact, Equation 11.92 shows that as the source intensity increases the fractional population of state 2 approaches but does not exceed 1/2. The population of state 2 is said to 'saturate' at that limit. We can define a parameter S that indicates the degree of saturation as

$$S = \frac{\frac{1}{2}|\Omega_0|^2}{\Delta\omega^2 + \left(\frac{\Gamma}{2}\right)^2} \tag{11.96}$$

and rewrite the force expressions, Equations 11.93 and 11.94, in terms of the Rabi frequency and the saturation parameter:

$$F_{dg} = -\frac{\hbar\Delta\omega}{6}\nabla S \cdot \frac{1}{1+S} \tag{11.97}$$

and

$$F_r = \frac{\hbar k}{3}\left(\frac{\Gamma}{2}\right) \cdot \frac{S}{1+S} \tag{11.98}$$

If we set $S = 1$ as the on-resonance tuning condition, $\Delta\omega = 0$, as a somewhat arbitrary definition of saturation, we have a saturation Rabi frequency:

$$\Omega_{0sat} = \frac{\Gamma}{\sqrt{2}} \tag{11.99}$$

Figure 11.5 shows how S grows as the excitation power increases. The saturation radiation pressure force becomes

$$F_r = \hbar k \frac{\Gamma}{12} \tag{11.100}$$

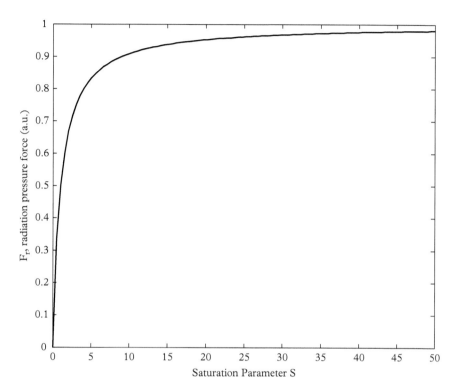

Figure 11.5 *Radiation pressure force F_r (arbitrary units) as a function of the saturation parameter S.*

The dipole gradient force is conservative, and integrating over the spatial extent of the field gradient determines an atom-optical potential U_T, where the subscript indicates 'trap':

$$U_T = -\int \mathbf{F_{dg}}d\mathbf{r} = \frac{\hbar\Delta\omega}{6}\ln(1 + S) \tag{11.101}$$

If the field is tuned relatively far from resonance so that $\Delta\omega \gg \Gamma$ and $S \ll 1$, then

$$U_T \simeq \frac{1}{12}\frac{\Omega_0^2}{\Delta\omega} \tag{11.102}$$

When the optical trap (usually one or more focused laser beams) is tuned to the red of resonance, U_T presents a negative potential and the atom is drawn to the field amplitude maximum. By far the most frequent realisation of the optical trap is a single-mode,

Gaussian-profile, focused laser beam. The two-level atoms then find the potential min-imum along the laser beam axis. As we shall see shortly, atoms with sufficiently low temperature can be confined in such an optical trap. If the detuning is to the blue, the potential becomes repulsive and the atoms are expelled from the laser beam centre. Both attractive and repulsive potentials can be used to manipulate cold atoms.

11.5.4   Atom Doppler cooling

Consider the atom moving in the $+z$ direction with velocity v_z and counter-propagating to the light wave detuned from resonance by $\Delta\omega_L$, the *effective* detuning from the point of view of the atom will be

$$\Delta\omega = \Delta\omega_L + kv_z \tag{11.103}$$

In the rest frame of the atom the light appears shifted to the blue, $\delta\omega = kv_z$, by the Doppler effect. The radiation pressure force acting on the atom as it scatters the light will be in the direction opposite to its motion. We have from Equations 11.94, 11.95, and 11.103:

$$\mathbf{F_{r\pm}} = \pm\frac{\hbar k}{6}|\Omega_0|^2\left[\frac{\Gamma/2}{(\Delta\omega \mp kv)^2 + \left(\frac{\Gamma}{2}\right)^2 + \frac{1}{2}|\Omega_0|^2}\right] \tag{11.104}$$

Suppose we have, as shown in Figure 11.6, two continuous wave Gaussian modes $\pm k$, tuned within the atomic natural linewidth, propagating in the $\pm z$ directions, and we take the amplitude of the net force to be the difference between the two components of the radiation pressure force when the atom is moving along z with v_z. Figure 11.7 shows that if $\Delta\omega$ is tuned red $\sim\Gamma/2$, then the Doppler blue shift arising from the light beam propagating right to left in Figure 11.6 increases the scattering probability, thereby increasing the opposing radiation pressure force. The light beam propagating left to right is shifted further into the wing of the absorption profile. If kv_z is small compared to $\Delta\omega$, we can expand the force expression in a Taylor expansion, and the net force acting on the atom will be

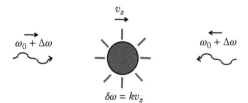

Figure 11.6 *Schematic situation for 1-D atom-optical cooling. Two light beams tuned $\Delta\omega$ to the red of the resonance frequency ω_0 counter-propagate along z and scatter from the atom moving along $+z$ with velocity v_z. The Doppler light shift experienced by the atom of the light beam propagating from the right is $+\delta\omega = kv_z$; from the left, $-\delta\omega$.*

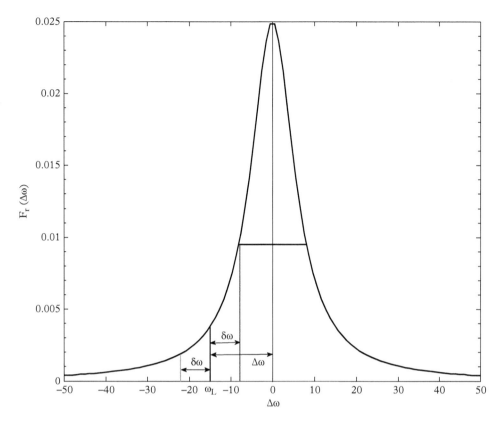

Figure 11.7 *Absorption profile for the atom moving with velocity $+v_z$ in the presence of two counter-propagating light beams tuned $\omega_L = \omega_0 - \Delta\omega \simeq -\Gamma/2$. The blue Doppler shift $+\delta\omega$ (vertical line right of ω_L) shifts absorption closer to line centre. The red Doppler shift $-\delta\omega$ (shortest vertical line to the left of ω_L) shifts absorption further into the absorption wing.*

$$F_{net} \simeq \frac{\hbar k}{6} |\Omega_0|^2 \left(\frac{\Gamma}{2}\right) \left[\frac{4kv_z \Delta\omega}{\left(\Delta\omega^2 + \left(\frac{\Gamma}{2}\right)^2 + \frac{1}{2}|\Omega_0|^2\right)^2} \right] \qquad (11.105)$$

This expression can be simplified somewhat by setting $\Delta\omega \simeq \Gamma/2$, which in practical experiments is close to reality, and defining an on-resonant saturation parameter as

$$S_0 = \frac{|\Omega_0|^2}{\Gamma^2/2} \qquad (11.106)$$

so that

$$F_{net} \simeq \frac{\hbar k}{3} S_0 \left(\frac{1}{\Gamma}\right) \left[\frac{8kv_z \Delta\omega}{(2 + S_0)^2} \right] \qquad (11.107)$$

Furthermore, in practical cases atoms subject to this radiation pressure force are spin-polarised so that we can drop the averaging over the transition dipole orientation:

$$F_{net} \simeq \hbar k \left(\frac{8kv_z \Delta \omega}{\Gamma} \right) \left[\frac{S_0}{(2 + S_0)^2} \right] \qquad (11.108)$$

We see that when the detuning is negative (red detuned), the Doppler-modified radiation pressure force acts in a direction opposite to the atom propagation and is proportional to the atomic velocity. Figure 11.8 shows how, within a relatively narrow range of atom motion kv, the radiation pressure force acts in a direction opposite to the motion, while Figure 11.7 shows this restoring force also increases the dissipative light scattering. The atom undergoes an oscillatory motion along $\pm z$ subject to a velocity-dependent restoring force. This situation corresponds to the damped harmonic oscillator expressed by Equation 11.13. The damping constant β is given by

$$\beta = 8\hbar k^2 \frac{\Delta \omega}{\Gamma} \frac{S_0}{(2 + S_0)^2} \qquad (11.109)$$

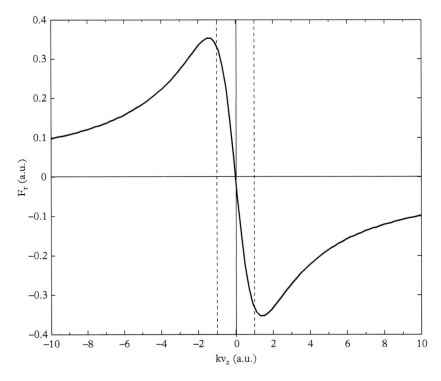

Figure 11.8 *Restoring force plotted against kv_z. Note the approximate linear dependence of F_r when kv_z does not deviate too far from zero such that the Doppler shift remains within an interval $\simeq \Gamma/2$. This linear range is indicated by the vertical dashed lines.*

Assuming a typical red detuning of $\Gamma/2$, we have

$$\beta = 4\hbar k^2 \frac{S_0}{(2 + S_0)^2} \tag{11.110}$$

and further assuming that the transition is far from saturation,

$$\beta \simeq \hbar k^2 S_0 \tag{11.111}$$

The time to damp the atom motion of m is

$$\tau = \frac{m}{2\beta} \tag{11.112}$$

The frictional force removes kinetic energy from the atom and dissipates it as spontaneous emission. The atom can be said to be 'cooled' by this dissipative light force, the cooling rate being the inverse of the damping time:

$$\rho_{cool} \simeq \frac{2\hbar k^2}{m} S_0 \tag{11.113}$$

The atom does not cool indefinitely however. The cooling rate must be balanced by the heating rate arising from the recoil impulsion each time the atom absorbs a photon. This heating rate is related to the fluctuation coefficient of the radiation pressure force[2], $\mathscr{D}$,

$$\mathscr{D} = 3\hbar^2 k^2 \Gamma S_0 \tag{11.114}$$

When the heating and cooling rates are balanced, we find the effective steady-state temperature for the atom:

$$T_{eff} = \frac{\mathscr{D}}{3k_B m \rho_{cool}} \tag{11.115}$$

$$T_{eff} = \frac{1}{2} \frac{\hbar \Gamma}{k_B} \tag{11.116}$$

where k_B is the Boltzmann constant. This effective temperature is called the 'Doppler cooling limit' and corresponds to a temperature of several hundred micro-Kelvin for alkali atoms such as sodium or rubidium. In fact, because these atoms are not true two-level atoms, other cooling mechanisms involving the hyperfine levels can drive the steady-state temperature much lower than the Doppler cooling limit. Discussion of these sub-Doppler cooling mechanisms is, however, beyond the scope of this chapter.

---

[2] The fluctuation force is discussed at length in the thesis of J. Dalibard, referenced in *Further reading* in Section 11.8.

11.6 Summary

The chapter begins with the atom as a damped harmonic oscillator. The next step is to consider the more realistic model of the semiclassical two-level atom where we find the transition rate between levels and from which we can calculate the Einstein B_{12} coefficient, the spontaneous emission loss, and develop expressions from the polarisation and susceptibilities (real and imaginary parts). Light-atom forces (dipole gradient and radiation pressure) are then expressed in terms of the susceptibilities and field or intensity gradients (dipole-gradient force) or phase gradient (radiation pressure force). The question is again examined from the perspective of the optical Bloch equations and the saturation parameter S. The chapter ends with a discussion of the simplest atom-optical cooling process–Doppler cooling.

11.7 Exercises

1. Given that the spontaneous emission lifetime $\tau = 1/\Gamma$ of the sodium atom $Na(3p)\,{}^2P_{3/2}$ level is 16 ns, calculate the magnitude of the transition dipole moment of the $Na(3s)\,{}^2S_{1/2} \to Na(3p)\,{}^2P_{3/2}$ transition.

2. The wavelength of the resonant $Na(3s)\,{}^2S_{1/2} \to Na(3p)\,{}^2P_{3/2}$ transition in atomic sodium is 589 nm. Calculate the transverse dipole gradient force on a Na atom illuminated by a Gaussian mode laser focused to a beam waist of $1\,\mu m$ and detuned to the red of the resonance frequency by $10\,\Gamma$. Take the light source to be a single-mode, continuous-wave laser with an average power of 100 mW.

3. Calculate the trap depth transverse potential energy well (in units of $k_B T$) for Exercise 2. What is the maximum atom temperature that would still permit atom confinement?

4. Calculate the restoring force on a Na atom moving in the $+z$ direction with velocity $v = 10\,\mathrm{ms}^{-1}$. Assume a light detuning $\Delta\omega = -\Gamma/2$ and a saturation parameter $S_0 = 1$.

5. Calculate the Doppler cooling limiting temperature for a Na atom and the time required to cool the atom from ambient temperature to this limit.

11.8 Further reading

1. J.-P Pérez, R. Carles, and R. Fleckinger, *Électromagnétism, Fondements et Applications, 3ème édition*, Masson (1997).

2. C. Cohen-Tannoudji, B. Diu, and F. Laloé, *Quantum Mechanics*, Wiley (1977).

3. J. Weiner, P.-T. Ho, *Light-Matter Interaction: Fundamentals and Applications*, Wiley-Interscience (2003).

4. John Weiner, *Cold and Ultracold Collisions in Quantum Microscopic and Mesoscopic Systems*, Cambridge (2003).

5. Jean Dalibard, *Le Role des Fluctuations dans la Dynamique d'un Atome Couplé au Champ Électromagnétique*, Thèse de Doctorat d'Etat, Université Pierre et Marie Curie (1986).

12

Radiation in Classical and Quantal Atoms

12.1 Introduction

The hydrogen atom is the simplest atomic structure interacting with light in the optical regime. We know from the earliest days of atomic spectroscopy that the hydrogen atom emits and absorbs dipole radiation. But the emitted and radiated light only appears at certain frequencies; the hydrogen atom does not exhibit a continuous spectrum as the classical dipole antenna. The reason is that the atom is a quantum object, not a classical dipole. Nevertheless, a 'semiclassical' picture of the light–atom interaction is possible. In this picture, the electron, with dynamics defined by the quantal nature of atomic internal states, still acts as a receiving antenna or a source of classical (usually dipole) radiation. This hybrid picture of quantal matter and classical electrodynamics captures a great deal of the atom-optical physics. Non-relativistic quantum mechanics and the classical radiation field combine to explain almost all the features of the hydrogen atom spectrum. We will use these tools to develop an explanation for the quantisation of the energy levels of the atom and the 'selection rules' for dipole transitions between them.

12.2 Dipole emission of an atomic electron

Here for convenience we summarise some of the discussion in Section 11.2.

In Section 2.8 we considered a classical electric dipole aligned along the z-axis and wrote the dipole as

$$\boldsymbol{p}(t) = q_0 a \cos \omega t \, \hat{\boldsymbol{z}}$$

If we consider an atom as an electron bound to a massive proton, subject to a radial restoring force centred at the proton position, then we can write down a similar dipole expression,

$$\boldsymbol{p}(t) = -e\boldsymbol{r}(t) \qquad \boldsymbol{p}_0 = -e\boldsymbol{r}_0 \qquad (12.1)$$

Light-Matter Interaction. Second Edition. John Weiner and Frederico Nunes.

where $-e$ is the electron charge, r is the radial distance from the atom centre, and r_0 is the maximum radial distance. The electron is pictured as vibrating radially through the proton position and in this simple model exhibits no angular momentum from orbital motion about the charge centre. Using the result for power emitted by a classical dipole, Equation 2.157, we can write the optical-cycle-averaged power emitted by the atom as

$$\langle W \rangle = \frac{1}{4\pi\varepsilon_0} \frac{\omega_0^4 p_0^2}{3c^3} = \frac{e^2}{4\pi\varepsilon_0} \frac{\omega_0^4 r_0^2}{3c^3} \tag{12.2}$$

For harmonic oscillation the maximum acceleration is $a_0 = -\omega_0^2 r_0$ and

$$\langle W \rangle = \frac{e^2}{4\pi\varepsilon_0} \frac{a_0^2}{3c^3} \tag{12.3}$$

Writing the cycle-averaged acceleration as $\langle a \cdot a \rangle = \langle a^2 \rangle = (1/2)a_0^2$, we have

$$\langle W \rangle = \frac{e^2}{4\pi\varepsilon_0} \frac{2\langle a^2 \rangle}{3c^3} \tag{12.4}$$

and we see that the average power emitted by the oscillating electron is proportional to the average square of the electron acceleration. A charged particle must be accelerating to emit radiation; an electron moving at constant velocity, not subject to external forces, does not emit.

The energy (cycle-averaged) emitted by the oscillator must come from mechanical (kinetic + potential) energy, $\mathscr{E}_{\text{mech}}$:

$$\mathscr{E}_{\text{mech}} = \frac{1}{2}m_e\omega_0^2 r_0^2 \tag{12.5}$$

where m_e is the electron mass. The average power emitted by the oscillator can be equated to the diminution of mechanical energy with time:

$$\langle W \rangle = -\frac{\mathscr{E}_{\text{mech}}}{\tau} \tag{12.6}$$

From Equations 12.2 and 12.5, and 12.6 we see that

$$\tau = \left(\frac{e^2}{4\pi\varepsilon_0}\right)^{-1} \frac{3m_e c^3}{2\omega_0^2} \tag{12.7}$$

The 'lifetime' of the oscillator τ can be written a little more simply by introducing the *classical radius* of the electron,

$$r_c = \frac{e^2}{4\pi\varepsilon_0} \frac{1}{m_e c^2} \simeq 2.8 \times 10^{-15} \quad \text{m} \tag{12.8}$$

and

$$\tau = \frac{3c}{2r_c\omega_0^2} \qquad (12.9)$$

In differential form Equation 12.6 is written as

$$\langle W \rangle = -\frac{d\mathscr{E}_{mech}}{dt}$$

so

$$\frac{d\mathscr{E}_{mech}}{\mathscr{E}_{mech}} = -\frac{dt}{\tau} \qquad (12.10)$$

and

$$\mathscr{E}_{mech}(t) = \mathscr{E}_{mech}(0)e^{-t/\tau} \qquad (12.11)$$

We see that the internal energy of the atom, modelled as a classical electron oscillating around a positive charge centre, decreases exponentially with time as the energy is radiated away.

A specific example provides some feeling for the relevant time and energy scales. Consider a sodium atom, structured with 10 electrons in a 'closed shell' and one electron outside. This exposed or 'active' electron is subject to an effective positive charge at the nucleus roughly equal to that of one proton. The potential and electric field experienced by the active electron is radial and the simple dipole oscillator model might therefore be appropriate. Emission from the first excited state to the ground state is concentrated in a narrow band of wavelengths centred at 590 nm, the well-known 'D line' of atomic spectroscopy. Considering this emission as emanating from a dipole oscillator with resonant frequency ω_0,

$$\omega_0 = \frac{2\pi c}{\lambda} = 3.2 \times 10^{15} \quad s^{-1}$$

and

$$\tau = \frac{3c}{2r_c\omega_0^2} \simeq 16 \times 10^{-9} \quad s$$

This result is actually quite close to the measured lifetime of the first excited state of the sodium atom and lends credence to the harmonic oscillator model. In reality, however, the atom must be treated quantum mechanically because there is no lower limit to the classical radiative emission, and the classical electron would fall into the nucleus as its mechanical energy diminishes to zero. Furthermore, the first excited state with a quantum-mechanically 'allowed' transition possesses one quantum of orbital angular momentum, and this angular momentum does not appear in the simple oscillator model.

12.3   Radiative damping and electron scattering

Nevertheless, the damped harmonic oscillator model provides some insight into the physics of atomic emission, absorption of radiation, and atomic scattering. We can write down the equation of motion of the classical, radiatively damped, externally driven harmonic oscillator, moving along the z-axis, as

$$\frac{d^2z}{dt^2} + \frac{1}{\tau}\frac{dz}{dt} + \omega_0^2 z = -\frac{eE_z}{m_e}e^{-i\omega t} \tag{12.12}$$

where as before ω_0 is the characteristic frequency of the oscillator, e, m_e are the electron charge and mass, respectively, τ the classical oscillator lifetime, and E_z is the amplitude of the external driving field oscillating at frequency ω. The solution to this equation of motion is

$$z = \frac{-eE_z e^{-i\omega t}}{m_e\left[\left(\omega_0^2 - \omega^2\right) - i\omega/\tau\right]} \tag{12.13}$$

From this result we can write down the frequency-dependent dipole moment,

$$p = -ez\hat{z} = \frac{-eE_z e^{-i\omega t}}{m_e\left[\left(\omega_0^2 - \omega^2\right) - i\omega/\tau\right]}\hat{z} \tag{12.14}$$

From Equation 12.2 we have for the average dissipated power,

$$\langle W \rangle = \frac{e^4 E_z^2 \omega^4}{12\pi\,\varepsilon_0 m_e^2\left[\left(\omega_0^2 - \omega^2\right)^2 + (\omega/\tau)^2\right]} \tag{12.15}$$

Using the classical electron radius, Equation 11.9, we can write this expression for the dissipated power as the product of an energy flux and a scattering cross section,

$$\langle W \rangle = \underbrace{\left(\frac{\varepsilon_0 E_z^2}{2}\right) c}_{\text{energy flux}} \cdot \underbrace{\left\{\left[\frac{8\pi\,r_c^2}{3}\right]\left[\frac{\omega^4}{\left(\omega_0^2 - \omega^2\right)^2 + (\omega/\tau)^2}\right]\right\}}_{\text{cross section}} \tag{12.16}$$

where the scattering cross section is defined as

$$\sigma(\omega) = \left[\frac{8\pi\,r_c^2}{3}\right]\left[\frac{\omega^4}{\left(\omega_0^2 - \omega^2\right)^2 + (\omega/\tau)^2}\right] \tag{12.17}$$

12.4   The Schrödinger equation for the hydrogen atom

The basic expression for the Schrödinger equation of any atomic system is described by an operator equation,

$$\hat{H}\Psi_n(r,t) = i\hbar\frac{\partial \Psi_n(r,t)}{\partial t} \qquad (12.18)$$

where $\Psi_n(r,t)$ represents the state n of the atom, and $\hat{H}$ is the Hamiltonian operator, $\hat{H} = i\hbar\partial/\partial t$. According to the postulates of quantum mechanics, a characteristic or 'eigen' state of a system must be represented by a wave function that is single-valued, normalised, and orthogonal to all the other eigenstates of the system. An arbitrary system state may be expressed as a linear combination of these eigenstates that form a 'complete set' in a linear, orthogonal vector space. If the state is 'stationary' or time independent, as it is for the isolated atom, we may write the state function as a product of a spatial term $\psi(r)$ and a time-varying phase, $\exp(-i\omega t)$:

$$\Psi(r,t) = \psi(r)e^{-i\omega t} \qquad (12.19)$$

The frequency of the phase is related to the energy E_n of the stationary state by the Planck relation, $E_n = \hbar\omega$. The spatial, time-independent part can factor, and be separated by, substitution of Equation 12.19 into Equation 12.18 with the result

$$\hat{H}\,[\psi(r) - E_n] = 0 \qquad (12.20)$$

where $\psi(r)$ is the spatial wave function that is the solution to the time-independent Schrödinger equation, Equation 12.20. Now the Hamiltonian operator $\hat{H}$ is constructed from the *Bohr correspondence principle* that permits us to identify certain mathematical operators with their dynamical variable counterparts in the Hamiltonian formulation of classical mechanics. The classical Hamiltonian H expresses the energy of a system. For example,

$$H = T + V \qquad (12.21)$$

where T is the kinetic energy and V the potential energy. The correspondence principle is a rule that indicates the equivalence of these classical functions to the operators representing them in quantum mechanics. For example,

$$H \rightarrow \hat{H} \qquad T \rightarrow \hat{T} \qquad V \rightarrow \hat{V} \qquad (12.22)$$

The kinetic and potential energies themselves are functions of position and momentum, which correspond to the operators $\hat{r}$ and $\hat{p}$, respectively. The 'operator' for the position is the position itself,

$$\hat{r} \rightarrow r \qquad (12.23)$$

and the operator for the momentum is

$$\hat{p} \rightarrow -i\hbar\nabla \tag{12.24}$$

Therefore the operator expression for the kinetic energy of a particle with mass m becomes

$$T = \frac{p^2}{2m} \rightarrow \hat{T} = -\frac{\hbar^2}{2m}\nabla^2 \tag{12.25}$$

and if the potential energy is a function only of position, then

$$V(r) \rightarrow \hat{V}(\hat{r}) \tag{12.26}$$

From Equations 12.25 and 12.26 we can construct the Hamiltonian operator and the Schrödinger operator equation for the energy of the system,

$$\left[\hat{H} - E_n\right]\psi = 0 \rightarrow \left[-\frac{\hbar^2}{2m}\nabla^2 + V(r) - E_n\right]\psi = 0 \tag{12.27}$$

The physically allowed solutions to the time-independent Schrödinger equation, Equation 12.27, constitute the 'states' of the physical system described by the energy terms of the Hamiltonian. In the case of a hydrogen atom, the system is quite simple, consisting of a single proton and a single electron bound together by the electrostatic Coulomb potential. Strictly speaking, this situation is a two-particle problem, and the kinetic energy term in Equation 12.27 should be the sum of two terms for the kinetic energies of the proton and electron. However, the rest mass of the proton is greater than that of the electron by a factor of nearly 2000, and it is convenient to consider the 'reduced mass' μ of the equivalent single-particle problem:

$$\mu = \frac{m_e m_p}{m_e + m_p} \simeq m_e \tag{12.28}$$

and the centre-of-mass (cm) coordinate,

$$r_{cm} = \frac{m_p r_p + m_e r_e}{m_p + m_e} \simeq r_p \tag{12.29}$$

So to a very good approximation we can consider a model system where the reduced mass can, essentially, be considered that of the electron and the centre-of-mass coordinate considered that of the proton. By placing the origin of the coordinate system at the centre-of-mass position we simplify the problem to a single electron moving about a stationary positive charge centred at the origin. For all practical purposes, therefore, we can write the time-independent Schrödinger equation for the hydrogen atom as

$$-\frac{\hbar^2}{2m_e}\nabla^2\psi(x,y,z) - \frac{1}{4\pi\varepsilon_0}\frac{e^2}{\left(x^2+y^2+z^2\right)^{1/2}}\psi(x,y,z) - E_n\psi(x,y,z) = 0 \qquad (12.30)$$

where the Laplacian operator ∇^2 in Cartesian coordinates is

$$\nabla^2 = \frac{\partial^2}{\partial x^2} + \frac{\partial^2}{\partial y^2} + \frac{\partial^2}{\partial z^2} \qquad (12.31)$$

and the Coulomb potential is

$$V = -\frac{1}{4\pi\varepsilon_0}\frac{e^2}{\left(x^2+y^2+z^2\right)^{1/2}} \qquad (12.32)$$

Since the Coulomb potential term exhibits spherical symmetry, Equation 12.30 can be solved far more easily in spherical coordinates, r, θ, ϕ, where,

$$x = r\sin\theta\cos\varphi \qquad (12.33)$$
$$y = r\sin\theta\sin\varphi \qquad (12.34)$$
$$z = r\cos\theta \qquad (12.35)$$
$$r = \left(x^2+y^2+z^2\right)^{1/2} \qquad (12.36)$$

and the Laplacian operator ∇^2 is expressed as

$$\nabla^2 = \frac{1}{r^2}\frac{\partial}{\partial r}\left(r^2\frac{\partial}{\partial r}\right) + \frac{1}{r^2\sin^2\theta}\frac{\partial^2}{\partial\varphi^2} + \frac{1}{r^2\sin\theta}\frac{\partial}{\partial\theta}\left(\sin\theta\frac{\partial}{\partial\theta}\right) \qquad (12.37)$$

The Schrödinger equation in spherical coordinates takes the form

$$-\frac{\hbar^2}{2m_e}\nabla^2\psi(r,\theta,\varphi) - \frac{1}{4\pi\varepsilon_0}\frac{e^2}{r}\psi(r,\theta,\varphi) - E_n\psi(r,\theta,\varphi) = 0 \qquad (12.38)$$

The relation between Cartesian and spherical coordinates and the representation of the Laplacian operator in spherical coordinates is discussed in Appendix D. In spherical coordinates the wave function solution to Equation 12.30 factors into three functions of each coordinate,

$$\psi(r,\theta,\varphi) = R(r)P(\theta)\Phi(\varphi) \qquad (12.39)$$

Substituting this form back into Equation 12.38 results in the separation into three differential equations in r, θ, φ, respectively:

$$\frac{1}{r^2}\frac{d}{dr}\left(r^2\frac{dR_{nl}}{dr}\right) + \frac{2\mu}{\hbar^2}[E - V(r)]R_{nl} = l(l+1)\frac{R_{nl}}{r^2} \tag{12.40}$$

$$-\frac{1}{\sin\theta}\frac{d}{d\theta}\left(\sin\theta\frac{dP_l}{d\theta}\right) + \frac{P_l}{\sin^2\theta} = \frac{l(l+1)}{m_l^2}P_l \tag{12.41}$$

$$\frac{d^2\Phi}{d\varphi^2} = -m_l^2\Phi \tag{12.42}$$

The expressions $l(l+1)$ and $-(m_l)^2$ are constants of separation.

12.4.1 Radial equation

The independent variable of the radial equation, Equation 12.40, is r, the distance of the electron from the origin. The physically admissible solutions are labelled by the quantum number integers n, l. By 'physically admissible' we mean solutions so that $R_{nl} \to 0$ as $r \to \infty$ and R_{nl} remains finite as $r \to 0$. There are in fact a family of admissible solutions, each member of which is identified by a unique n, l label. The n quantum number is called the principal quantum number and is directly identified with the total energy of the atom:

$$E_n = \frac{\mu e^4}{(4\pi\varepsilon_0)^2\,2\hbar^2 n^2} \simeq -\frac{m_e e^4}{(4\pi\varepsilon_0)^2\,2\hbar^2 n^2} \tag{12.43}$$

where n can take on the integer values $n = l, l+1, l+2, \ldots$ and l can be an integer $l = 0, 1, 2, \ldots$. The radial solutions take the form of three factors: a normalisation factor, a polynomial, and an exponential. In fact, Equation 12.40 can be rearranged into a form that bears a striking resemblance to the differential equation whose solutions are the 'modified Laguerre functions'. The rearranged form is obtained by multiplying Equation 12.40 by $-2\mu/\hbar^2$ and transposing two terms:

$$-\frac{\hbar^2}{2\mu}\frac{1}{r^2}\frac{d}{dr}\left(r^2\frac{dR_{nl}}{dr}\right) - \frac{e^2}{4\pi\varepsilon_0 r}R_{nl} + \frac{\hbar^2}{2\mu}\frac{l(l+1)}{r^2}R_{nl} = E_n R_{nl} \tag{12.44}$$

The differential equation for the modified Laguerre functions and its solutions, Equations F.18 and F.19, are discussed in Appendix F. Now if we make the following substitutions, Equation 12.44 can be cast into the form of Equation F.19:

$$\rho = \alpha r \quad \text{and} \quad \alpha = \sqrt{-\frac{8\mu E}{\hbar^2}} \tag{12.45}$$

where it is understood that $E < 0$ since we only consider the allowed bound-state energies of the hydrogen atom, and the reference zero of energy is at the ionisation limit $r \to \infty$. The quantity α has units of m^{-1} so ρ is unitless. We make an additional substitution to consolidate many constants,

$$\lambda = \frac{\mu e^2}{2\pi \varepsilon_0 \alpha \hbar^2} \qquad (12.46)$$

In order for the λ substitution to be consistent with the physically admissible values of the energy E_n, λ must be restricted to integers, $\lambda = 1, 2, 3, \ldots$, and defining $\chi(\rho) = R(\rho/\alpha)$, we can recast Equation 12.44 as,

$$\frac{1}{\rho^2} \frac{d}{d\rho} \left(\rho^2 \frac{d\chi(\rho)}{d\rho} \right) + \left(\frac{\lambda}{\rho} - \frac{1}{4} - \frac{l(l+1)}{\rho^2} \right) \chi(\rho) = 0 \qquad (12.47)$$

Using one last substitution trick,

$$\frac{1}{\rho^2} \frac{d}{d\rho} \rho^2 \frac{d\chi}{d\rho} = \frac{1}{\rho} \frac{d^2}{d\rho^2} (\rho \chi) \qquad (12.48)$$

we have, finally,

$$\frac{d^2(\rho\chi)}{d\rho^2} + \left(-\frac{1}{4} + \frac{\lambda}{\rho} - \frac{l(l+1)}{\rho^2} \right) \rho\chi = 0 \qquad (12.49)$$

Now we compare this last result to Equation F.19 rewritten here for convenience,

$$\frac{d^2 \Phi_n^k(x)}{dx^2} + \left(-\frac{1}{4} + \frac{2n+k+1}{2x} - \frac{k^2-1}{4x^2} \right) \Phi_n^k(x) = 0$$

We see that Equation 12.49 becomes equivalent to Equation F.19 if

$$\rho \leftrightarrow x \qquad l \leftrightarrow \frac{k-1}{2} \qquad \lambda \to n+l+1 \qquad (12.50)$$

Then the solutions to Equation 12.49 are

$$\rho\chi(\rho) = e^{-\rho/2} \rho^{l+1} L_{\lambda-l-1}^{2l+1}(\rho) \qquad (12.51)$$

equivalent to the solutions for Equation F.19:

$$\Phi_n^k(x) = e^{-x/2} x^{(k+1)/2} L_n^k(x) \qquad (12.52)$$

where $L_n^k(x)$ are the associated Laguerre polynomials. Thus, Equation 12.51 expresses the solutions to Equation 12.49 that in turn is just the hydrogen radial equation, Equation 12.40, in a different guise. Now we recover from the scaling parameters α and ρ

terms of physically meaningful quantities. From the expression for the allowed bound energies E_n of the hydrogen atom, Equation 12.43, we can write α and ρ in terms of n the principal quantum number:

$$\alpha = \frac{2}{na_0} \quad \text{and} \quad \rho = \frac{2}{na_0}r \quad\quad (12.53)$$

where $a_0 = 4\pi\varepsilon_0\hbar^2/m_ee^2$, the well-known atomic radius of the Bohr model. Finally, we can write out the solutions to the hydrogen radial equation, Equation 12.40,

$$R_{nl}(r) = (\alpha r)^l L_{n-l-1}^{2l+1}(\alpha r)e^{-\alpha r/2} \quad\quad (12.54)$$

In quantum mechanics the solutions to the Schrödinger equation play an analogous role to force fields in the Maxwell equations. Although the wave function is not a vector function, it is, in general, complex. Just as in electrodynamics, the force field F itself is not observed but the *intensity*, FF^*, guaranteed to be a real quantity, is the detected signal, so in quantum mechanics it is the *probability density*, $\psi\psi^*$ that connects theory to experiment. Note that EE^* and HH^* are directly proportional to the energy density of an electromagnetic field. It is therefore plausible to assign $\psi\psi^*$ the role of probability density. When ψ is a function only of space and is the solution to a single-particle Schrödinger equation, then $\psi(r)\psi^*(r)$ is interpreted as the probability density of finding the particle at position r. Evidently then,

$$\int_{\text{all space}} \psi(r)\psi^*(r)\, dr = 1 \quad\quad (12.55)$$

and when this condition is fulfilled, the wave function is said to be normalised. In order to be consistent with this probability density interpretation, the functions $R_{nl}(r)$ must be prefixed by a constant normalising factor $\sqrt{N_{R_{nl}}}$ such that

$$\frac{1}{N_{R_{nl}}} \int_0^\infty R_{nl}^*(r)R_{nl}(r)r^2\, dr = 1 \qu\quad (12.56)$$

It can be shown from the normalisation condition of the associated Laguerre functions, Equation F.16 in Appendix F, that

$$N_{R_{nl}} = \left(\frac{na_0}{2}\right)^3 \frac{2n(n+l)!}{(n-l-1)!} \quad\quad (12.57)$$

so that, finally, the normalised radial solutions to the hydrogen atom Schrödinger equation are:

$$R_{nl}(r) = \left[\left(\frac{2}{na_0}\right)^3 \frac{(n-l-1)!}{2n(n+l)!}\right]^{1/2} \left(\frac{2}{na_0}r\right)^l L_{n-l-1}^{2l+1} e^{-(2/na_0)r/2} \quad\quad (12.58)$$

Table 12.1 *Hydrogen radial wave functions R_{nl}, $n=1-3;l=0$*

n	l	R_{nl}
1	0	$\left(\frac{1}{a_0}\right)^{3/2} 2e^{-r/a_0}$
2	0	$\left(\frac{1}{2a_0}\right)^{3/2}\left[2-\frac{r}{a_0}\right]e^{-r/2a_0}$
3	0	$\left(\frac{1}{3a_0}\right)^{3/2} 2\left[1-2\left(\frac{r}{a_0}\right)+\frac{2}{27}\left(\frac{r}{a_0}\right)^2\right]e^{-r/3a_0}$

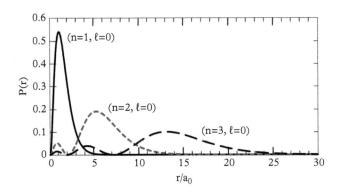

Figure 12.1 *Normalised radial probability density as a function of the radial coordinate (in units of a_0) for three radial wave functions.*

Figure 12.1 shows the normalised radial probability density functions,

$$P(r) = R_{nl}^* R_{nl}\, r^2 \tag{12.59}$$

for the hydrogen atom ground state $n = 1, l = 0$ and two excited states. Table 12.1 lists the radial functions, R_{nl} used to plot the probability densities of Figure 12.1.

12.4.2 Angular equations

12.4.2.1 *Functions of the azimuthal angle φ*

The solutions to Equation 12.41 are a family of functions of the polar angle θ and are usually denoted $P_l(\cos\theta)$ with $\cos\theta$ as the independent variable. This family is called the associated Legendre functions, and their properties are discussed in Appendix G. The solutions to Equation 12.42, the functions of the azimuthal angle φ, can be obtained by inspection. It is easy to verify that

$$\Phi_{m_l}(\varphi) = e^{im_l\varphi} \tag{12.60}$$

According to the postulates of quantum mechanics, the solution to a differential equation must be single-valued in the independent variable in order for the function to suitably represent a physical state. Therefore, as φ cycles through integer multiples of 2π, increasing from $0 \to 2\pi$, $2\pi \to 4\pi$, ... the value of the function Φ_{m_l} should remain single-valued. Thus, $\Phi_{m_l}(\varphi)$ must have the same value as $\Phi_{m_l}(\varphi')$ if $\varphi' = \varphi + n2\pi$ where $n = 1, 2, 3, \dots$. In order to assure this behaviour, the value of m_l must be an integer, $m_l = \pm 1, \pm 2, \pm 3, \dots$. Furthermore, Φ_{m_l} is subject to a normalisation requirement analogous to that of Equation 12.56:

$$\frac{1}{N} \int_0^{2\pi} \Phi_{m_l}^*(\varphi) \Phi_{m_l}(\varphi) \, d\varphi = 1 \tag{12.61}$$

The normalised azimuthal wave function is therefore

$$\Phi_{m_l}(\varphi) = \frac{1}{\sqrt{2\pi}} e^{i(\pm m_l)\varphi} \tag{12.62}$$

12.4.2.2 Functions of the polar angle θ

The solution to Equation 12.41 constitutes a family of functions, $P_l^{m_l}(\cos\theta)$ called the associated Legendre functions. The properties of the Legendre functions and the Legendre polynomials, the relevant functions when $m_l = 0$, are discussed in some detail in Appendix G. These functions describe the polar-angle dependence of the hydrogen atom wave function, $\psi(r, \theta, \varphi) = R_{nl} P_l^{m_l}(\cos\theta) \Phi_{m_l}(\varphi)$. Just as R_{nl} and Φ_{m_l} must be normalised, so must the functions $P_l^{m_l}(\cos\theta)$. The normalisation requirement specifies that

$$\frac{1}{N} \int_0^{\pi} P_l^{*m_l}(\cos\theta) P_l^{m_l}(\cos\theta) \sin\theta \, d\theta = 1 \tag{12.63}$$

with N expressed by

$$N = \frac{2}{2l + 1} \cdot \frac{(l + m_l)!}{(l - m_l)!} \tag{12.64}$$

The family of functions that are the product of $P_l(\cos\theta)$ and $\Phi(\varphi)$ is called the *spherical harmonics*, $Y_l^{m_l}(\theta, \varphi)$. The normalised spherical harmonics are

$$Y_l^{m_l}(\theta, \varphi) = (-1)^{m_l} \sqrt{\frac{2l + 1}{4\pi} \cdot \frac{(l - m_l)!}{(l + m_l)!}} P_l^{m_l}(\cos\theta) e^{im_l\varphi} \tag{12.65}$$

and

$$\int_0^{\pi} \int_0^{2\pi} Y_l^{*m_l}(\theta, \varphi) Y_l^{m_l}(\theta, \varphi) \sin\theta \, d\theta \, d\varphi = 1 \tag{12.66}$$

In order to maintain physically admissible, well-behaved functions there are certain restrictions on the function labels, n, l, m_l. We summarise them here without proof:

$$n = 1, 2, 3, \ldots \tag{12.67}$$
$$l = 0, 1, 2, \ldots n-1 \tag{12.68}$$
$$m_l = -l, -l+1, -l+2, \ldots 0 \ldots, l-2, l-1, l \tag{12.69}$$

Table 12.2 lists the spherical harmonic functions in spherical coordinates up to $n = 3$.

Putting the spherical harmonic angular wave functions together with the radial wave functions gives us, finally, the full three-dimensional wave functions that specify that quantum state of the hydrogen atom:

$$\psi(r, \theta, \varphi)_{nlm_l} = R_{nl}(r) Y_{lm_l}(\theta, \varphi) \tag{12.70}$$

The expression $\psi^*_{nlm_l}(r, \theta \varphi) \psi_{nlm_l}(r, \theta, \varphi)$ is the probability density of finding the electron at r.

$$r = r\hat{r} \qquad \text{polar coordinates} \tag{12.71}$$
$$r = x\hat{x} + y\hat{y} + z\hat{z} \qquad \text{Cartesian coordinates} \tag{12.72}$$

The probability dP of finding the electron in a volume interval between V and $V + dV$ is

$$dP = \psi^*_{nlm_l}(r, \theta \varphi) \psi_{nlm_l}(r, \theta, \varphi)\, dV \tag{12.73}$$
$$dP = \psi^*_{nlm_l}(r, \theta \varphi) \psi_{nlm_l}(r, \theta, \varphi)\, r^2 \sin\theta\, d\theta\, d\varphi\, dr \tag{12.74}$$

Table 12.2 *Angular functions, spherical harmonics $Y_l^{m_l}$.*

n	l	m_l	$Y_l^{m_l}$
1	0	0	$\frac{1}{2}\sqrt{\frac{1}{\pi}}$
2	1	-1	$\frac{1}{2}\sqrt{\frac{3}{2\pi}} \sin\theta e^{-\varphi}$
2	1	0	$\frac{1}{2}\sqrt{\frac{3}{2\pi}} \cos\theta$
2	1	1	$-\frac{1}{2}\sqrt{\frac{3}{2\pi}} \sin\theta e^{\varphi}$
3	2	-2	$\frac{1}{4}\sqrt{\frac{15}{2\pi}} \sin^2\theta e^{-2\varphi}$
3	2	-1	$\frac{1}{2}\sqrt{\frac{15}{2\pi}} \sin\theta \cos\theta e^{-\varphi}$
3	2	0	$\frac{1}{4}\sqrt{\frac{5}{\pi}} \left(\cos^2\theta - 1\right)$
3	2	1	$-\frac{1}{2}\sqrt{\frac{15}{2\pi}} \sin\theta \cos\theta e^{i\varphi}$
3	2	2	$\frac{1}{4}\sqrt{\frac{15}{2\pi}} \sin^2\theta e^{i2\varphi}$

Table 12.3 *Hydrogen wave functions* $R_{nl}Y_l^{m_l}$, *n=1 – 3.*

n	l	m_l	R_{nl}	$Y_l^{m_l}$
1	0	0	$\frac{1}{\sqrt{\pi}}\left(\frac{1}{a_0}\right)^{3/2}e^{-r/a_0}$	1
2	0	0	$\frac{1}{4\sqrt{2\pi}}\left(\frac{1}{a_0}\right)^{3/2}\left[2-\left(\frac{r}{a_0}\right)\right]e^{-r/2a_0}$	1
2	1	0	$\frac{1}{4\sqrt{2\pi}}\left(\frac{1}{a_0}\right)^{3/2}\left(\frac{r}{a_0}\right)e^{-r/2a_0}$	$\cos\theta$
2	1	±1	$\frac{1}{8\sqrt{\pi}}\left(\frac{1}{a_0}\right)^{3/2}\left(\frac{r}{a_0}\right)e^{-r/2a_0}$	$\sin\theta e^{\pm i\varphi}$
3	0	0	$\frac{1}{81\sqrt{3\pi}}\left(\frac{1}{a_0}\right)^{3/2}\left[27-18\left(\frac{r}{a_0}\right)+2\left(\frac{r}{a_0}\right)^2\right]e^{-r/3a_0}$	1
3	1	0	$\frac{1}{81}\sqrt{\frac{2}{\pi}}\left(\frac{1}{a_0}\right)^{3/2}\left(6-\frac{r}{a_0}\right)\left(\frac{r}{a_0}\right)e^{-r/3a_0}$	$\cos\theta$
3	1	±1	$\frac{1}{81\sqrt{\pi}}\left(\frac{1}{a_0}\right)^{3/2}\left(6-\frac{r}{a_0}\right)\left(\frac{r}{a_0}\right)e^{-r/3a_0}$	$\sin\theta e^{\pm i\varphi}$
3	2	0	$\frac{1}{81\sqrt{6\pi}}\left(\frac{1}{a_0}\right)^{3/2}\left(\frac{r^2}{a_0}\right)^2 e^{-r/3a_0}$	$3\cos^2\theta-1$
3	2	±1	$\frac{1}{81\sqrt{\pi}}\left(\frac{1}{a_0}\right)^{3/2}\left(\frac{r^2}{a_0}\right)^2 e^{-r/3a_0}$	$\sin\theta\cos\theta e^{\pm i\varphi}$
3	2	±2	$\frac{1}{162\sqrt{\pi}}\left(\frac{1}{a_0}\right)^{3/2}\left(\frac{r^2}{a_0}\right)^2 e^{-r/3a_0}$	$\sin^2\theta e^{\pm i2\varphi}$

and

$$\int_{\text{all space}} dP = \int_0^\infty \int_0^\pi \int_0^{2\pi} \psi^*_{nlm_l}\psi_{nlm_l}r^2\sin\theta\, d\theta\, d\varphi\, dr = 1 \qquad (12.75)$$

Table 12.3 lists the normalised ψ_{nlm_l} solutions to the hydrogen atom Schrödinger equation for $n = 1 - 3$.

12.4.3   Orthogonality

If ψ_{n,l,m_l} represents an eigenstate of the hydrogen atom with an eigen energy and angular momentum, then according to the postulates of quantum mechanics it must be not only single-valued and normalised, but also orthogonal to all other states with different energies or angular momenta. More succinctly,

$$\int_{\text{all space}} \psi^*_{n,l,m_l}\psi_{n',l',m'_l}\, d\tau = \delta_{nn'}\delta_{ll'}\delta_{m_l m'_l} \qquad (12.76)$$

where the δ function takes the value of unity when n, l, m_l are equal to their primed counterparts but is null otherwise. Wave functions that exhibit this property are said to be 'orthonormal' because they are orthogonal and normalised. In polar coordinates the wave functions of the hydrogen atom are products of radial and angular functions,

$$\psi_{n,l,m_l}(r,\theta,\varphi) = R_{nl}(r)\,Y_{lm_l}(\theta,\varphi) = R_{nl}(r)P_l^{m_l}(\cos\theta)\,\Phi_{m_l}(\varphi) \tag{12.77}$$

and therefore these radial and angular functions must be orthonormal as well.

12.5   State energy and angular momentum

Now that we have the functional form of the radial and angular solutions to the Schrö-dinger equation for the hydrogen atom, we can examine their properties in more detail.

12.5.1   Eigen energies

Considering the time-independent Schrödinger equation as an operator equation, Equation 12.20, we see that an eigenstate, when operated upon by the energy operator $\hat{H}$, returns the energy eigenvalue of that state,

$$\hat{H}\psi_{n,l,m_l} = E_n\psi_{n,l,m_l} \tag{12.78}$$

The eigen energies E_n are labelled by the principal quantum number n and are char-acterised by the analytical expression given in Equation 12.43. The constants in this expression can be collected together to facilitate numerical calculation in convenient units:

$$E_n = -\frac{2.18 \times 10^{-18}}{n^2}\text{J} = -\frac{13.6}{n^2}\text{eV} = -\frac{1.097 \times 10^5}{n^2}\text{cm}^{-1} \tag{12.79}$$

The allowed values of n are $n = 1, 2, 3, \ldots$. The lowest lying (ground) state of the atom corresponds to $n = 1$, and $E_1 = -13.6$ eV. The first excited state corresponds to $n = 2$, and $E_2 = -3.40$ eV, and so forth. As $n \to \infty$, $E_n \to 0$, and the zero-reference energy corresponds to the ionisation limit where the electron is no longer bound by the Coulomb potential. Figure 12.2 shows a diagram of the hydrogen atom energy levels bound within the Coulomb potential.

12.5.2   Angular momentum

In addition to energy, the states of the hydrogen atom may exhibit orbital angular mo-mentum. In the classical Bohr model, the electron rotated around the proton in various orbital configurations reminiscent of the planets around a star. The uncertainty prin-ciple no longer permits us to make such a literal interpretation, but the Schrödinger equation does show us how to calculate the orbital angular momentum of the hydro-gen atom states. We can obtain the quantum mechanical operator corresponding to the classical angular momentum by applying the 'correspondence principle' discussed in Section 12.4. From classical mechanics we have, for the angular momentum of the

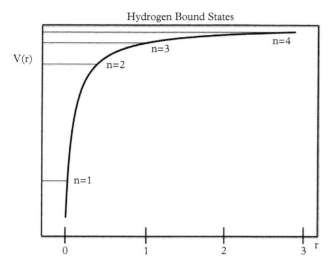

Figure 12.2 *Coulomb potential to which the bound electron in the hydrogen atom is subject. The lowest four bound states (n = 1 – 4) are shown.*

electron L, from the position r, and the linear momentum p,

$$L = r \times p \tag{12.80}$$

and the three Cartesian components:

$$L_x = yp_z - zp_y \tag{12.81}$$
$$L_y = zp_x - xp_z \tag{12.82}$$
$$L_z = xp_y - yp_x \tag{12.83}$$

Then from the correspondence principle,

$$r \to \hat{r} \tag{12.84}$$
$$p \to -i\hbar \nabla \tag{12.85}$$

The operator equivalents of three angular momentum components are therefore

$$\hat{L}_x = -i\hbar \left(y \frac{\partial}{\partial z} - z \frac{\partial}{\partial y} \right) \tag{12.86}$$

$$\hat{L}_y = -i\hbar \left(z \frac{\partial}{\partial x} - x \frac{\partial}{\partial z} \right) \tag{12.87}$$

$$\hat{L}_z = -i\hbar \left(x \frac{\partial}{\partial y} - y \frac{\partial}{\partial x} \right) \tag{12.88}$$

and the operator corresponding to the square of the total orbital angular momentum is

$$\hat{L}^2 = \left(\hat{L}_x^2 + \hat{L}_y^2 + \hat{L}_z^2 \right) \tag{12.89}$$

In polar coordinates, as discussed in Appendix D, we have

$$\hat{L}_x = i\hbar \left[\sin\varphi \frac{\partial}{\partial\theta} + \left(\frac{1}{\tan\theta} \right) \cos\varphi \frac{\partial}{\partial\varphi} \right] \tag{12.90}$$

$$\hat{L}_y = i\hbar \left[-\cos\varphi \frac{\partial}{\partial\theta} + \left(\frac{1}{\tan\theta} \right) \sin\varphi \frac{\partial}{\partial\varphi} \right] \tag{12.91}$$

$$\hat{L}_z = -i\hbar \frac{\partial}{\partial\varphi} \tag{12.92}$$

and Equation 12.89 transforms to

$$\hat{L}^2 = -\hbar^2 \left[\frac{1}{\sin\theta} \frac{\partial}{\partial\theta} \left(\sin\theta \frac{\partial}{\partial\theta} \right) + \frac{1}{\sin^2\theta} \frac{\partial^2}{\partial\varphi^2} \right] \tag{12.93}$$

From Equations 12.41 and 12.42 and the definition of the spherical harmonic functions, Equation 12.65, we see that

$$\hat{L}^2 P_l^0(\theta) = \hbar^2 l(l+1) P_l^0(\theta) \tag{12.94}$$

$$\hat{L}_z Y_{l,m_l}(\theta,\varphi) = \hbar m_l Y_{l,m_l}(\theta,\varphi) \tag{12.95}$$

The value of the orbital angular momentum for a given eigenstate of the hydrogen atom is therefore:

$$L = \hbar\sqrt{l(l+1)} \tag{12.96}$$

and the L_z component is given by

$$L_z = \hbar m_l \tag{12.97}$$

Thus, we see that the ground state exhibits no orbital angular momentum, with $n = 1$, $l = 0$, and $L, L_z = 0$. The first excited state, $n = 2$, can have values of $l = 0, 1$. Therefore, the $n = 2$ level exhibits two states of orbital angular momentum, $L = 0$ and $L = \hbar\sqrt{2}$. The level with $l = 1$ possesses three possible values of L_z: $\hbar, 0, -\hbar$. It is already clear from these results that the Schrödinger equation yields a physical interpretation of the internal atomic structure quite different from that of the Bohr orbits. Firstly, since the ground state has no angular momentum, we cannot imagine the electron as orbiting the nucleus. The electron is restricted to radial motion. Secondly, even when the orbital state does have non-zero angular momentum, the projection L on the z-axis can only take on discrete values. In the case of $l = 1$, only three L_z values are allowed. In a classical orbit, there would be no restriction on L_z from $-L$ to L. A third nonclassical feature of the angular momentum is the peculiar status of the two other components, L_x, L_y. A consequence of the uncertainty principle is that the position and momentum of a particle cannot be known (measured) simultaneously. If all three components of the angular momentum were measurable *and* the magnitude of L was known as well, then the position and momentum of the electron in the atom would be determined simultaneously in violation of this fundamental postulate of the quantum theory.

12.6 Real orbitals

It is sometimes useful to consider linear combinations of the spherical harmonics, which of course, are themselves solutions to the angular part of the Schrödinger equation. In particular, linear combinations can be chosen so that resulting solutions are real. These real one-electron functions are termed 'orbitals' and they play an important role in molecular structure. These real orbitals give spatial directionality to the angular wave functions and can help to interpret the geometry of chemical bonding when atoms combine to form molecules. Linear combinations of $l = 1$ solutions to the Schrödinger equation are called p-orbitals. They exhibit electron probability density localised along the three orthogonal x, y, z coordinate axes, as shown in the first row of Figure 12.3. Combinations of the five $l = 2$ solutions result in d-orbitals. The electron probability density of three of these orbitals are indicated in the second row of Figure 12.3. Table 12.4 lists the linear combinations of spherical harmonics for $l = 1$ and $l = 2$. The labels

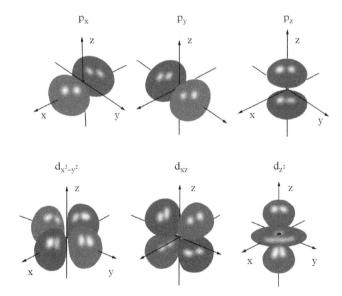

Figure 12.3 *Plots of the p-orbitals and three of the d-oribitals showing their spatial directions and orientations.*

Table 12.4 *Real orbitals from linear combinations of spherical harmonics.*

n	l	real orbital	linear combination	Cartesian coordinates
2	1	p_x	$\sqrt{\frac{1}{2}}\left(Y_1^{-1} - Y_1^1\right)$	$\sqrt{\frac{3}{4\pi}} \cdot \frac{x}{r}$
2	1	p_y	$i\sqrt{\frac{1}{2}}\left(Y_1^{-1} + Y_1^1\right)$	$\sqrt{\frac{3}{4\pi}} \cdot \frac{y}{r}$
2	1	p_z	Y_1^0	$\sqrt{\frac{3}{4\pi}} \cdot \frac{z}{r}$
3	2	$d_{x^2-y^2}$	$\sqrt{\frac{1}{2}}\left(Y_2^{-2} + Y_2^2\right)$	$\frac{1}{4}\sqrt{\frac{15}{\pi}} \cdot \frac{x^2-y^2}{r^2}$
3	2	d_{xy}	$i\sqrt{\frac{1}{2}}\left(Y_2^{-2} - Y_2^2\right)$	$\frac{1}{2}\sqrt{\frac{15}{\pi}} \cdot \frac{xy}{r^2}$
3	2	d_{xz}	$\sqrt{\frac{1}{2}}\left(Y_2^{-1} - Y_2^1\right)$	$\frac{1}{2}\sqrt{\frac{15}{\pi}} \cdot \frac{xz}{r^2}$
3	2	d_{yz}	$i\sqrt{\frac{1}{2}}\left(Y_2^{-1} + Y_2^1\right)$	$\frac{1}{2}\sqrt{\frac{15}{\pi}} \cdot \frac{yz}{r^2}$
3	2	d_{z^2}	Y_2^0	$\frac{1}{4}\sqrt{\frac{5}{\pi}} \cdot \frac{2z^2-x^2-y^2}{r^2}$

for the real orbitals arise from the functional form of these combinations when expressed in Cartesian coordinates as listed in the right-most column. Highly symmetric molecular structures can be explained on the basis of the linear combination of orbitals. For example, the methane CH_4 exhibits a spatial structure with the carbon atom at the centre and the four hydrogen atoms at the vertices of an equilateral tetrahedron. This structure is *not* explained by the p-orbitals of the central carbon atom alone but is due to a further orbital mixing. The $l = 0$ s-orbital combines with the three p-orbitals to produce the tetrahedral structure. This further mixing of s- and p-orbitals is called *hybridisation* because it involves linear combinations of solutions to the Schrödinger equation with *different* angular momenta. As such, these 'hybridised' orbitals are not pure quantum angular momentum states, but they nevertheless appear in nature because the hybridisation lowers the energy of the molecule. Highly symmetric molecules like methane CH_4 or sulphur hexafluoride SF_6 do not exhibit dipole absorption or emission between rotational or vibrational states because they have no permanent dipole moment. However, a simple substitution of a chlorine atom for one of the hydrogen atoms in methane, for example, yields chloromethane CH_3Cl, which does have a permanent dipole moment. Rotation and vibration of the molecular structure results in oscillation of the dipole, which in turn permits rotation-vibration transitions through a dipolar interaction with incident light. We will examine in more detail this type of molecular spectra for simple diatomic molecules when we discuss the rigid-rotor, harmonic oscillator model of molecular spectra in Section 12.9.

12.7   Interaction of light with the hydrogen atom

12.7.1   Semiclassical absorption and emission

We will first consider a two-level system is order to keep the notation simple and the ideas clear. Once we have expressions for the rates of absorption and emission we will apply the result to the transitions between the hydrogen atom ground state and the first excited state.

We start by returning to the time-dependent Schrödinger equation, Equation 12.18, rewritten here for convenience:

$$\hat{H}\Psi_n(r, t) = i\hbar\frac{\partial\Psi_n(r, t)}{\partial t} \tag{12.98}$$

The Hamiltonian operator now consists of two terms: the first term is the Hamiltonian of the atom itself that we developed earlier in the chapter, $\hat{H}_{atom}$. To this atomic Hamiltonian, we add a time-dependent term, $\hat{V}(t)$, which will be closely related to a driving electromagnetic field, with frequency ω_0. The total Hamiltonian is the sum of these two terms. Now we write the stationary state n of the atom as

$$\Psi_n(\mathbf{r}, t) = e^{-iE_nt/\hbar}\psi_n(\mathbf{r}) = e^{-i\omega_nt}\psi_n(\mathbf{r}) \tag{12.99}$$

where we have used

$$E_n = \hbar\omega_n \tag{12.100}$$

the Planck relation between quantised energy and frequency (see Chapter 6). For the case of the hydrogen atom, the label n can be considered the principal quantum number n discussed in Section 12.4.1. The time-independent Schrödinger equation becomes

$$\hat{H}_{atom}\psi_n(\mathbf{r}) - E_n\psi_n(\mathbf{r}) = 0 \tag{12.101}$$

For the moment we consider the hydrogen atom as a two-level system and write

$$\hat{H}_{atom}\psi_1 = E_1\psi_1 = \hbar\omega_1\psi_1$$
$$\hat{H}_{atom}\psi_2 = E_2\psi_1 = \hbar\omega_1\psi_2$$

with

$$\hbar\omega_0 = \hbar(\omega_2 - \omega_1)) = E_2 - E_1$$

We now add the $\hat{V}(t)$ term, the interaction energy of the atom with the driving field with frequency close to ω_0:

$$\hat{H} = \hat{H}_{atom} + \hat{V}(t) \tag{12.102}$$

With the field turned on the state of the system is no longer stationary and becomes a coupled, time-dependent linear combination of the two stationary states:

$$\Psi(\mathbf{r}, t) = C_1(t)\psi_1e^{-i\omega_1t} + C_2(t)\psi_2e^{-i\omega_2t} \tag{12.103}$$

with C_1, C_2 the coupling coefficients. We require $\Psi(\mathbf{r}, t)$ to be normalised:

$$\int |\Psi(\mathbf{r}, t)|^2 \, dV = |C_1(t)|^2 + |C_2(t)|^2 = 1$$

Now, if we substitute the time-dependent solution, Equation 12.103, back into the time-dependent Schrödinger equation, Equation 12.98, multiply from the left with the complex conjugate, $\psi_1^* e^{i\omega t}$, and integrate over all space, we obtain

$$C_1 \int \psi_1^* \hat{V} \psi_1 \, d\mathbf{r} + C_2 e^{-i\omega_0 t} \int \psi_1^* \hat{V} \psi_2 \, d\mathbf{r} = i\hbar \frac{dC_1}{dt} \qquad (12.104)$$

To simplify notation we write the integral 'matrix elements' $\int \psi_1^* \hat{V} \psi_1 \, d\mathbf{r}$ and $\int \psi_1^* \hat{V} \psi_2 \, d\mathbf{r}$ as V_{11} and V_{12}. So we have,

$$C_1 V_{11} + C_2 e^{-i\omega_0 t} V_{12} = i\hbar \frac{dC_1}{dt} \qquad (12.105)$$

and similarly for C_2 we get

$$C_1 e^{i\omega_0 t} V_{21} + C_2 V_{22} = i\hbar \frac{dC_2}{dt} \qquad (12.106)$$

These two equations define the time evolution of the atom-plus-field system as the atom undergoes a stimulated change between states 1 and 2. We need to solve this pair of equations for $C_1(t)$ and $C_2(t)$. The squares of the absolute values of these coefficients can be interpreted as the probability of finding the atom in either stationary state 1 or 2, as a function of time. Therefore, we are really more interested in $|C_1(t)|^2$, $|C_2(t)|^2$ than the coefficients themselves.

12.7.1.1 Atom-field coupling operator

We now turn our attention to the details of $\hat{V}(t)$. This term is an interaction energy operator, and from the classical theory of electrostatics we know that the leading interaction V between an electric field and neutral matter is the dipole term:

$$V = -\mathbf{p} \cdot \mathbf{E} \qquad (12.107)$$

where $\mathbf{p}$ is the dipole moment and $\mathbf{E}$ is the electric field of, say, a propagating electromagnetic wave through a polarisable material. The classical dipole itself is defined as

$$\mathbf{p} = \sum_i \mathbf{p}_i = \sum_i q_i \mathbf{r}_i \qquad (12.108)$$

where q_i is the charge at position $\mathbf{r}_i$ with respect to the origin of the coordinate system (see Chapter 5). In the case of a classical atom with one electron orbiting a massive proton, the instantaneous classical dipole can be taken as

$$\mathbf{p} = -e\mathbf{r} \tag{12.109}$$

where e is the charge on the electron and $\mathbf{r}$ is the radial position of the electron with respect to the origin, centred at the proton. If an external E-field is now applied to the classical atom, the interaction energy is given by

$$V = -\mathbf{p} \cdot \mathbf{E} = e\mathbf{r} \cdot \mathbf{E} \tag{12.110}$$

Passing to the quantal atom, we use the correspondence principle to write the $\mathbf{r}$ and $\mathbf{p}$ operators:

$$\mathbf{r} \to \hat{\mathbf{r}} \tag{12.111}$$
$$\mathbf{p} \to \hat{\mathbf{p}} \tag{12.112}$$

but we do not quantise the E-field. The rate of atomic absorption and emission we will obtain is based on this *semiclassical* approximation in which the atom is quantised but the field remains classical. We write the dipole interaction operator as

$$\hat{V} = -\hat{\mathbf{p}}(\mathbf{r}) \cdot \mathbf{E} \tag{12.113}$$
$$= e\hat{\mathbf{r}} \cdot \mathbf{E} \tag{12.114}$$

where $\hat{\mathbf{r}} = \mathbf{r}$ is the electron radial coordinate centred at the proton. The operator $\hat{V}$ exhibits odd parity with respect to this electron coordinate so that the matrix elements V_{11} and V_{22} in Equations 12.105 and 12.106 will vanish. Only atomic states of opposite parity can be coupled by the dipole interaction. For the E-field we write the real form,

$$\mathbf{E} = \hat{\mathbf{e}}E_0 \cos(\omega t - kz)$$

were $\hat{\mathbf{e}}$ is the polarisation unit vector, E_0 the field amplitude, ω the frequency, and k the propagation parameter. We have chosen the field to be propagating along z. The interaction takes place over the spatial extent of the atom, $\sim a_0$, which is only about 0.5 nm. For electromagnetic fields ranging in wavelength from the near ultraviolet to the near infrared, the wavelength is hundreds of nanometres. Therefore, the kz term in the field argument is negligible and the E-field can be considered effectively constant over atomic dimensions. The absolute interaction matrix element can therefore be written as

$$|V_{12}(t)| = |V_{12}| \cos \omega t$$
$$= \left[\int \psi_1^*(\mathbf{r}) e\mathbf{r}\psi_2(\mathbf{r}) \, dV \right] E_0 \cos \omega t$$
$$= p_{12} E_0 \cos \omega t \tag{12.115}$$

Since the dipole-field interaction is an energy, we can also write it as

$$|V_{12}(t)| = \hbar\Omega_0 \cos \omega t$$

with the *Rabi frequency* given by

$$\Omega_0 = \frac{p_{12}E_0}{\hbar} \tag{12.116}$$

12.7.2   Selection rules for hydrogen atom transitions

In classical electromagnetic theory, an oscillating charged dipole can radiate a continuous spectrum. We found in Chapter 2, Equation 2.157 that the power emitted by a classical dipole varies as the fourth power of the oscillation frequency ω, on which there is no restriction. In contrast, the absorption and emission of radiation by atoms only occurs at certain well-defined frequencies, giving rise to the characteristic atomic line spectrum. The rules that govern these allowed transition frequencies are called selection rules. The quantisation of the emission and absorption of radiation is one of the principal differences between classical and quantal light–matter interaction. We can understand the origin of the selection rules by studying the quantal dipole transition moment $\mathbf{p}_{12}$:

$$\mathbf{p}_{12} = \int \psi^*_{nlm_l}(r,\theta,\varphi)e\mathbf{r}\psi_{n'l'm'_l}(r,\theta,\varphi)r^2 \sin\theta \, d\theta \, d\varphi \, dr \tag{12.117}$$

where the primes on the quantum number labels indicate a final state different from the initial state. This dipole matrix element can be simplified by factoring the wave functions into their r, θ, φ constituents and resolving the vector dipole moment into its x, y, z components:

$$\mathbf{p}_{12} = e \int \mathbf{r} R^*_{nl}(r)R_{n'l'}r^2 drP^*_l(\cos\theta)P_{l'}(\cos\theta)e^{-im_l\varphi}e^{im_{l'}\varphi} \sin\theta \, d\theta \, d\varphi \tag{12.118}$$

$$p_z = r\cos\theta \tag{12.119}$$

$$p_x = r\sin\theta\cos\varphi \tag{12.120}$$

$$p_y = r\sin\theta\sin\varphi \tag{12.121}$$

First, considering p_z, we see that this component is independent of φ. Therefore, the dipole transition matrix element will have a factor

$$p_z \sim \int_0^{2\pi} e^{i(m_l - m_{l'})\varphi} \, d\varphi \tag{12.122}$$

Since the integral over the period of any oscillatory exponential function with imaginary argument is zero, the only way this factor can be non-zero is if the argument of the exponential itself is zero. Therefore, we have the first selection rule:

$$m_l - m_{l'} = \Delta m_l = 0 \qquad (12.123)$$

Next, still considering p_z, we see that we will have another factor in the transition moment integral,

$$p_z \sim \int_0^\pi P_l^*(\cos\theta) P_{l'}(\cos\theta) \cos\theta \sin\theta \, d\theta \qquad (12.124)$$

Now, as we show in Appendix G, the Legendre functions $P_l(\cos\theta)$ have definite parity. Functions labelled with even l are unchanged (even) with respect to the parity operation and functions labelled with odd l exhibit odd parity (change sign). The parity operation is sign reversal of the function variable. In the case of the Legendre functions the variable is $\cos\theta$. The parity operation is

$$\cos\theta \rightarrow \cos(\pi - \theta) = -\cos\theta \qquad (12.125)$$

and the parity property of the Legendre functions is that

$$\begin{aligned}
P_l[\cos(\pi - \theta)] &\rightarrow P_l(\cos\theta) & l &= 0, 2, 4, \ldots \\
P_l[\cos(\pi - \theta)] &\rightarrow -P_l(\cos\theta) & l &= 1, 3, 5, \ldots
\end{aligned}$$

The integrand of the integral in Equation 12.124 must be even with respect to θ over the limits of integration. Since the factor $\cos\theta\sin\theta$ exhibits odd parity, the product $P_l P_{l'}$ must also be odd so as to render the overall integrand even. Therefore, we see that $\Delta l = 0$ is a 'forbidden' transition and $\Delta l = \pm 1$ is allowed, or at least not forbidden, by the integrand symmetry. Turning now to p_x, p_y, we will have factors

$$p_x \sim \int_0^{2\pi} e^{-im_l\varphi} \cos\varphi \, e^{im_{l'}\varphi} \, d\varphi \qquad (12.126)$$

$$p_y \sim \int_0^{2\pi} e^{-im_l\varphi} \sin\varphi \, e^{im_{l'}\varphi} \, d\varphi \qquad (12.127)$$

and since $\cos\varphi$ and $\sin\varphi$ are linear combinations of $e^{i\varphi}$ and $e^{-i\varphi}$, it is clear that in order to avoid a null integral, another selection rule for these two components of the transition moment is

$$\Delta m_l = \pm 1 \qquad (12.128)$$

More generally we have to consider the overall parity operation that transforms the dipole vector $e\mathbf{r}$ to $-e\mathbf{r}$. In polar coordinates this parity inversion is accomplished by

$$\begin{aligned}
r &\rightarrow -r & (12.129) \\
\theta &\rightarrow \pi - \theta & (12.130) \\
\varphi &\rightarrow \varphi \pm \pi & (12.131)
\end{aligned}$$

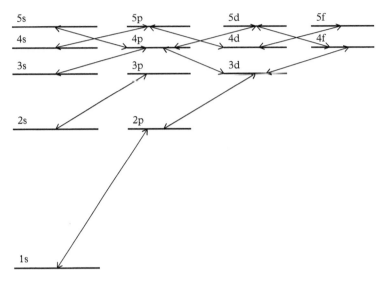

Figure 12.4 *Grotrian diagram for nl energy levels of the hydrogen atom. The arrows indicate some of the allowed transitions. Even with the restrictive selection rule $\Delta l = \pm 1$, the complicated cascading transition from high-lying energy states to lower-lying states is evident.*

We see from Equation 12.118 that the radial part of the integrand has odd parity, and therefore, the product of the Legendre functions must show odd parity as well. Therefore, l must differ from l' by ± 1 or we have the selection rule

$$\Delta l = \pm 1 \tag{12.132}$$

Summarising, the selection rules for allowed dipole transitions in the hydrogen atom are

$$\Delta l = \pm 1 \tag{12.133}$$
$$\Delta m_l = \pm 1, 0 \tag{12.134}$$

The various allowed atomic transitions are often summarised in a schematic chart called a Grotrian diagram. Figure 12.4 shows a Grotrian diagram for some of the transitions in the hydrogen atom. The notation for the levels derives from atomic spectroscopy. The principal quantum number n is indicated as $n = 1, 2, 3, 4, \ldots$ but the l quantum number is assigned a letter $s, p, d, f, \ldots$. The letters indicate progressions of spectroscopic lines that have some common property. The quantum number $l = 1$ is labelled s for 'sharp' lines. The $l = 2$ levels are labelled p for 'principal' lines, $l = 2 \rightarrow d$ for 'diffuse' lines.

12.7.2.1   *Calculation of atomic absorption and emission rates*

Now let us go back to the coupled equations, Equations 12.105 and 12.106, and re-write them in terms of the Rabi frequency and ω_0, the frequency difference between the ground and excited states:

$$\Omega_0 \cos \omega t \, e^{-i\omega_0 t} C_2 = i\frac{dC_1}{dt} \tag{12.135}$$

$$\Omega_0^* \cos \omega t \, e^{i\omega_0 t} C_1 = i\frac{dC_2}{dt} \tag{12.136}$$

We take the initial conditions (at $t = 0$) to be unit probability of finding the two-state system in state 1 and null probability for finding it in the excited state 2. At later times the probability for finding the system in state 2 is $\mathscr{P} = |C_2(t)|^2$. At early times the rate of increase of this probability is

$$\frac{d\mathscr{P}}{dt} = \frac{|C_2(t)|^2}{t} \tag{12.137}$$

so our task is to calculate $C_2(t)$ at times just later than $t = 0$. Applying the initial conditions, it is not difficult to show that

$$C_2(t) = \frac{\Omega_0^*}{2}\left[\frac{1 - e^{i(\omega_0+\omega)t}}{\omega_0 + \omega} + \frac{1 - e^{i(\omega_0-\omega)t}}{\omega_0 - \omega}\right] \tag{12.138}$$

If the frequency ω of the driving wave approaches the two-state difference frequency ω_0 it is quite clear that the first term in Equation 12.138 will be negligible compared to the second. Near resonance we can therefore write to good approximation:

$$C_2(t) \simeq \frac{\Omega_0^*}{2}\left[\frac{1 - e^{i(\omega_0-\omega)t}}{\omega_0 - \omega}\right] \tag{12.139}$$

We have, therefore,

$$|C_2(t)|^2 = |\Omega_0|^2 \frac{\sin^2\left[(\omega_0 - \omega)(t/2)\right]}{(\omega_0 - \omega)^2} \tag{12.140}$$

and when $\omega \to \omega_0$, around resonance, we obtain a limiting expression:

$$|C_2(t)|^2 = \frac{1}{4}|\Omega_0|^2 t^2 \tag{12.141}$$

Thus, with ω near ω_0 and at early times under weak-field conditions, the probability rate for finding the system in the excited states increases linearly with time:

$$\frac{d\mathscr{P}}{dt} = \frac{1}{4}|\Omega_0|^2 t \tag{12.142}$$

12.7.3 Finite spectral width of absorption

The incident driving field always has some spectral width. The source might be, for example, a broad band arc lamp or the output from a monochromator coupled to a synchrotron. Therefore we need to write the field energy as an integral over the *spectral energy density* of the excitation source in the neighbourhood of the transition frequency:

$$\frac{1}{2}\varepsilon_0 E_0^2 = \int_{\omega_0 - \frac{1}{2}\Delta\omega}^{\omega_0 + \frac{1}{2}\Delta\omega} \rho_\omega \, d\omega \qquad (12.143)$$

where ρ_ω is the spectral energy density of the excitation source with a line width of $\Delta\omega$. Note that the MKS units of spectral energy density are $Jm^{-3}s$. Now from Equation 12.140 we can write

$$|C_2(t)|^2 = \left(\frac{p_{12}E_0}{\hbar}\right)^2 \frac{\sin^2\left[(\omega_0 - \omega)t/2\right]}{(\omega_0 - \omega)^2} \qquad (12.144)$$

and substituting from Equation 12.143 we have

$$|C_2(t)|^2 = \frac{2p_{12}^2}{\varepsilon_0 \hbar^2} \int_{\omega_0 - \frac{1}{2}\Delta\omega}^{\omega_0 + \frac{1}{2}\Delta\omega} \rho_\omega \, d\omega \qquad (12.145)$$

For conventional broad-band excitation sources we can safely assume that the spectral density is constant over the line width of the atomic transition and replace $\rho(\omega)$ in the integrand with $\rho(\omega_0)$ outside it. Assuming a continuous wave (cw) excitation source, the integral can be evaluated:

$$\int_{\omega_0 - \frac{1}{2}\Delta\omega}^{\omega_0 + \frac{1}{2}\Delta\omega} \frac{\sin^2\left[(\omega_0 - \omega)t/2\right]}{(\omega_0 - \omega)^2} \, d\omega = \frac{\pi t}{2} \qquad (12.146)$$

So finally, the expression for the time rate of increase of the probability of finding the system in state 2 (i.e. the radiation absorption rate), integrated over the spectral width of the source, is

$$\frac{d\mathscr{P}}{dt} = \frac{|C_2(t)|^2}{t} = \frac{\pi p_{12}^2}{\varepsilon_0 \hbar^2} \rho(\omega_0) \qquad (12.147)$$

12.7.4 Comparison to Einstein *B* coefficient

Now, from Chapter 6 we know that the Einstein *A* and *B* coefficients represent spontaneous emission and stimulated absorption under conditions of equilibrium between a radiation field and a collection of material absorber-emitters. These coefficients were introduced into phenomenological rate equations and originally had no underlying interpretation of just *how* the light was absorbed or emitted. However, by recognising that the Einstein absorption rate, $B12\rho_\omega$, is equivalent to the semiclassical $d\mathscr{P}/dt$ we can write the *B* coefficient in terms of the quantum mechanical dipole matrix element p_{12}. More succinctly:

$$\frac{d\mathscr{P}}{dt} = \frac{\pi p_{12}^2}{\varepsilon_0 \hbar^2} \rho(\omega_0) = B_{12}\rho(\omega_0) \tag{12.148}$$

or

$$B_{12} = \frac{\pi p_{12}^2}{\varepsilon_0 \hbar^2} \tag{12.149}$$

The p_{12} that appears in Equation 12.148 has a specific orientation with respect to the polarisation of the exciting field. Equation 12.115 shows the angle θ of p_{12} with respect to the E-field of the exciting light, which is zero. However the Einstein B coefficient assumes isotropic radiation, so in order to really make the comparison of B and p_{12} meaningful we have to average the right-hand side of Equation 12.149 over all angles. The average value of the square of the dipole moment matrix element is given by

$$\langle p_{12}^2 \rangle = p_{12}^2 \langle \cos^2 \theta \rangle = \frac{1}{3} p_{12}^2 \tag{12.150}$$

So, finally we have

$$B_{12} = \frac{\pi p_{12}^2}{3\varepsilon_0 \hbar^2} \tag{12.151}$$

We have restricted this development to the consideration of an artificial atom, the two-level system without any substructure or energy-level degeneracy. No real atom corresponds to this simple scheme. Even the simplest atom, hydrogen, with one electron and one proton, exhibits orbital angular momentum degeneracy in the excited p level. However, it is a simple matter to take degeneracy into account in writing the expressions for B_{12}, B_{21}, and A_{21}, the stimulated absorption, stimulated emission, and spontaneous emission coefficients, respectively. The stimulated absorption and emission coefficients are related to the degeneracies g_1, g_2 of states 1, 2, respectively, by

$$g_1 B_{12} = g_2 B_{21} \tag{12.152}$$

The spontaneous emission rate A_{21} is independent of the upper state degeneracy but is related to B_{21} by

$$A_{21} = B_{21} \frac{\hbar \omega_0^3}{\pi^2 c^3} \tag{12.153}$$

Therefore, we have

$$B_{21} = \frac{g_1}{g_2} B_{12} = \frac{g_1}{g_2} \frac{\pi^2 c^3}{\hbar \omega_0^3} \tag{12.154}$$

and, using Equation 12.151,

$$A_{21} = \frac{g_1}{g_2} \frac{\omega_0^3 p_{12}^2}{3\pi \varepsilon_0 \hbar c^3} \tag{12.155}$$

Thus, if we can calculate or determine experimentally the transition dipole moment of a quantal atom we can obtain the spontaneous emission rate, and the stimulated emission and absorption rate coefficients.

12.7.5   Interpretation of dipole radiation in quantal atoms

We saw in Chapter 2, Section 2.8, Equation 2.138 that the leading source of classical electromagnetic radiation is the harmonically oscillating charge dipole,

$$\mathbf{p}(t) = p_0 \cos(\omega t)\hat{\mathbf{z}} = q_0 a \cos(\omega t)\hat{\mathbf{z}}$$

where q_0 is a point charge oscillating over a length a aligned along $\hat{\mathbf{z}}$ with frequency ω. In classical electrodynamics an oscillating dipole is a tangible, physical entity with observable properties. In quantum mechanics, the dipole, like all observables, is cast as an abstract mathematical operator, $\hat{\mathbf{p}}(t) = q\hat{\mathbf{r}}(t)$. How then does one connect the classical picture with the quantal formulation? The answer is through the dipole 'matrix element' or expression for the 'average value' of the dipole operator. The conventional form for the matrix element is to sandwich the operator between the wave function representing the state of the dipole (on the right) and its complex conjugate (on the left). The 'sandwich' is then integrated over all space:

$$\langle \hat{\mathbf{p}} \rangle = \int \Psi^*(\mathbf{r}, t)\hat{\mathbf{p}}\Psi(\mathbf{r}, t)\, dV \tag{12.156}$$

The rationale for this expression comes from the interpretation of the product of a wave function and its complex conjugate as a probability density $\mathscr{P}$:

$$\mathscr{P} = \Psi^*(\mathbf{r}, t)\Psi(\mathbf{r}, t)$$

Assuming the wave function is normalised,

$$\int \mathscr{P}\, dV = \int \Psi^*(\mathbf{r}, t)\Psi(\mathbf{r}, t)\, dV = 1$$

In classical statistics the average value of any quantity is calculated by integrating the probability distribution of values over the space of all possible values. A familiar example is the average speed v of a gas molecule subject to the Maxwell–Boltzmann distribution at fixed temperature T:

$$\langle v \rangle = \int \mathscr{P}(v)v\, dv = \left(\frac{m}{2\pi k_{mb} T}\right)^{3/2} \int 4\pi v^2 e^{-\frac{1}{2}mv^2/(k_{mb} T)} \cdot v\, dv$$

where k_{mb} is the Maxwell–Boltzmann constant. Similarly, in quantum mechanics

$$\langle \hat{\mathbf{p}}(t) \rangle = \int \Psi^*(\mathbf{r}, t) \Psi(\mathbf{r}, t) \hat{\mathbf{p}}(t) \, dV \tag{12.157}$$

but because $\hat{\mathbf{p}}$ is an operator, it is conventionally placed to the left of Ψ so that it can 'operate' on it, left to right. Now, the state of the atom, when subject to a harmonic driving field, is *not* one of the stationary states but a time-dependent linear combination of them. In the two-level atom we have

$$\Psi(\mathbf{r}, t) = C_1(t) \psi_1(\mathbf{r}) e^{-i\omega_1 t} + C_2(t) \psi_2(\mathbf{r}) e^{-i\omega_2 t} \tag{12.158}$$

with mixing coefficients C_1, C_2 subject to the normalisation condition

$$|C_1|^2 + |C_2|^2 = 1 \tag{12.159}$$

The differential probability of finding the dipole $\mathbf{p}(\mathbf{r}) = -e\mathbf{r}$ between $\mathbf{r}$ an $\mathbf{r} + d\mathbf{r}$ is then

$$\mathscr{P}\mathbf{p}\, dV = \Psi^*\Psi\mathbf{p}\, dV = -e\Psi^*\Psi\mathbf{r}\, dV \tag{12.160}$$

and the average value of the dipole moment is the integral of this differential probability over all space:

$$\begin{aligned}
\langle \mathbf{p}(t) \rangle = -e\langle \mathbf{r}(t) \rangle &= \int \left[C_1^* \psi_1^*(\mathbf{r}) e^{i\omega_1 t} + C_2^* \psi_2 e^{i\omega_2 t} \right] (-e\mathbf{r}) \cdot \\
&\quad \left[C_1 \psi_1(\mathbf{r}) e^{-i\omega_1 t} + C_2 \psi_2(\mathbf{r}) e^{-i\omega_2 t} \right] dV \\
&= 2\cos(\omega_0 t) \int C_1 C_2^* \psi_1^*(\mathbf{r})(-e\mathbf{r}) \psi_2(\mathbf{r}) \, dV \\
&= 2 C_{12} p_0 \cos(\omega_0 t)
\end{aligned} \tag{12.161}$$

with $\omega_0 = \omega_2 - \omega_1$ and $C_{12} = C_1^* C_2 = C_2^* C_1$ and

$$p_0 = \int \psi_1^*(-e\mathbf{r}) \psi_2 \, dV \tag{12.162}$$

We see that the dipole matrix element oscillates in time at frequency $\omega_0 = \omega_2 - \omega_1$ and with an amplitude controlled by the coefficients C_1, C_2. When the two stationary states are strongly coupled to form the time-dependent oscillating states, the dipole transition is strong.

12.8   The fourth quantum number: intrinsic spin

Up to this point we have considered only the quantum numbers n, l, m_l. We have used these indices to identify properties of the stationary states of the hydrogen atom. The number n indexes the energy of the state, the number l the magnitude of its orbital angular momentum, and m_l the projection of the angular momentum on the axis of quantisation; specifying, within the limits of the uncertainty principle, the direction in space to which the orbital angular momentum orients. In addition to these three, a fourth quantum number m_s exists. This quantum labels the orientation of the *intrinsic spin* angular momentum $\mathbf{S}$ of the electron. This spin angular momentum has no classical analogue. It does not correspond to the classical magnetic moment of a spinning electron and it does not correspond to the magnetic field of an electron circulating in a Bohr orbit. It arises naturally from the theory of quantum electrodynamics (QED), which is, unfortunately, beyond the scope of this book. Every electron carries a quantity of intrinsic spin $S = \hbar\sqrt{s(s+1)}$ where the spin quantum number $s = 1/2$. Therefore, the magnitude of spin angular momentum intrinsic to every electron is $S = (\sqrt{3}/2)\hbar$. Analogous to the orbital angular momentum, the spin direction is given by the projection of S on the quantisation axis. For the electron spin there are only two possible values, $m_s = \pm 1/2$, and the spin state is labelled by this quantum number. A complete specification of a stationary state of the hydrogen atom is therefore given by four quantum numbers: n, l, m_l, m_s. The quantum charge-dipole operator does not couple states of different intrinsic spin, and so the quantum number m_s remains unchanged during electric dipole transitions in the hydrogen atom.[1] Our last selection rule is therefore $\Delta m_s = 0$.

In fact, stationary states of the hydrogen atom with $l \neq 0$ do exhibit a magnetic moment that can be interpreted as arising from the electron orbital motion. The orbiting electron is a current, and Ampère's law tells us that a magnetic field is always produced at right angles to this loop current. This orbital magnetic is *not* the intrinsic spin, but in multi-electron atoms the two magnetic dipoles, orbital and spin, can couple. The results of this coupling can be observed in the atomic transition spectra, but these considerations are the proper domain of atomic structure and spectroscopy and will not be considered further here.

12.9   Other simple quantum dipolar systems

12.9.1   Rigid rotor model of a diatomic molecule

A heteronuclear diatomic molecule has an asymmetric charge density localised near the two nuclei comprising the molecule. This charge asymmetry produces a net charge separated by internuclear distance of the molecule, and therefore a permanent dipole. In the

---

[1] In multi-electron atoms the orbital and spin angular momenta may couple strongly. When this 'spin-orbit coupling' becomes significant, the individual angular momenta $\mathbf{L}$ and $\mathbf{S}$ are no longer conserved, and only the total angular momentum $\mathbf{J} = \mathbf{L} + \mathbf{S}$ specifies the state of the system. Further discussion of angular momentum coupling in multi-electron systems is, however, beyond the scope of what we can discuss here.

rigid rotator model, we ignore the molecular vibration and consider only the molecular rotation in space. The projection of a rotating dipole on any axis contained in the plane of rotation reveals an effective dipole oscillation, and therefore, the possibility of absorption and emission of dipole radiation. Note that this dipolar oscillation is due to *nuclear* motion, not the electron orbital motion. The nuclei of molecules move much more slowly than the electrons in the charge density binding the nuclei together. Because of this great difference between effective nuclear and electron velocities, their motions are, in many cases, effectively decoupled. The energy of the molecule can then be calculated as the sum of the rotational, vibrational, and electronic energy; and we can write down a Schrödinger equation for each type of motion, solve them separately, and write the total wave function as a product of three. Thus:

$$E_{molecule} \simeq E_{rot} + E_{vib} + E_{elec} \tag{12.163}$$

$$\psi_{molecule} \simeq \psi_{rot}\psi_{vib}\psi_{elec} \tag{12.164}$$

This motional decoupling is called the Born-Oppenheimer approximation.

Figure 12.5 shows the setup for modelling the rotational motion. The classical rotational energy is usually expressed in terms of the angular frequency ω and the moment of inertia, I:

$$E_{rot} = \frac{1}{2}I\omega^2 \tag{12.165}$$

The moment of inertia is given by

$$I = m_1 r_1^2 + m_2 r_2^2 \tag{12.166}$$

where

$$r_1 = \frac{m_1}{m_1 + m_2}r \quad \text{and} \quad r_2 = \frac{m_2}{m_1 + m_2}r \tag{12.167}$$

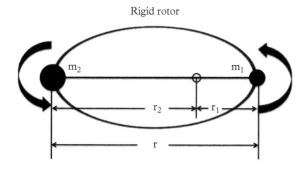

Rigid rotor

Figure 12.5 *Rigid rotor model of a heteronuclear diatomic molecule. Internuclear separation r is considered constant. Mass $m_2 > m_1$. Permanent electric dipole is aligned along r and rotates with the molecule.*

Now substitute Equation 12.167 into Equation 12.166. The result is

$$I = \frac{m_1 m_2}{m_1 + m_2} r^2 \tag{12.168}$$

and identify the *reduced mass* μ as

$$\mu = \frac{m_1 m_2}{m_1 + m_2} \tag{12.169}$$

We now have reduced the problem to one mass and one coordinate,

$$I = \mu r^2 \tag{12.170}$$

and the energy of rotation can be written in terms of the rigid rotor angular momentum, J:

$$J = I\omega \tag{12.171}$$

and

$$E_{\text{rot}} = \frac{J^2}{2I} \tag{12.172}$$

The quantum rigid rotor problem is setup using the correspondence principle between classical dynamical variables and their quantum operator counterparts. From Equation 12.172 we draw the following correspondences,

$$E_{\text{rot}} \rightarrow \hat{H}_{\text{rot}} \tag{12.173}$$

$$J \rightarrow \hat{J} = -\hbar^2 \nabla^2 \tag{12.174}$$

$$r \rightarrow \hat{r} = r \tag{12.175}$$

Then we write the time-independent Schrödinger equation,

$$\left[\hat{H}_{\text{rot}} - E_{\text{rot}} \right] \psi_{\text{rot}} = 0 \tag{12.176}$$

$$\left[\nabla^2 + \frac{2I}{\hbar^2} \right] \psi_{\text{rot}} = 0 \tag{12.177}$$

The solutions to Equation 12.177 are analogous to the orbital angular momentum of the electron in the hydrogen atom problem. There is a family of solutions:

$$\psi_{\text{rot}} = P_J(\cos \theta) e^{im J\varphi} = Y_J^{m_j}(\theta, \varphi) \tag{12.178}$$

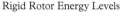

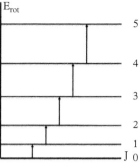

Figure 12.6 *Energy level diagram of the rigid rotor eigenstates.*

where $P_J(\cos\theta)$ are again the associated Legendre functions and $Y_J^{m_j}$ are the spherical harmonics. The eigen energies are

$$E_{\text{rot}} = \frac{\hbar^2 J(J+1)}{2I}; \qquad J = 0, 1, 2, \dots \tag{12.179}$$

where J are the rigid-rotor angular momentum quantum numbers (analogous to l for the hydrogen atom). The selection rules on rigid rotor transitions are similar to the selection rules in hydrogen:

$$\Delta m_J = 0, \pm 1; \qquad \Delta J = \pm 1 \tag{12.180}$$

Figure 12.6 shows a diagram of the energy levels and pure rotational transitions in the rigid rotor model.

In the realm of molecular spectroscopy the energy levels are often classified according to *terms*, by which is meant that the transition energy is expressed in units of reciprocal length. Thus, Equation 12.179 is expressed as

$$F(J) = \frac{E_{\text{rot}}(J)}{hc} = \frac{h}{8\pi^2 cI}J(J+1) = BJ(J+1) \tag{12.181}$$

where h, c have their usual meanings and B is called the *rotational constant*. It charac-terises the rotational energy in units of reciprocal length. Traditionally in spectroscopy the conventional unit is cm^{-1}, called a 'wavenumber'. Pure rotational transitions typically have rotational constants of some tens of cm^{-1}. For example, the rotational constant of hydrogen iodide, HI, is about $6.5\ \text{cm}^{-1}$ and that of H_2 is about $61\ \text{cm}^{-1}$.

12.9.2   Harmonic oscillator model of a diatomic molecule

For heteronuclear diatomic molecules, a permanent dipole moment exists, oriented along the internuclear axis. Harmonic molecular vibrations along this axis will result

Harmonic Oscillator

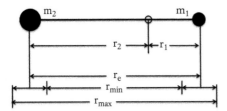

Figure 12.7 *Harmonic oscillator model of molecular vibration along the internuclear axis r. If $m_1 \neq m_2$ then a charge dipole exists along the axis, the amplitude of which will vary with the oscillation of r.*

in a dipole oscillation at the same frequency, and therefore quantised dipole emission and absorption can take place. Figure 12.7 shows the harmonic oscillator model of the heteronuclear diatomic molecule. The energy of a harmonic oscillator is the sum of its kinetic and potential energies. From elementary treatments we know that the kinetic energy is $p^2/2\mu$, where p is the linear momentum of the reduced mass μ, and the potential energy $V(r)$ is given by

$$V(r) = \frac{1}{2} k r^2 \tag{12.182}$$

where k is the *force constant* given by

$$k = \mu \omega^2 \tag{12.183}$$

with ω the angular frequency of oscillation. Once again we use the correspondence principle to form the Hamiltonian operator:

$$p \to \hat{p} = -i\hbar \frac{\partial}{\partial r} \tag{12.184}$$

$$r \to \hat{r} = r \tag{12.185}$$

$$E_{vib} \to \hat{H}_{vib} = \frac{\hat{p}^2}{2\mu} + \frac{1}{2} k \hat{r}^2 \tag{12.186}$$

The harmonic oscillator Schrödinger equation then becomes

$$\left[\hat{H}_{vib} - E_{vib} \right] \psi_{vib} = 0 \tag{12.187}$$

The eigen energies take on the form

$$E_{vib} = \hbar \omega \left(v + \frac{1}{2} \right); \qquad v = 0, 1, 2, \ldots \tag{12.188}$$

and the eigenfunctions are closely related to the Hermite polynomials $H_v(\sqrt{\alpha} r)$ discussed in Appendix H. The physically admissible solutions to Equation 12.187 are

Harmonic Oscillator Energy Levels

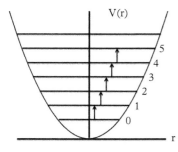

Figure 12.8 *Energy level diagram showing allowed energies of the quantum harmonic oscillator. Note that the ground state energy* $E_0 = 1/2\hbar\omega$.

$$\psi_{\text{vib}} = N_{norm}H_v(\sqrt{\alpha}r)e^{-\mu\omega/\hbar x^2} \tag{12.189}$$

where $N_{norm} = 1/\sqrt{2^v\pi^{1/2}v!}$ is a normalisation factor and $\alpha = \mu\omega/\hbar$ is a scaling factor for the coordinate r. Figure 12.8 shows the equally spaced ladder of energies $E_{\text{vib}} = \hbar\omega$ starting from the ground state with *zero-point* energy $1/2\hbar\omega$.

In spectroscopy, the energy of vibrational transitions are expressed as vibrational *terms* analogous to the spectroscopy of pure rotations. The vibrational terms are expressed as

$$G(v) = \frac{E_{\text{vib}}}{hc} = \omega\left(v + \frac{1}{2}\right); \qquad v = 0, 1, 2, \dots \tag{12.190}$$

where ω means the vibrational energy, usually reported in units of cm^{-1}. This vibrational ω should not be confused with the angular frequency of some wave or transition. Unfortunately, spectroscopic notation became entrenched long before the modern association of ω with a unit of angular frequency. The vibrational transitions of diatomic molecules exhibit much greater energy than rotational transitions. For example, the vibrational term $G(v)$ in HI is $2309\,\text{cm}^{-1}$, a factor of about 350 greater than the rotational term.

12.9.3 Electronic states of molecules

In addition the rotational and vibrational transitions, arising from the motion of permanent dipoles, molecules can undergo electronic transitions, somewhat analogous to the transitions we have studied in the hydrogen atom. Diatomic molecules have lower symmetry (cylindrical) than atoms (spherical), and therefore the great simplification of variable separation is not so easy to carry out. Nevertheless, one can characterise molecular orbitals and assign symmetry quantum number to electronic states. The proper study of molecular structure and spectroscopy is outside the scope of this book, but suffice it to state that molecular electronic transitions do occur and their energies are again much greater than pure vibrational or rotational transitions. Continuing with the example of HI, the first electronic transition between molecular bound states is the $X \rightarrow B$ transition. Here, X indicates a molecular ground electronic state and B labels a bound

Table 12.5 *Molecular transitions, term energies, and frequencies.*

Transition Type	Term Energy (cm^{-1})	Frequency Range
Rotation	3–30	EHF Radio
Vibration	1000–3000	THz-IR
Electronic	10 000–30 000	Vis-UV

excited electronic state. The electronic spectroscopic term for this transition is about 5.6×10^3 cm^{-1}, about 25 times greater than the vibrational term. Hierarchy of Transitions Rotation, vibration, and electronic transitions follow an energy hierarchy from relatively low to high. Table 12.5 summarises the typical energies for these dipole transitions. Rotational spectroscopy can be studied with microwave technology, but pure vibrational transitions (at least for 'typical' diatomic molecules) fall in an awkward part of the electromagnetic spectrum. Radiation sources and detectors in the THz (10^{12} Hz and far infrared (IR) regimes are not, at present, a well-developed and readily available technology. Electronic spectroscopy, in contrast, can be studied with a vast array of sources, optical dispersion instrumentation, and detectors. Of course, when a molecule absorbs or emits light in the visible-ultraviolet range it undergoes a 'rovibronic' transition in which all three types of internal molecular motion change. Figure 12.9 shows, schematically, how these rovibronic transitions couple lower states to higher states. The rotation-vibration motion gives rise to underlying manifolds of transition lines, riding on top of the electronic transition. The amplitude and spacing of the transition lines within the manifold provides information on vibrational and rotational motion.

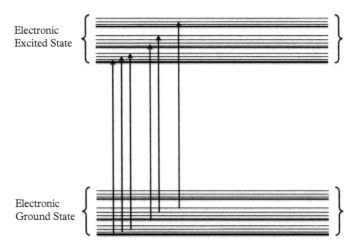

Figure 12.9 *Schematic of rovibronic transitions where radiation connects a ground-state manifold of vibration-rotation states to an excited-state manifold. Separations between states are not to scale.*

12.10 Summary

This chapter again starts with the most elementary models of harmonic oscillation and radiative damping as a warm-up to the main event, the Schrödinger equation for the hydrogen atom. The full 3-D Hamiltonian is presented, the equation solved in polar coordinates, and the solutions for radial space coordinate r, and polar (θ), azimuthal (φ) angular coordinates established. The final 3-D wave function is then the product of the three solutions. The nature of the bound states leading to the electron continuum is discussed together with the significance of the n, l, m_l quantum numbers. Then real orbitals, constructed from linear combinations of the complex product solutions (hybridisation), are developed and shown to be useful in theories of molecular structure and chemical binding. Dipole radiative emission and absorption is the next topic where the symmetries of the radial and angular solutions, together with the symmetry of the dipole coupling, are used to determine transition quantum number 'selection rules'. The chapter ends with a brief discussion of the 'fourth quantum number', magnetic dipole spin, and the structures and spectroscopy (rotational, vibrational and electronic) of diatomic molecules.

12.11 Exercises

1. Calculate the average value of the electron radius for the hydrogen atom in its ground state, $\mathcal{R}_{10}$. Use the expression from Table 12.1.

2. For the same state, calculate the maximum probability for finding the electron at some distance from the proton.

3. Explain why the average value for the electron radius and the most probable value for the electron radius are not equal.

4. Given the ground state and first excited states of the hydrogen as

$$\psi_{100} = \frac{1}{\sqrt{\pi}} \frac{1}{a_0^{3/2}} e^{-r/a_0}$$

and

$$\psi_{210} = \frac{1}{\sqrt{\pi}} \frac{1}{(2a_0)^{5/2}} r e^{-r/2a_0} \cos \theta$$

show that

$$p_{12} = \frac{2^{15/2} e a_0}{3^5}$$

where e is the charge on the electron and a_0 is the Bohr radius.

5. The energy-level spacing for this first transition in hydrogen is called the Lyman α transition. The wavelength at line centre is $\lambda = 121.7\,\text{nm}$. Calculate the spontaneous emission rate A_{21} (s^{-1}) for this transition.

12.12   Further reading

1. J.-P. Pérez, R. Carles, and R. Fleckinger, *Électromagnétism, Fondements et applications*, 3ème édition, Masson (1997).

2. C. Cohen-Tannoudji, B. Diu, and F. Laloë, *Quantum Mechanics*, Wiley (1977).

3. G. B. Arfken and H. J. Weber, *Mathematical Methods for Physicists*, 6th edition, Elsevier Academic Press (2005).

4. R. Loudon, *The Quantum Theory of Light*, 3rd edition, Oxford University Press (2003).

5. A. Corney, *Atomic and Laser Spectroscopy*, Clarendon Press, Oxford (1977).

6. J. Weiner and P.-T. Ho, *Light-Matter Interaction: Fundamentals and Applications*, Wiley-Interscience (2003).

7. G. Herzberg, *Molecular Spectra and Molecular Structure; I Spectra of Diatomic Molecules*, Van Nostrand Reinhold (1950).

Appendix A
Numerical Constants and Dimensions

A.1 Numerical values of some fundamental constants

Permittivity of free space ε_0

$$\varepsilon_0 = 8.854 \times 10^{-12} \quad \text{farad m}^{-1} \tag{A.1}$$

Permeability of free space μ_0

$$\mu_0 = 4\pi \times 10^{-7} = 1.257 \times 10^{-6} \quad \text{henry m}^{-1} \tag{A.2}$$

Speed of light

$$c = \frac{1}{\sqrt{\mu_0 \varepsilon_0}} = 2.998 \times 10^{8} \quad \text{metres s}^{-1} \tag{A.3}$$

Resistance of free space Z

$$Z = \sqrt{\frac{\mu_0}{\varepsilon_0}} = 376.7 \quad \text{ohms} \tag{A.4}$$

Boltzmann constant k_B

$$k_B = 1.38065 \times 10^{-23} \quad \text{J K}^{-1} \tag{A.5}$$

Stefan–Boltzmann constant σ

$$\sigma = 5.670 \times 10^{-8} \quad \text{W m}^{-2} \text{ K}^{-4} \tag{A.6}$$

Atomic mass unit amu

$$amu = 1.660 \times 10^{-27} \quad \text{kg} \tag{A.7}$$

Planck constant $\hbar$

$$\hbar = 1.054572 \times 10^{-34} \quad \text{J s} \tag{A.8}$$

Elementary charge e

$$e = 1.602176 \times 10^{-19} \quad \text{C} \tag{A.9}$$

Bohr radius a_0

$$a_0 = 0.529177 \times^{-10} \quad \text{m} \tag{A.10}$$

Bohr magneton μ_B

$$\mu_B = 927.401 \times 10^{-26} \quad \text{J T}^{-1} \tag{A.11}$$

Electron volt eV

$$eV = 1.602176 \times 10^{-19} \quad \text{J} \tag{A.12}$$

A.2   Dimensions of electromagnetic quantities

Fundamental quantities are chosen to be mass M, length L, time T, and charge Q. The units are the SI units of kg, m, s, and C, respectively.

Quantity	Symbol	Dimensions	SI unit
Force	F	MLT^{-2}	Newton
Energy	$E, \mathscr{E}$	ML^2T^{-2}	Joule
Power	W	ML^2T^{-3}	Watt
Field energy flux	S	MT^{-3}	Watt per square metre
Field momentum flux	g	$ML^{-2}T^{-1}$	Joule per cubic metre · metre per second
Charge	q	Q	Coulomb
Current	I	QT^{-1}	Ampère
Charge density	ρ	QL^{-3}	Coulomb per cubic metre
Current density	J	$QT^{-1}M^{-2}$	Ampère per square metre
Resistance	R	$ML^2T^{-1}Q^{-2}$	Ohm
Conductivity	σ	$M^{-1}L^{-3}TQ^2$	(Ohm·metre)$^{-1}$
Electric potential	V	$ML^2T^{-2}Q^{-1}$	Volt
Electric field	E	$MLT^{-2}Q^{-1}$	Volt per metre
Capacitance	C	$M^{-1}L^{-2}T^2Q^2$	Farad
Displacement field	D	QL^{-2}	Coulomb per square metre
Permittivity	ε	$M^{-1}L^{-3}T^2Q^2$	Farad per metre
Electric dipole moment	p	LQ	Coulomb·metre
Magnetic flux	Φ	$ML^2T^{-1}Q^{-1}$	Weber
Magnetic induction field	B	$MT^{-1}Q^{-1}$	Weber per square metre
Magnetic field	H	$QT^{-1}L^{-1}$	Ampère per metre
Inductance	L	ML^2Q^{-2}	Henry
Permeability	μ	MLQ^{-2}	Henry per metre
Magnetic dipole moment	m	$L^2T^{-1}Q$	Ampère·square metre

Appendix B
Systems of Units in Electromagnetism

B.1 General discussion of units and dimensions

We conventionally decide to take the three mechanical units of **mass** m, **length** l, and **time** t, as three fundamental, basic units. However, this choice is just for convenience. The number of units need not be three and they need not be these three. For example, we could *increase* the number of units by defining the kinetic energy as

$$E = k \cdot m \cdot \left(\frac{l}{t}\right)^2 = kv^2 \tag{B.1}$$

where k is a constant of proportionality with some dimensions, say, the ratio of charge to mass of an electron, (e/m_e). We could define

$$k = 2\frac{e}{m_e} \tag{B.2}$$

Then kinetic energy would be expressed by

$$E = \left(\frac{e}{m_e}\right)v^2 \tag{B.3}$$

and would have units of l^2/qt^2, where q is some unit of electrical charge. The number of basic units is now four: mass, length, time, and electric charge. We could also *decrease* the number of units by setting some constants equal to unity and without dimensions. For example, setting $m_e = e = \hbar = 1$ is such a choice, sometimes called *atomic units*, and convenient in atomic and molecular quantum mechanical expressions.

Here is another example: the volume can be defined as function of the length unit,

$$V = kl^3 \tag{B.4}$$

where k is an arbitrary constant, usually set equal to unity. But suppose we had some problem dealing only with spherical volumes,

$$V = k\frac{4}{3}\pi r^3 \tag{B.5}$$

We could define k to be $\frac{3}{4\pi}$ which would make the volume read

$$V = r^3 \tag{B.6}$$

in some system of 'rationalised' units.

Now electromagnetic (E-M) theory differs from classical mechanics in that a fundamental constant, 'the speed of light', c appears explicitly. This constant has dimensions, l/t, and if these units are changed, the appearance of E-M expression will change as well. We will see how the speed of light makes its appearance by comparing electrostatic and magnetostatic forces.

B.2   Coulomb's law

Suppose we start with Coulomb's law, where 'force' has been defined in terms of mass, length, and time but the constant of proportionality, K_1, or the units of electrical charge, q, have not yet been fixed:

$$F = K_1 \cdot \left(\frac{q_1 q_2}{r^2}\right) \tag{B.7}$$

We now choose to set $K_1 = 1$, which then determines the dimensionality of the charge to be $l^{3/2} m^{1/2}/t$. This choice of proportionality constant in Coulomb's law is used in the esu (electrostatic units) system of units. The usual units of mass, length, and time in this system are the gram, centimetre, and second. The unit of charge is called the 'statcoulomb'.

We can use Coulomb's law to write the electric field as 'force per unit charge',

$$\frac{F}{q_1} = E = \frac{q_2}{r^2} \tag{B.8}$$

Or more properly in vector form,

$$E = q\frac{\hat{r}}{r^3} \rightarrow \left[\int_V \rho(x)d^3x\right]\frac{\hat{r}}{r^3} \tag{B.9}$$

where $\hat{r}$ is the unit vector pointing in the E direction, and $\rho(x)$ is the charge density. Now from Gauss's law we know that

$$\oint_S E \cdot n\, da = 4\pi \int_V \rho(x)d^3x \tag{B.10}$$

and from the divergence theorem,

$$\oint_S E \cdot n\, da = \int_V \nabla \cdot E\, d^3x \tag{B.11}$$

from which we get the differential form of Gauss's law,

$$\nabla \cdot E = 4\pi\rho \tag{B.12}$$

But instead of choosing $K_1 = 1$ in Equation B.7 we could have chosen $K_1 = \frac{1}{4\pi}$, which would make the 4π factor in Equations B.10 and B.12 disappear. Such a choice of scaled units might be called 'rationalised' units.

Up to now we have only considered E-M fields in a vacuum, but when we generalise to dielectric media, the electric vector field **E** becomes the displacement vector **D** and is expressed as

$$\mathbf{D} = \varepsilon_0 \mathbf{E} + \kappa \mathbf{P} \tag{B.13}$$

The vector field **P** is called the 'polarisation' and represents the density of dipole moments in the material. The constants ε_0 and κ are proportionality constants. These constants are not the same in esu and MKS units. In the esu system the displacement vector field is

$$\mathbf{D} = \mathbf{E} + 4\pi \mathbf{P} \tag{B.14}$$

In this system ε_0 is chosen to be a pure number with unit value. This choice implies that **D**, **E**, and **P** all have the same dimensions.

In the MKS system

$$\mathbf{D} = \varepsilon_0 \mathbf{E} + \mathbf{P} \tag{B.15}$$

As usual, the 4π factor is eliminated, but the factor ε_0 appears explicitly. Note from Equation B.7 and the choice of $K_1 = 1/(4\pi\varepsilon_0)$ that this factor in the MKS system has dimensions of $q^2 t^2/ml^3$ and units of 'Coulomb squared per Joule metre' $(C^2/J \cdot m)$, so the electric field vector **E** does *not* have the same dimensions as the displacement vector **D** and the polarisation vector **P**. To be consistent with the displacement field in polarisable materials, the displacement field in free space is just written as

$$\mathbf{D_0} = \varepsilon_0 \mathbf{E_0} \tag{B.16}$$

and **P** is zero since we do not consider free space to be polarisable (at least in classical electrodynamics). Now, going back to the differential form of Gauss's law (Equation B.12) and inserting ε_0 on both sides, we have the generalisation to the displacement field form that gives us

$$\nabla \cdot \mathbf{D_0} = 4\pi \varepsilon_0 \rho \tag{B.17}$$

But again, instead of choosing $K_1 = 1$ as we did in Equation B.7, we could have chosen $K_1 = \frac{1}{4\pi\varepsilon_0}$, which would result in the disappearance of messy constants in the final expression for Gauss's law:

$$\nabla \cdot \mathbf{D_0} = \rho \quad \text{and} \quad \nabla \cdot \mathbf{E_0} = \frac{1}{\varepsilon_0}\rho \tag{B.18}$$

This choice for K_1 forms the basis for the 'rationalised MKS' system of units.

B.3   Ampère's law

Currents I_1, I_2 moving in two parallel wires separated by some known distance d also give rise to a force. The force equation, analogous to the Coulomb force in electrostatics, is

$$\frac{dF}{dl} = 2K_2 \frac{I_1 I_2}{d} \qquad (B.19)$$

where K_2 is a proportionality constant whose value depends on the choice of units. Note that here the force is 'per unit length' of the infinitely long wires. Now, from Equations B.7 and B.19 we can do a dimensional analysis of the ratio, K_1/K_2. The result is that the ratio of the two constants has units of velocity squared $(l/t)^2$. It turns out from *measurement* that for equal electrostatic and magnetostatic forces the value of this ratio is equivalent to the speed of light in vacuum squared, c^2. So we have $K_1/K_2 = c^2$.

The magnetic induction field **B** is defined as the 'force per unit current' analogous to the 'force per unit charge' that defines the electric field **E**:

$$\mathbf{B} = 2K_2 \frac{I}{d} \qquad (B.20)$$

However, Equation B.20 is not quite the whole story. A defining relation between **E** and **B** would imply that **B** is just *proportional* to the 'force per unit current' so that, strictly speaking, we should write

$$\mathbf{B} = 2K_2 K_3 \frac{I}{d} \qquad (B.21)$$

and then make some convenience argument about the choice of K_3. In order to really set K_3 correctly we need the equation that relates **E** to **B**. Such an equation exists. It is Faraday's law of induction:

$$\nabla \cdot \mathbf{E} = -K_3 \frac{\partial \mathbf{B}}{\partial t} \qquad (B.22)$$

Table B.1 *Common systems of units in electromagnetism.*

Name	Units	K_1	K_2	Remarks
Electrostatic	cgs	1	$1/c^2$	charge unit statcoulomb
Electromagnetic	cgs	c^2	1	current unit abampere
Heaviside–Lorentz	cgs	$1/4\pi$	$1/(4\pi c^2)$	rationalized emu
Gaussian	cgs	1	$1/c^2$	convenient formally
SI	m,kg,s	$1/(4\pi\varepsilon_0)$	$\mu_0/4\pi$	convenient numerically

Notice that the Faraday relates a spatial derivative of $\mathbf{E}$ to a time derivative of $\mathbf{B}$. Suppose, just to keep things simple, that $\mathbf{E}$ and $\mathbf{B}$ are the field components of a plane wave. Then we have immediately the scalar relation,

$$ikE = K_3\omega B \quad \text{or} \quad E = K_3\left(\frac{\omega}{k}\right)B \tag{B.23}$$

The factor ω/k has units of l/t or velocity. We have, therefore, two easy choices for K_3. If we set $K_3 = 1$, then $\mathbf{B} = (1/v)\,\mathbf{E}$, and in free space $\mathbf{B} = (1/c)\,\mathbf{E}$. We could also set $K_3 = 1/c$, in which case $\mathbf{B} = \mathbf{E}$.

We see that in the esu system of units the choice of $K_1 = 1$ means that $K_2 = c^2$. In the rationalised MKS system $K_1 = 1/(4\pi\varepsilon_0)$. The constant K_2 has to be chosen so that the ratio is c^2. The choice is the following:

$$K_1 = \frac{1}{4\pi\varepsilon_0} = 10^{-7}c^2 \quad \text{and} \quad K_2 = \frac{\mu_0}{4\pi} = 10^{-7} \tag{B.24}$$

where the 'permeability of free space' μ_0 has been introduced. The permeability serves the same function for magnetic fields that permittivity serves for electric fields. The difference between the Electrostatic system and the Gaussian system is in the choice of K_3. In the Electrostatic system of units $K_3 = 1$ while in the Gaussian system $K_3 = 1/c$. The

Table B.2 *Electromagnetic symbols and units*

Quantity	Symbol	Units
Electric charge	q	C
Electric charge density	ρ	Cm^{-3}
Electric current density	J	Am^{-2}
Electric current	I	A
Electric conductivity	σ	Sm^{-1}
Electric permittivity	ε	Fm^{-1}
Electric susceptibility	χ_e	unitless
Electric field	E	Vm^{-1}
Displacement field	D	Cm^{-2}
Magnetic permeability	μ	Hm^{-1}
Magnetic susceptibility	χ_m	unitless
Magnetic field	H	Am^{-1}
Magnetic induction field	B	Vsm^{-2}

Heaviside–Lorentz system is similar to the Gaussian except it is 'rationalised' thereby eliminating factors of 4π that often appear in the field equations of the Gaussian, Electrostatic, and Electromagnetic systems.

The magnetic fields in matter, analogous to Equations B.14 and B.15, are

$$\mathbf{H} = c^2\mathbf{B} - 4\pi\mathbf{M} \text{ (esu)} \qquad \text{and} \qquad \mathbf{H} = \frac{1}{\mu_0}\mathbf{B} - \mathbf{M} \text{ (MKS)} \qquad \text{(B.25)}$$

The vector field $\mathbf{H}$ is, strictly speaking, the *magnetic field*, while $\mathbf{B}$ should always be referred to as the *magnetic induction field*. Obviously, the magnetic field $\mathbf{H}$ plays an analogous role to the displacement field $\mathbf{D}$ in the macroscopic equations of electrostatics. The vector field $\mathbf{M}$ is called the 'magnetisation' and plays a role analogous to $\mathbf{P}$, the polarisation. Note that neither in the esu system nor in the MKS system does $\mathbf{B}$ have the same units as $\mathbf{H}$.

Note finally that the choice of $K_2 = \mu_0/4\pi$ in the MKS system implies that

$$\frac{1}{\varepsilon_0\mu_0} = c^2 \qquad \text{(B.26)}$$

Many of the quantities, symbols, and units (in the rationalised MKS system) are compiled in Table B.2.

Appendix C
Review of Vector Calculus

C.1 Vectors

A spatial vector is essentially a geometric entity comprising a magnitude (or length) and a direction. A scalar is some entity characterised only by magnitude. Many physical quantities can be represented by scalars and vectors. As an example, Figure C.1 shows an object of mass m moving in a curved trajectory with velocity $\mathbf{v}$ and subject to a force $\mathbf{F}$. The acceleration $\mathbf{a}_c$ points toward the origin of the central force. The distance from the object to the force centre is R. In this example m and R are scalars with only one number (amplitude) associated with these physical quantities. The vector quantities, $\mathbf{v}, \mathbf{F}, \mathbf{a}_c$ have two associated properties: the amplitude and the direction. The geometric entity of the vector can be *represented* by the coordinates of the two end points. If the coordinate system is transformed to some other one, the vector remains unchanged but its representation changes with the coordinate values of the endpoints. For example, the direction can be represented by two numbers along a line in one-dimensional (1-D) space, two pairs of numbers in two-dimensional (2-D) space, and a pair of three numbers in a three-dimensional (3-D) space. Very often the 'tail' of the vector is referred to the origin of the coordinate system. The 3-D velocity vector, for example, in Cartesian coordinates takes on the form

$$\mathbf{v} = v_x\hat{\mathbf{x}} + v_y\hat{\mathbf{y}} + v_z\hat{\mathbf{z}} \tag{C.1}$$

where the unit vectors in the x, y, z directions are $\hat{\mathbf{x}}, \hat{\mathbf{y}}, \hat{\mathbf{z}}$ and the magnitude of the vector components in the x, y, z directions are v_x, v_y, v_z. The overall length or amplitude of the vector is $|v| = \sqrt{v_x^2 + v_y^2 + v_z^2}$. A vector can also be expressed by a column of numbers, each element corresponding to the component along a Cartesian coordinate axis:

$$\mathbf{v} = \begin{bmatrix} v_x \\ v_y \\ v_z \end{bmatrix} \tag{C.2}$$

A vector can be rotated within a coordinate space, or alternatively, the coordinate axes themselves can be rotated (transformed so that the vector is represented in a new coordinate space). The rotation operation can be represented by a matrix. For a 2-D anticlockwise rotation we have, for example,

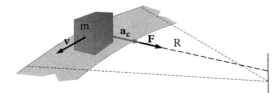

Figure C.1 *Examples of vector quantities and scalar quantities. Velocity v, force F, and acceleration a_c are vectors. The distance to the origin R and the mass m are scalars.*

$$\mathbf{v'} = \begin{bmatrix} v'_x \\ v'_y \end{bmatrix} = \begin{bmatrix} \cos\theta & -\sin\theta \\ \sin\theta & \cos\theta \end{bmatrix} \cdot \begin{bmatrix} v_x \\ v_y \end{bmatrix} \tag{C.3}$$

Writing out the matrix multiplication explicitly we have

$$v'_x = \cos\theta\, v_x - \sin\theta\, v_y \tag{C.4}$$
$$v'_y = \sin\theta\, v_x + \cos\theta\, v_y \tag{C.5}$$

and Figure C.2 shows the coordinate rotation. This type of coordinate transformation of a vector is called a *unitary* transformation because the length of the vector is preserved. This property is easy to see by taking the absolute value of the original vector and the transformed vector. The length of the original vector is given by

$$|\mathbf{v}| = \left(v_x^2 + v_y^2\right)^{1/2} \tag{C.6}$$

and the length of the rotated vector is

$$|\mathbf{v'}| = \left(v'^2_x + v'^2_y\right)^{1/2} = \left[\left(\cos^2\theta + \sin^2\theta\right)\left(v_x^2 + v_y^2\right)\right]^{1/2}$$
$$= \left(v_x^2 + v_y^2\right)^{1/2} = |\mathbf{v}| \tag{C.7}$$

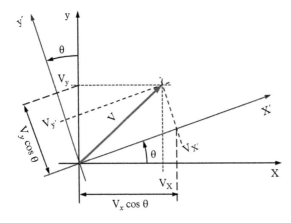

Figure C.2 *Coordinate rotation of coordinate axes $x-y$ to coordinate axes $x'-y'$ by angle θ anticlockwise around vector V. Projections onto $x-y$ and $x'-y'$ coordinate axes are shown.*

In 3-D we can represent the rotation matrix about the three Cartesian coordinate axes as

$$R_x(\theta) = \begin{bmatrix} 1 & 0 & 0 \\ 0 & \cos\theta & -\sin\theta \\ 0 & \sin\theta & \cos\theta \end{bmatrix} \tag{C.8}$$

$$R_y(\theta) = \begin{bmatrix} \cos\theta & 0 & \sin\theta \\ 0 & 1 & 0 \\ -\sin\theta & 0 & \cos\theta \end{bmatrix} \tag{C.9}$$

$$R_z(\theta) = \begin{bmatrix} \cos\theta & -\sin\theta & 0 \\ \sin\theta & \cos\theta & 0 \\ 0 & 0 & 1 \end{bmatrix} \tag{C.10}$$

These matrices rotate the vectors anticlockwise about each axis, assuming that the rotation axis considered is pointing towards the reader.

C.2   Axioms of vector addition and scalar multiplication

Vectors can undergo algebraic operations such as addition and scalar multiplication. The axioms that define these operations in linear algebra are summarised in Table C.1.

Table C.1 *Axiomatic properties of vector addition and scalar multiplication.*

Axiomatic property		Vector expression
association		$v_1 + (v_2 + v_3) =$ $(v_1 + v_2) + v_3$
commutation		$v_1 + v_2 = v_2 + v_1$
sum identity (existence of null vector)		$v + 0 = v$
inversion (existence of −v)		$v + (-v) = 0$
distribution	scalar times vector sum	$s((v_1 + v_2) = sv_1 + sv_2$
	scalar sum times vector	$(s_1 + s_2)v = s_1 v + s_2 v$
	scalar times vector	$s_1 (s_2 v) = (s_1 s_2)v$
multiplication identity (existence of $s = 1$)		$1(v) = v$

C.3 Vector multiplication

C.3.1 Scalar product

The projection of some vector $\mathbf{A}$ onto its Cartesian coordinates provides a special case of the definition of the scalar or 'dot' product from which we can get a general definition between two arbitrary vectors $\mathbf{A}$ and $\mathbf{B}$. As shown in Figure C.3 we project $\mathbf{A}$ onto the coordinate axes x, y, z and define this projection as the 'dot' product of $\mathbf{A}$ and the unit vectors $\hat{\mathbf{x}}, \hat{\mathbf{y}}, \hat{\mathbf{z}}$:

$$A_x = A \cos \alpha \equiv \mathbf{A} \cdot \hat{\mathbf{x}}$$
$$A_y = A \cos \beta \equiv \mathbf{A} \cdot \hat{\mathbf{y}}$$
$$A_z = A \cos \gamma \equiv \mathbf{A} \cdot \hat{\mathbf{z}}$$

We posit that the scalar product is associative and distributive:

$$\mathbf{A} \cdot (y\mathbf{B}) = (y\mathbf{A}) \cdot \mathbf{B} = y\mathbf{A} \cdot \mathbf{B} \tag{C.11}$$
$$\mathbf{A} \cdot (\mathbf{B} + \mathbf{C}) = \mathbf{A} \cdot \mathbf{B} + \mathbf{A} \cdot \mathbf{B} \tag{C.12}$$

We use the associative property to get a generalisation of the dot product between two vectors $\mathbf{A}$ and $\mathbf{B}$. We can write $\mathbf{B}$ in terms of its components and the unit vectors along the Cartesian axes:

$$\mathbf{B} = B_x\hat{\mathbf{x}} + B_y\hat{\mathbf{y}} + B_z\hat{\mathbf{z}}$$

Then,

$$\mathbf{A} \cdot \mathbf{B} = \mathbf{A} \cdot \left(B_x\hat{\mathbf{x}} + B_y\hat{\mathbf{y}} + B_z\hat{\mathbf{z}} \right) \tag{C.13}$$

But then invoking the associative property,

$$\mathbf{A} \cdot \mathbf{B} = B_x\mathbf{A} \cdot \hat{\mathbf{x}} + B_y\mathbf{A} \cdot \hat{\mathbf{y}} + B_z\mathbf{A} \cdot \hat{\mathbf{z}}$$
$$= B_xA_x + B_yA_y + B_zA_z$$

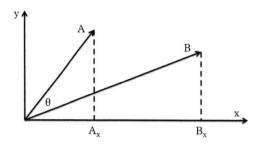

Figure C.3 *Vector scalar product.* $A \cdot B = AB \cos \theta$.

Therefore:

$$\mathbf{A} \cdot \mathbf{B} = \sum_i B_i A_i = \sum_i A_i B_i = \mathbf{B} \cdot \mathbf{A} \tag{C.14}$$

and we see that the scalar product between two vectors is commutative as well as associative. In the same way we defined $A_x = \mathbf{A} \cdot \hat{\mathbf{x}} = A \cos\alpha$, we could define the projection of $\mathbf{A}$ onto the $\mathbf{B}$ direction as

$$A_B = A\cos\theta \equiv \mathbf{A} \cdot \hat{\mathbf{B}} = \mathbf{A} \cdot \mathbf{B}/B$$

Then multiplying this last equivalency by B,

$$BA\cos\theta = \mathbf{A} \cdot \mathbf{B}$$

Similarly, we could specify the projection of $\mathbf{B}$ onto the $\mathbf{A}$ direction. Then $B_A = B\cos\theta \equiv \mathbf{B} \cdot \mathbf{A}/A$. Then

$$AB\cos\theta = \mathbf{B} \cdot \mathbf{A}$$

Once again we see that the scalar product between any two vectors is commutative and equal to the product of the two vector 'lengths' and the cosine of the angle between them.

The distributive property, Equation C.12, is illustrated in Figure C.4. The figure shows geometrically that the sum of the projections of $\mathbf{B}$ and $\mathbf{C}$ onto $\mathbf{A}$ is equivalent to the projection of their vector sum $\mathbf{B} + \mathbf{C}$ onto $\mathbf{A}$:

$$\mathbf{B} \cdot \mathbf{A} + \mathbf{C} \cdot \mathbf{A} = (\mathbf{B} + \mathbf{C}) \cdot \mathbf{A}$$

C.3.2 Vector product

The vector or 'cross' product of two vectors $\mathbf{A}$ and $\mathbf{B}$ has meaning in 3-D and is defined as

$$\mathbf{A} \times \mathbf{B} = \mathbf{C} \tag{C.15}$$

The vector product of $\mathbf{A}$ and $\mathbf{B}$ is another vector $\mathbf{C}$. The magnitude of $\mathbf{C}$ is related to those of $\mathbf{A}$ and $\mathbf{B}$ by

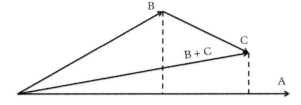

Figure C.4 *Distributive law of vector addition.* $A \cdot (B + C) = A \cdot B + A \cdot C.$

$$|\mathbf{C}| = |\mathbf{A}| \cdot |\mathbf{B}| \sin\theta \tag{C.16}$$

where θ is the angle between vectors $\mathbf{A}$ and $\mathbf{B}$. The direction of $\mathbf{C}$ is governed by the 'right-hand screw rule'. The vectors $\mathbf{A}$ and $\mathbf{B}$, separated by angle θ, form a plane. As $\mathbf{A}$ rotates towards $\mathbf{B}$, the positive direction of $\mathbf{C}$ points in the same direction as would a right-hand threaded screw. A consequence of this convention is that

$$\mathbf{A} \times \mathbf{B} = -\mathbf{B} \times \mathbf{A} \tag{C.17}$$

The cross product is *anticommutative*. From the definition we can directly obtain basic relations among the Cartesian unit vectors:

$$\hat{\mathbf{x}} \times \hat{\mathbf{y}} = \hat{\mathbf{z}} \qquad \hat{\mathbf{y}} \times \hat{\mathbf{z}} = \hat{\mathbf{x}} \qquad \hat{\mathbf{z}} \times \hat{\mathbf{x}} = \hat{\mathbf{y}} \tag{C.18}$$

and

$$\hat{\mathbf{y}} \times \hat{\mathbf{x}} = -\hat{\mathbf{z}} \qquad \hat{\mathbf{z}} \times \hat{\mathbf{y}} = -\hat{\mathbf{x}} \qquad \hat{\mathbf{x}} \times \hat{\mathbf{z}} = -\hat{\mathbf{y}} \tag{C.19}$$

It is also clear from the definition, Equation C.16 that

$$\hat{\mathbf{x}} \times \hat{\mathbf{x}} = \hat{\mathbf{y}} \times \hat{\mathbf{y}} = \hat{\mathbf{z}} \times \hat{\mathbf{z}} = 0 \tag{C.20}$$

Notice that these relations can be generated by a cyclic commutation of $\hat{\mathbf{x}}, \hat{\mathbf{y}}, \hat{\mathbf{z}}$ starting from any one of them. In analogy to the scalar product we posit that the cross product operation is distributive and associative. Thus:

$$\mathbf{A} \times (\mathbf{B} + \mathbf{C}) = \mathbf{A} \times \mathbf{B} + \mathbf{A} \times \mathbf{C} \tag{C.21}$$
$$(\mathbf{A} + \mathbf{B}) \times \mathbf{C} = \mathbf{A} \times \mathbf{C} + \mathbf{B} \times \mathbf{C} \tag{C.22}$$
$$\mathbf{A} \times (s\mathbf{B}) = s\mathbf{A} \times \mathbf{B} = (s\mathbf{A}) \times \mathbf{B} \tag{C.23}$$

Now let us take $\mathbf{A} \times \mathbf{B}$ and decompose this expression into its Cartesian components:

$$\mathbf{A} \times \mathbf{B} = \left(A_x\hat{\mathbf{x}} + A_y\hat{\mathbf{y}} + A_z\hat{\mathbf{z}}\right) \times \left(B_x\hat{\mathbf{x}} + B_y\hat{\mathbf{y}} + B_z\hat{\mathbf{z}}\right) \tag{C.24}$$
$$= (A_yB_z - A_zB_y)\hat{\mathbf{x}} - (A_xB_z - A_zB_x)\hat{\mathbf{y}} + \left(A_xB_y - A_yB_x\right)\hat{\mathbf{z}}$$
$$= C_x\hat{\mathbf{x}} + C_y\hat{\mathbf{y}} + C_z\hat{\mathbf{z}} = \mathbf{C} \tag{C.25}$$

We see that the Cartesian components of $\mathbf{C}$ are constructed from combining products of the Cartesian components of $\mathbf{A}$ and $\mathbf{B}$. Note also that these combinations commute cyclically. The easiest way to generate the components of the vector product is by writing a determinant and expanding it. The first row of the determinant consists of the Cartesian unit vectors, $\hat{\mathbf{x}}, \hat{\mathbf{y}}, \hat{\mathbf{z}}$. The Cartesian components of $\mathbf{A}$ composes the second row and the Cartesian components of $\mathbf{B}$, the third:

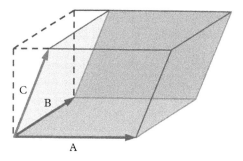

A

B

C

Figure C.5 *Geometrical interpretation of the triple scalar product, $A \cdot (B \times C)$.*

$$C = \begin{vmatrix} \hat{\mathbf{x}} & \hat{\mathbf{y}} & \hat{\mathbf{z}} \\ A_x & A_y & A_z \\ B_x & B_y & B_z \end{vmatrix} \qquad \text{(C.26)}$$

Expansion of this determinant results in an expression identical to the second line of Equation C.24, obtained from the cross product definition applied to the Cartesian unit vectors.

C.3.3 Triple products

Because the cross product of two vectors is itself a vector we can 'dot' this result and 'cross' it with yet another vector. Since the dot operation results in a scalar quantity, this triple product is called the scalar triple product. Figure C.5 shows that the scalar triple product can be interpreted geometrically as the volume of a parallelepiped. Furthermore, the scalar triple product is cyclically commutative:

$$\mathbf{A} \cdot (\mathbf{B} \times \mathbf{C}) = \mathbf{B} \cdot (\mathbf{C} \times \mathbf{A}) = \mathbf{C} \cdot (\mathbf{A} \times \mathbf{B}) \qquad \text{(C.27)}$$

Because of the anticommutivity of the cross product operation,

$$\mathbf{A} \cdot (\mathbf{B} \times \mathbf{C}) = -\mathbf{A} \cdot (\mathbf{C} \times \mathbf{B})$$

The vector triple product can be simplified by the 'BAC-CAB' rule:

$$\mathbf{A} \times (\mathbf{B} \times \mathbf{C}) = \mathbf{B}(\mathbf{A} \cdot \mathbf{C}) - \mathbf{C}(\mathbf{A} \cdot \mathbf{B}) \qquad \text{(C.28)}$$

which can be verified by direct expansion in Cartesian coordinates.

C.4 Vector fields

The idea of a *field* is the assignment of some value or function to all points in a region of geometric space. If the entity assigned is a scalar, then we have a scalar field. The distribution of temperature throughout the solar system is an example of a scalar field. In the theory of electromagnetism, the concept of a *vector* field is essential. A vector field is an assignment of a vector to each point in some region of geometrical space. The electric **E** and magnetic **B** force fields produced by an oscillating dipole source or propagating in a plane wave are examples of vector force fields.

Associated with vector and scalar fields are three differential operations important in physics and engineering: the gradient, the divergence, and (in 3-D) the curl operations.

C.4.1   Gradient

If $\varphi(x, y, z)$ is a scalar field then the *gradient* of φ is defined as

$$\nabla \varphi(x, y, z) = \frac{\partial \varphi}{\partial x} \hat{\mathbf{x}} + \frac{\partial \varphi}{\partial y} \hat{\mathbf{y}} + \frac{\partial \varphi}{\partial z} \hat{\mathbf{z}} \tag{C.29}$$

Notice that the gradient of scalar field is a vector field consisting of the partial derivatives of the scalar field pointing in the x, y, z directions. If we just considered the total differential of φ in terms of its partial derivatives we have

$$d\varphi = \frac{\partial \varphi}{\partial x} dx + \frac{\partial \varphi}{\partial y} dy + \frac{\partial \varphi}{\partial z} dz \tag{C.30}$$

In fact, we can write this scalar differential as the scalar product of $\nabla \varphi$ and the vector coordinate differential $d\mathbf{r} = dx\,\hat{\mathbf{x}} + dy\,\hat{\mathbf{y}} + dz\,\hat{\mathbf{z}}$:

$$d\varphi = \nabla \varphi \cdot d\mathbf{r} = |\nabla \varphi||d\mathbf{r}| \cos \theta \tag{C.31}$$

where the angle θ is the angle between the gradient and the coordinate vector. Clearly, in order for Equation C.31 to be in accord with Equation C.30, the angle $\theta = 0$. If θ were any other value, then the dot product in Equation C.31 would be less than $d\varphi$. So $\nabla \varphi$ as defined in Equation C.29 points in the direction of maximum change in φ and its magnitude $|\nabla \varphi|$ yields the greatest rate of change (slope) along this direction.

The $\nabla = \partial/\partial x\,\hat{\mathbf{x}} + \partial/\partial y\,\hat{\mathbf{y}} + \partial/\partial z\,\hat{\mathbf{z}}$ is a vector operator which acts on a scalar field to produce a vector field. The most familiar example of the use of the gradient in electromagnetism is the relation between an electrostatic field **E** and an electrical potential $V(x, y, z)$:

$$\mathbf{E} = -\nabla V$$

The gradient operator is sometimes called 'grad' or 'del' and the gradient operation on a scalar field φ is termed 'grad φ' or 'del φ'. The ∇ symbol itself is called the 'nabla'.

C.4.2   Divergence

The divergence operation is the dot product of the vector gradient operator with a *vector* field. Suppose we have a vector field $\boldsymbol{\varphi}$. The result of the divergence operation is

$$\nabla \cdot \boldsymbol{\varphi} = \left[\frac{\partial}{\partial x}\hat{\mathbf{x}} + \frac{\partial}{\partial y}\hat{\mathbf{y}} + \frac{\partial}{\partial z}\hat{\mathbf{z}} \right] \cdot \left[\varphi_x \hat{\mathbf{x}} + \varphi_y \hat{\mathbf{y}} + \varphi_z \hat{\mathbf{z}} \right]$$

$$= \frac{\partial \varphi_x}{\partial x} + \frac{\partial \varphi_y}{\partial y} + \frac{\partial \varphi_z}{\partial z} \qquad (C.32)$$

Notice that the divergence operation results in a scalar field because it is a 'dot product' operation between the vector operator ∇ and the vector field $\boldsymbol{\varphi}$. The divergence of a vector field is a measure of how much the field increases or diverges along the field coordinates. A vector field radiating from a point in all directions has a very high divergence. A vector field whose magnitude is invariant along some coordinate has a null divergence. A positive (negative) divergence denotes an increasing (decreasing) field. This geometrical representation of divergence is shown in Figure C.6. The divergence of vector field is sometimes written as *div* $\boldsymbol{\varphi}$. The most familiar example of the use of the divergence in electromagnetism is the differential form of Gauss's law,

$$\nabla \cdot \mathbf{E} = \frac{\rho}{\varepsilon_0} \qquad (C.33)$$

where $\mathbf{E}$ is the electric field and ρ the electric charge density.

C.4.3   Curl

The vector product of ∇ and a vector field $\boldsymbol{\varphi}$ is another vector field $\boldsymbol{\chi}$, called the curl of $\boldsymbol{\varphi}$:

$$\boldsymbol{\chi} = \nabla \times \boldsymbol{\varphi} \qquad (C.34)$$

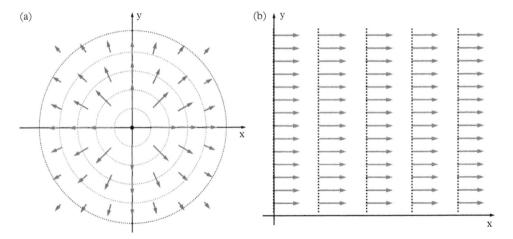

Figure C.6 *Panel (a) shows a vector field with very high negative divergence. Vector amplitudes decrease with distance from origin. Panel (b) shows a constant vector field along x which exhibits null divergence.*

From the determinant rule for the cross product of two vectors we can write

$$\chi = \begin{vmatrix} \hat{x} & \hat{y} & \hat{z} \\ \frac{\partial}{\partial x} & \frac{\partial}{\partial y} & \frac{\partial}{\partial z} \\ \varphi_x & \varphi_y & \varphi_z \end{vmatrix} \tag{C.35}$$

Expanding the determinant, we have explicitly,

$$\chi = \left[\frac{\partial \varphi_z}{\partial y} - \frac{\partial \varphi_z}{\partial y}\right]\hat{x} - \left[\frac{\partial \varphi_z}{\partial x} - \frac{\partial \varphi_x}{\partial z}\right]\hat{y} + \left[\frac{\partial \varphi_y}{\partial x} - \frac{\partial \varphi_x}{\partial y}\right]\hat{z} \tag{C.36}$$

The curl of a vector field is measure of its 'rotation' around an axis perpendicular to the plane of rotation. A vortex field has a very large curl while a field emanating radially from a point in all directions has no curl. Figure C.7 shows a field of all curl and null divergence. The explicit expression for this vector field is

$$V(x, y) = -\frac{y}{\sqrt{x^2 + y^2}}\hat{i} + \frac{x}{\sqrt{x^2 + y^2}}\hat{j} \tag{C.37}$$

Figure C.8 shows a vector field with finite divergence but null curl along the $x = y$ diagonals. The expression for this vector field is

$$V(x, y) = \frac{y}{\sqrt{x^2 + y^2}}\hat{i} + \frac{x}{\sqrt{x^2 + y^2}}\hat{j} \tag{C.38}$$

One can show that the divergence of the curl of a vector field is null:

$$\nabla \cdot [\nabla \times (\nabla \varphi)] = 0 \tag{C.39}$$

and the curl of the gradient of a scalar field is null:

$$\nabla \times (\nabla \varphi) = 0 \tag{C.40}$$

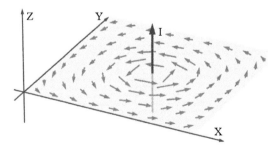

Figure C.7 *A vector field circulating anti-clockwise in the x–y plane producing a curl field along z but no divergence.*

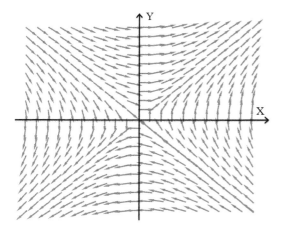

Figure C.8 *A vector field with a null curl at the origin and along the x = y diagonals but with finite divergence for all x, y except at the point x = 0, y = 0 where the divergence is null.*

The most celebrated curl equation in electromagnetism is Faraday's law:

$$\nabla \times \mathbf{E} = -\frac{\partial \mathbf{B}}{\partial t}$$

In this discussion we have described the grad, div, and curl operations in terms of Cartesian coordinates. But many other curvilinear coordinate systems are a more convenient choice, depending on the symmetry of the problem. Two commonly encountered curvilinear coordinate systems in physics are cylindrical and polar coordinates. The expressions for grad, div, and curl in these coordinates is discussed in Appendix D.

C.5 Integral theorems for vector fields

C.5.1 Integral of the gradient

We saw in Equation C.31 that the differential of a scalar field $d\varphi$ is equal to the dot product of the gradient of the field $\nabla\varphi$ and the differential of coordinate length d**r**. If we integrate the differential $d\varphi$ from $\varphi(x_0, y_0, z_0)$ to $\varphi(x_1, y_1, z_1)$ we have

$$\int_0^1 d\varphi = \varphi(x_1, y_1, z_1) - \varphi(x_0, y_0, z_0) = \int_0^1 \nabla\varphi \cdot d\mathbf{r} \tag{C.41}$$

Equation C.41 says that the line integral of the vector field $\nabla\varphi$ using the path d**r** is equal to the difference between the scalar field values at the two end points. Since this difference does not depend on the path, the result of the line integral is independent of the path taken; and in particular,

$$\oint \nabla\varphi \cdot d\mathbf{r} = 0 \tag{C.42}$$

for any closed loop since for these the ending point is identical to the starting point. The integral over a path of the gradient of a scalar field is commonly encountered in electrostatic problems where the task is to calculate the potential difference between, say, two conductors that form an electric field between them. The potential difference is given by

$$V_{21} = V_2 - V_1 = -\int_1^2 \mathbf{E} \cdot d\mathbf{r} = -\int_1^2 \nabla V \cdot d\mathbf{r} \qquad (C.43)$$

C.5.2 Integral of the divergence

The expression for the divergence theorem or Gauss's theorem is

$$\int_{\mathrm{vol}} (\nabla \cdot \boldsymbol{\chi}) \, dV = \oint_{\mathrm{surf}} \boldsymbol{\chi} \cdot d\mathbf{S} \qquad (C.44)$$

where dV is the differential volume element and $d\mathbf{S}$ is a differential element of the surface surrounding (bounding) the volume. Here $\boldsymbol{\chi}$ is a vector field, in contrast to φ of Equation C.31, a scalar field. Note that the surface element $d\mathbf{S}$ is itself a differential vector with the unit vector direction perpendicular to the surface and conventionally pointing outwards. A common application of the divergence theorem is the integral form of Gauss's law, discussed in Chapter 2, Equation 2.124:

$$\int_{\mathrm{vol}} \nabla \cdot \mathbf{D} \, dV = \varepsilon_0 \int_{\mathrm{vol}} \nabla \cdot \mathbf{E} \, dV = \int_{\mathrm{vol}} \rho \, dV = \varepsilon_0 \int_{\mathrm{surf}} \mathbf{E} \cdot d\mathbf{S}$$

where $\mathbf{D}$ is the displacement vector field and $\mathbf{E}$ is the electric field emanating from the charge density ρ enclosed by the surface boundary S of volume V. Note that the surface differential element is a vector differential with direction pointing outwards from the boundary and perpendicular to it.

C.5.3 Integral of the curl

The surface integral over the curl of a vector is called Stokes' theorem:

$$\int_{\mathrm{surf}} (\nabla \times \boldsymbol{\chi}) \cdot d\mathbf{S} = \oint \boldsymbol{\chi} \cdot d\mathbf{r} \qquad (C.45)$$

The line integral on the right is taken around the boundary of the surface integral on the left. A convenient way to remember and visualise the geometric interpretation of the curl integral is to think of Ampère's law. In differential form it is

$$\nabla \times \mathbf{H} = \mathbf{J}_{\mathrm{free}}$$

where $\mathbf{H}$ is the magnetic field and $\mathbf{J}_{\text{free}}$ is a free charge current density. The integral form of this equation is, taking into account the curl theorem,

$$\int_{\text{surf}} (\nabla \times \mathbf{H}) \cdot d\mathbf{S} = \oint \mathbf{H} \cdot d\mathbf{r} = \int_{\text{surf}} \mathbf{J}_{\text{free}} \cdot d\mathbf{S} = I \tag{C.46}$$

where I is the charge current running in a wire of cross section S. This relation can be pictured as a loop of H-field surrounding a straight wire and generated by a current I running through it.

C.6 Useful identities of vector calculus

$$a \cdot b \times c = b \cdot c \times a - c \cdot a \times b \tag{C.47}$$
$$a \times (b \times c) = (a \cdot c)b - (a \cdot b)c \tag{C.48}$$
$$(a \times b) \cdot (c \times d) = a \cdot b \times (c \times d) \tag{C.49}$$
$$= a \cdot (b \cdot dc - b \cdot cd) \tag{C.50}$$
$$= (a \cdot c)(b \cdot d) - (a \cdot d)(b \cdot c) \tag{C.51}$$
$$(a \times b) \times (c \times d) = (a \times b \cdot d)c - (a \times b \cdot c)d \tag{C.52}$$
$$\nabla(\varphi + \psi) = \nabla\varphi + \nabla\psi \tag{C.53}$$
$$\nabla(\varphi\psi) = \varphi\nabla\psi + \psi\nabla\varphi \tag{C.54}$$
$$\nabla \cdot (a + b) = \nabla \cdot a + \nabla \cdot b \tag{C.55}$$
$$\nabla \times (a + b) = \nabla \times a + \nabla \times b \tag{C.56}$$
$$\nabla \cdot (\varphi a) = a \cdot \nabla\varphi + \varphi\nabla \cdot a \tag{C.57}$$
$$\nabla \times (\varphi a) = \nabla\varphi \times a + \varphi\nabla \times a \tag{C.58}$$
$$\nabla(a \cdot b) = (a \cdot \nabla)b + (b \cdot \nabla)a + a \times (\nabla \times b) + b \times (\nabla \times a) \tag{C.59}$$
$$\nabla \cdot (a \times b) = b \cdot \nabla \times a - a \cdot \nabla \times b \tag{C.60}$$
$$\nabla \times (a \times b) = a\nabla \cdot b - b\nabla \cdot a + (b \cdot \nabla)a - (a \cdot \nabla)b \tag{C.61}$$
$$\nabla \times \nabla \times a = \nabla\nabla \cdot a - \nabla^2 a \tag{C.62}$$
$$\nabla \times \nabla\varphi = 0 \tag{C.63}$$
$$\nabla \cdot \nabla \times a = 0 \tag{C.64}$$

Appendix D
Gradient, Divergence, and Curl in Cylindrical and Polar Coordinates

Although Cartesian coordinates are the most familiar and serve many purposes, they are not the only orthogonal coordinate system that can be used to define a set of basis vectors. Often we can take advantage of the natural symmetry of a problem to express vectors in some other orthogonal curvilinear coordinate system. The most common of these are the cylindrical and polar coordinates because they are appropriate for many practical problems.

In general we can expand a vector V in basis vectors of the Cartesian system or some other system with basis vectors $\{q\}$,

$$V = \hat{x}V_x + \hat{y}V_y + \hat{z}V_z \qquad \text{or} \qquad V = \hat{q}_1 V_1 + \hat{q}_2 V_2 + \hat{q}_3 V_3 \qquad (D.1)$$

However, we cannot treat the position vector with this general rule. In this special case

$$
\begin{aligned}
r &= x\hat{x} + y\hat{y} + z\hat{z} & \text{cartesian coordinates} & \qquad (D.2) \\
r &= \rho\hat{\rho} + z\hat{z} & \text{cylindrical coordinates} & \qquad (D.3) \\
r &= r\hat{r} & \text{spherical coordinates} & \qquad (D.4)
\end{aligned}
$$

In the general case we convert from the Cartesian system to another through the relations

$$
\begin{aligned}
x &= x(q_1, q_2, q_3) \\
y &= y(q_1, q_2, q_3) \\
z &= z(q_1, q_2, q_3)
\end{aligned}
$$

The complete differential in x can be written as

$$dx = \frac{\partial x}{\partial q_1} dq_1 + \frac{\partial y}{\partial q_2} dq_2 + \frac{\partial y}{\partial q_3} dq_3$$

and similarly for y and z. The differential of the position vector r in the Cartesian basis is

$$dr = d\hat{x} + d\hat{y} + d\hat{z}$$

and the square of the distance between two neighbouring points can then be written as

$$ds^2 = dr \cdot dr = dx^2 + dy^2 + dz^2$$

But the differential of dr can also be expanded in the curvilinear coordinate system:

$$dr = \frac{\partial r}{\partial q_1} dq_1 + \frac{\partial r}{\partial q_2} dq_2 + \frac{\partial r}{\partial q_3} dq_3 \tag{D.5}$$

and

$$ds^2 = dr \cdot dr = \sum_{ij} \frac{\partial r}{\partial q_i} \cdot \frac{\partial r}{\partial q_j} dq_i dq_j \tag{D.6}$$

Remembering that we have supposed that the curvilinear coordinates are orthogonal,

$$\hat{q}_i \cdot \hat{q}_j = \delta_{ij}$$

we write Equation D.6:

$$ds^2 = \sum_i \frac{\partial^2 r}{\partial q_i^2} dq_i^2 = g_{11} dq_1^2 + g_{22} dq_2^2 + g_{33} dq_3^2 = \sum_i (h_i dq_i)^2$$

where we have identified $g_{ii} = h_i^2$. We see from Equation D.6 that h_i is a scale factor such that the differential length segment ds_i can be written as

$$ds_i = h_i dq_i \qquad \text{and} \qquad \frac{\partial r}{\partial q_i} = h_i \hat{q}_i$$

and from Equation D.5 the position vector differential dr can be expanded in terms of these scale factors in the curvilinear basis as

$$dr = h_1 dq_1 \hat{q}_1 + h_2 dq_2 \hat{q}_2 + h_3 dq_3 \hat{q}_3 \tag{D.7}$$

From the line segment ds_i we can easily write the surface and volume elements σ, τ in the curvilinear system:

$$d\sigma_{ij} = ds_i ds_j = h_i h_j \, dq_i dq_j \qquad i \neq j$$

and

$$d\tau = ds_1 ds_2 ds_3 = h_1 h_2 h_3 \, dq_1 dq_2 dq_3$$

We can expand an area vector σ in the curvilinear basis set analogous to the position vector r in Equation D.7:

$$d\boldsymbol{\sigma} = ds_2 ds_3 d\hat{\boldsymbol{q}}_1 + ds_3 ds_1 \hat{\boldsymbol{q}}_2 + ds_1 ds_2 \hat{\boldsymbol{q}}_3$$
$$= h_2 h_3 \, dq_2 dq_3 \hat{\boldsymbol{q}}_1 + h_3 h_1 \, dq_3 dq_1 \hat{\boldsymbol{q}}_2 + h_1 h_2 \, dq_1 dq_2 \hat{\boldsymbol{q}}_3 \tag{D.8}$$

For example, we can take an ordinary vector quantity F and expand it in Cartesian coordinates or in spherical coordinates:

$$\boldsymbol{F} = F_x \hat{\boldsymbol{x}} + F_y \hat{\boldsymbol{y}} + F_z \hat{\boldsymbol{z}} \tag{D.9}$$
$$\boldsymbol{F} = F_r \hat{\boldsymbol{r}} + F_\theta \hat{\boldsymbol{\theta}} + F_\varphi \hat{\boldsymbol{\varphi}} \tag{D.10}$$

The unit vectors $\hat{r}, \hat{\theta}, \hat{\varphi}$ in terms of the Cartesian unit vectors are

$$\hat{r} = \sin\theta \, \cos\varphi \, \hat{\boldsymbol{x}} + \sin\theta \, \sin\varphi \, \hat{\boldsymbol{y}} + \cos\theta \, \hat{\boldsymbol{z}} \tag{D.11}$$
$$\hat{\theta} = \cos\theta \, \cos\varphi \, \hat{\boldsymbol{x}} + \cos\theta \, \sin\varphi \, \hat{\boldsymbol{y}} - \sin\theta \, \hat{\boldsymbol{z}} \tag{D.12}$$
$$\hat{\varphi} = -\sin\varphi \, \hat{\boldsymbol{x}} + \cos\varphi \, \hat{\boldsymbol{y}} \tag{D.13}$$

The differential length vector along the surface in spherical coordinates is then

$$d\boldsymbol{l} = dr\,\hat{r} + r\,d\theta\,\hat{\boldsymbol{\theta}} + r\sin\theta\,d\varphi\,\hat{\boldsymbol{\varphi}} \tag{D.14}$$

and the differential volume element in spherical coordinates is obtained from the product of $ds_1 \, ds_2 \, ds_3$ with

$$ds_1 = h_1 \, dq_1 = dr \tag{D.15}$$
$$ds_2 = h_2 \, dq_2 = r \, d\theta \tag{D.16}$$
$$ds_3 = h_3 \, dq_3 = r\sin\theta \, d\varphi \tag{D.17}$$

The differential volume element in spherical coordinates is therefore:

$$dV = ds_1 \, ds_2 \, ds_3 = dr \, r d\theta \, r\sin\theta \, d\varphi = r^2 \sin\theta \, d\theta \, d\varphi \, dr \tag{D.18}$$

The expressions for $d\boldsymbol{r}$ and $d\boldsymbol{\sigma}$ in Equations D.7 and D.8 give us the tools to write vector line and surface integrals in the curvilinear coordinates:

$$\int V \cdot d\boldsymbol{r} = \sum_i \int V_i h_i dq_i \tag{D.19}$$

$$\int V \cdot d\boldsymbol{\sigma} = \sum_i \int V_i h_j h_k \, dq_j dq_k \qquad j \neq k \neq i \tag{D.20}$$

D.1 The gradient in curvilinear coordinates

The gradient is a vector operator ∇ that operates on a scalar point function ψ. In Cartesian coordinates we write

$$\nabla\psi = \left(\frac{\partial}{\partial x}\hat{x} + \frac{\partial}{\partial y}\hat{y} + \frac{\partial}{\partial z}\hat{z}\right)\psi(x, y, z) \tag{D.21}$$

An alternative integral definition is more convenient for finding the expression for the gradient in some curvilinear coordinate system:

$$\nabla\psi = \lim_{\int d\tau \to 0}\frac{\int \psi\, d\boldsymbol{\sigma}}{\int d\tau} \tag{D.22}$$

The gradient, in general, is defined so that it yields the maximum spatial rate of change of the scalar function, and this maximum spatial change is independent of the coordinate system in which it is described. Therefore, we should be able to express it in coordinate systems other than Cartesian. Remembering that the length segments dx, dy, dz in Cartesian coordinates go over into ds_1, ds_2, ds_3 in our chosen curvilinear system, we can write the gradient in our new coordinates as

$$\nabla\psi(q_1, q_2, q_3) = \hat{q}_1\frac{\partial\psi}{\partial s_1} + \hat{q}_2\frac{\partial\psi}{\partial s_2} + \hat{q}_3\frac{\partial\psi}{\partial s_3} \tag{D.23}$$

and also remembering that $ds_i = h_i dq_i$ we have

$$\nabla\psi(q_1, q_2, q_3) = \hat{q}_1\frac{1}{h_1}\frac{\partial\psi}{\partial q_1} + \hat{q}_2\frac{1}{h_2}\frac{\partial\psi}{\partial q_2} + \hat{q}_3\frac{1}{h_3}\frac{\partial\psi}{\partial q_3} \tag{D.24}$$

D.2 The divergence in curvilinear coordinates

The divergence $\nabla \cdot V$ of some vector field V can be expressed as a differential in Cartesian coordinates as

$$\left(\frac{\partial}{\partial x}\hat{x} + \frac{\partial}{\partial y}\hat{y} + \frac{\partial}{\partial z}\hat{z}\right) \cdot \left(V_x\hat{x} + V_y\hat{y} + V_z\hat{z}\right) \tag{D.25}$$

Similar to the gradient operation, the divergence can be generalised to curvilinear coordinates by considering it as the result of taking the limit of a differential vector field surface integral divided by the differential volume enclosed by the surface as the volume approaches zero. Taking this limiting ratio operation as the definition of the divergence we have

$$\nabla \cdot V(q_1, q_2, q_3) = \lim_{\int d\tau \to 0}\frac{\delta\int V \cdot d\boldsymbol{\sigma}}{\int d\tau}$$

and the differential of the surface integral is

$$\delta \int V(q_1, q_2, q_3) \cdot d\boldsymbol{\sigma} =$$

$$\left[\frac{\partial(V_1 h_2 h_3)}{\partial q_1} + \frac{\partial(V_2 h_3 h_1)}{\partial q_2} + \frac{\partial(V_3 h_1 h_2)}{\partial q_3} \right] dq_1 dq_3 dq_3$$

and taking $\int d\tau$ in the limit as $ds_1 ds_2 ds_3 = h_1 dq_1 \, h_2 dq_2 \, h_3 dq_3$ we have for the limiting ratio

$$\nabla \cdot V(q_1, q_2, q_3) = \frac{1}{h_1 h_2 h_3} \left[\frac{\partial}{\partial q_1} (V_1 h_2 h_3) + \frac{\partial}{\partial q_2} (V_2 h_3 h_1) + \right.$$

$$\left. \frac{\partial}{\partial q_3} (V_3 h_1 h_2) \right] \tag{D.26}$$

Another useful general expression is the scalar Laplacian operator, $\nabla \cdot \nabla$, which we can get from combining the grad and div operations,

$$\nabla \cdot \nabla \psi = \frac{1}{h_1 h_2 h_3} \left[\frac{\partial}{\partial q_1} \left(\frac{h_2 h_3}{h_1} \frac{\partial \psi}{\partial q_1} \right) + \frac{\partial}{\partial q_2} \left(\frac{h_3 h_1}{h_2} \frac{\partial \psi}{\partial q_2} \right) + \right.$$

$$\left. \frac{\partial}{\partial q_3} \left(\frac{h_1 h_2}{h_3} \frac{\partial \psi}{\partial q_3} \right) \right] \tag{D.27}$$

D.3   The curl in curvilinear coordinates

The familiar differential expression for the curl operation in Cartesian coordinates is

$$\nabla \times V = \left(\frac{\partial V_z}{\partial y} - \frac{\partial V_y}{\partial z} \right) \hat{x} + \left(\frac{\partial V_x}{\partial z} - \frac{\partial V_z}{\partial x} \right) \hat{y} + \left(\frac{\partial V_y}{\partial x} - \frac{\partial V_x}{\partial y} \right) \hat{z}$$

and the easiest way to remember it, is by writing the expression as determinant:

$$\nabla \times V = \begin{vmatrix} \hat{x} & \hat{y} & \hat{z} \\ \frac{\partial}{\partial x} & \frac{\partial}{\partial y} & \frac{\partial}{\partial z} \\ V_x & V_y & V_z \end{vmatrix}$$

As with the gradient and divergence operations, the curl can also be written as the limiting integral operation,

$$\nabla \times V = \lim_{\int d\tau \to 0} \frac{\int d\boldsymbol{\sigma} \times V}{\int d\tau}$$

It can be shown that by using this integral limiting form, and applying Stokes' theorem, the curl operation can be written in curvilinear coordinates as

$$\nabla \times V = \frac{1}{h_1 h_2 h_3} \begin{vmatrix} \hat{q}_1 h_1 & \hat{q}_2 h_2 & \hat{q}_3 h_3 \\ \frac{\partial}{\partial q_1} & \frac{\partial}{\partial q_2} & \frac{\partial}{\partial q_3} \\ h_1 V_1 & h_2 V_2 & h_3 V_3 \end{vmatrix} \qquad (D.28)$$

D.4 Expressions for grad, div, curl in cylindrical and polar coordinates

D.4.1 Circular cylindrical coordinates

Circular cylindrical coordinates consist of three independent variables, ρ, φ, z and their associated unit vectors, $\hat{\rho}$, $\hat{\varphi}$, $\hat{z}$. The limits on these variables are,

$$0 \leq \rho \leq \infty \qquad 0 \leq \varphi \leq 2\pi \qquad -\infty < z < +\infty$$

Figure D.1 shows the relation between the Cartesian coordinates x, y, z and the cylindrical coordinates ρ, φ, z. The z coordinate is common to both systems. From Figure D.1 it is evident that

$$\rho = \sqrt{(x^2 + y^2)} \qquad x = \rho \cos \varphi \qquad y = \rho \sin \varphi \qquad z = z$$

The differential volume element dV is

$$dV = ds_\rho \, ds_\varphi \, ds_z = \rho d\rho \, \rho d\varphi \, dz$$

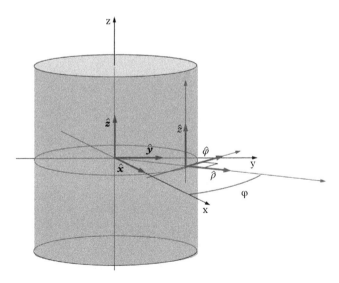

Figure D.1 *Cylindrical coordinates ρ, φ, z and unit vectors $\hat{\rho}, \hat{\varphi}, \hat{z}$ and their relations to Cartesian coordinates and unit vectors, x, y, z, $\hat{x}, \hat{y}, \hat{z}$.*

From which we identify

$$h_1 = h_\rho = 1 \qquad h_2 = h_\varphi = \rho \qquad h_3 = h_z = 1$$

Then from the general expressions for grad, div, curl, Equations D.24, D.26, and D.28 we write these operators in cylindrical coordinates as

$$\text{grad} \qquad \nabla\psi\,(\rho,\varphi,z) = \frac{\partial\psi}{\partial\rho}\hat{\rho} + \frac{1}{\rho}\frac{\partial\psi}{\partial\varphi}\hat{\varphi} + \frac{\partial\psi}{\partial z}\hat{z} \tag{D.29}$$

$$\text{div} \qquad \nabla\cdot V = \frac{1}{\rho}\frac{\partial(\rho V_\rho)}{\partial\rho} + \frac{1}{\rho}\frac{\partial V_\varphi}{\partial\varphi} + \frac{\partial V_z}{\partial z} \tag{D.30}$$

$$\text{curl} \qquad \nabla \times V = \frac{1}{\rho} \begin{vmatrix} \hat{\rho} & \rho\hat{\varphi} & \hat{z} \\ \frac{\partial}{\partial\rho} & \frac{\partial}{\partial\varphi} & \frac{\partial}{\partial z} \\ V_\rho & \rho V_\varphi & V_z \end{vmatrix} \tag{D.31}$$

We can also find the scalar Laplacian in cylindrical coordinates by applying the general expression Equation D.27:

$$\text{Laplacian} \qquad \nabla^2\psi = \frac{1}{\rho}\frac{\partial}{\partial\rho}\left(\rho\frac{\partial\psi}{\partial\rho}\right) + \frac{1}{\rho^2}\frac{\partial^2\psi}{\partial\varphi^2} + \frac{\partial^2\psi}{\partial z^2} \tag{D.32}$$

D.4.2   Polar coordinates

Polar coordinates consist of three independent variables r, θ φ and their associated unit vectors $\hat{r}$, $\hat{\theta}$, $\hat{\varphi}$. The limits on these variables are,

$$0 \le r \le \infty \qquad 0 \le \theta \le \pi \qquad 0 \le \varphi \le 2\pi$$

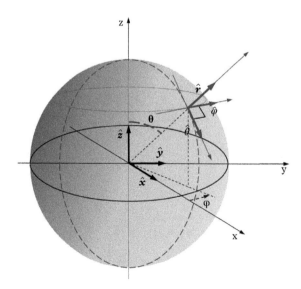

Figure D.2 *Polar coordinates r,θ,φ and unit vectors $\hat{r},\hat{\theta},\hat{\varphi}$ and their relations to Cartesian coordinates and unit vectors, $x,y,z,\hat{x},\hat{y},\hat{z}$.*

Figure D.2 shows the relation between the Cartesian coordinates x, y, z and the polar coordinates r, θ, φ. From Figure D.2 we see that

$$r = \sqrt{x^2 + y^2 + z^2} \qquad x = r\sin\theta\cos\varphi \qquad y = r\sin\theta\cos\varphi \qquad z = r\cos\theta$$

From the differential volume dV and inspection of Figure D.2 we can identify the scale factors h_1, h_2, h_3:

$$dV = ds_r\, ds_\theta\, ds_\varphi = dr\, rd\theta\, r\sin\theta\, d\varphi$$

$$h_r = 1 \qquad h_\theta = r \qquad h_\varphi = r\sin\theta$$

Then from Equations D.24, D.26, and D.28 we write expressions for the grad, div, and curl operators in polar coordinates:

$$\text{grad} \qquad \nabla\psi = \frac{\partial\psi}{\partial r}\hat{r} + \frac{1}{r}\frac{\partial\psi}{\partial\theta}\hat{\theta} + \frac{1}{r\sin\theta}\frac{\partial\psi}{\partial\varphi}\hat{\varphi} \tag{D.33}$$

$$\text{div} \qquad \nabla\cdot V = \frac{1}{r^2\sin\theta}\left[\sin\theta\frac{\partial r^2 V_r}{\partial r} + r\frac{\partial\sin\theta\, V_\theta}{\partial\theta} + r\frac{\partial V_\varphi}{\partial\varphi}\right] \tag{D.34}$$

$$\text{curl} \qquad \nabla\times V = \frac{1}{r^2\sin\theta}\begin{vmatrix} \hat{r} & r\hat{\theta} & r\sin\theta\,\hat{\varphi} \\ \frac{\partial}{\partial r} & \frac{\partial}{\partial\theta} & \frac{\partial}{\partial\varphi} \\ V_r & rV_\theta & r\sin\theta\, V_\varphi \end{vmatrix} \tag{D.35}$$

Finally, again by applying Equation D.27 we can obtain the Laplacian operator in polar coordinates:

$$\text{Laplacian} \qquad \nabla^2\psi = \frac{1}{r^2\sin\theta}\left[\sin\theta\frac{\partial}{\partial r}\left(r^2\frac{\partial\psi}{\partial r}\right) + \frac{\partial}{\partial\theta}\left(\sin\theta\frac{\partial\psi}{\partial\theta}\right) + \right.$$
$$\left. \frac{1}{\sin\theta}\frac{\partial^2\psi}{\partial\varphi^2}\right] \tag{D.36}$$

Or, expanding the radial terms,

$$\text{Laplacian} \qquad \nabla^2\psi = \frac{\partial^2\psi}{\partial r^2} + \frac{2}{r}\frac{\partial\psi}{\partial r} + \frac{1}{r^2\sin\theta}\left[\frac{\partial}{\partial\theta}\left(\sin\theta\frac{\partial\psi}{\partial\theta}\right) + \right.$$
$$\left. \frac{1}{\sin\theta}\frac{\partial^2\psi}{\partial\varphi^2}\right] \tag{D.37}$$

Appendix E
Properties of Phasors

E.1 Introduction

In Chapter 3, Section 3.1.1, we discussed the phasor form for electromagnetic propagating fields consisting of a single-frequency harmonic time-dependence factor $e^{-i\omega t}$ and the spatially dependent amplitude-modulation factor $e^{i(\mathbf{k}\cdot\mathbf{r}+\varphi)}$. The general form of the field is

$$A(\mathbf{r}, t) = A_0 e^{i(\mathbf{k}\cdot\mathbf{r}+\varphi)} e^{-i\omega t} \tag{E.1}$$

and we separated the field into a phasor factor, involving only the spatial dependence, and a harmonic time factor that was usually invariant.

In general, a 'phasor' is any function with a sinusoidal modulation, and in this Appendix, we examine some of the useful properties of phasors and their application to circuit theory.

E.1.1 Phasor addition

Since phasors can be represented as complex exponentials, multiplication of two phasors $Ae^{i\alpha}$ and $Be^{i\beta}$ is elementary:

$$Ae^{i\alpha} \cdot Be^{i\beta} = ABe^{i(\alpha+\beta)} = Ce^{i\gamma} \tag{E.2}$$

The product of two phasors is another phasor with a phase angle equal to the algebraic sum of the two individual arguments of the exponential factors.

The addition of two phasors is less obvious. In many cases we are interested in real sin and cos functions so let us examine the following sum of two real phasors, $V_1(t) = A\sin\omega t$ and $V_2(t) = B\sin(\omega t + \varphi)$:

$$\begin{aligned} V_3(t) &= V_1(t) + V_2(t) \\ &= A\sin\omega t + B\sin(\omega t + \varphi) \end{aligned} \tag{E.3}$$

On intuition we might suppose that the sum of two sin functions differing only in amplitude and relative phase would be another sin function. We posit therefore that the form of the sum is

$$V_3(t) = C\sin(\omega t + \delta) \tag{E.4}$$

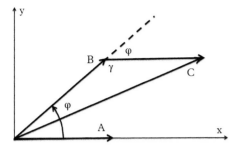

Figure E.1 *Relations among two phasors A and B that differ in amplitude and phase. Phase of A leads phase of B by φ. The interior angle γ is related to φ by γ = π − φ.*

where C and δ are to be determined. Expanding $V_3(t)$ in Equations E.3 and E.4, and setting coefficients of $\sin \omega t$ and $\cos \omega t$ equal, we find that

$$\delta = \tan^{-1} \left[\frac{B \sin \varphi}{A + B \cos \varphi} \right] \tag{E.5}$$

$$C = \left[A^2 + B^2 + 2AB \cos \varphi \right]^{1/2} \tag{E.6}$$

The expression for C, the amplitude of the phasor sum, is highly reminiscent of the result we would expect for the length of a vector $\mathbf{C}$ that is the vector sum of $\mathbf{A}$ and $\mathbf{B}$. In this case the length of the vector $\mathbf{C}$ would be determined by the law of cosines as indicated in Figure E.1. The interior angle γ between the two component vectors is closely related to φ, the relative angle between the two phasors. In fact, it is easy to show that

$$\gamma = \pi - \varphi \tag{E.7}$$

and substitution into Equation E.6 results in

$$C = \left[A^2 + B^2 - 2AB \cos \gamma \right]^{1/2} \tag{E.8}$$

confirming that two phasors add with amplitude equivalent to the resultant 'length' from the sum of two vectors.

E.2   Application of phasors to circuit analysis

As an illustration of the usefulness of phasors to harmonic circuit analysis, we consider a simple RLC circuit with

$$V(t) = Ri(t) + L\frac{di(t)}{dt} + \frac{1}{C} \int i(t)\,dt \tag{E.9}$$

where V, R, L, and C have their usual meanings of voltage, resistance, inductance, and capacitance, respectively, and $i(t)$ is the time-dependent current running through all the lumped circuit elements. We posit that the current source is harmonically oscillating at frequency ω and write $i(t)$ as a phasor,

$$i(t) = i_0 e^{-i\omega t} \tag{E.10}$$

Substituting the phasor form into Equation E.9 we find that

$$V(t) = Ri_0 e^{-i\omega t} + \omega L i_0 e^{-i(\omega t + \pi/2)} + \frac{1}{\omega C} i_0 e^{-(\omega t - \pi/2)} \tag{E.11}$$

$$= V_R + V_L + V_C \tag{E.12}$$

We see that the voltage drop across the resistor is in phase with the source, while the *inductive reactance*, ωL, shows a phase advance of $\pi/2$ and the *capacitive reactance*, $1/\omega C$, lags in phase by $\pi/2$. The resistive voltage drop is purely dissipative while the two reactances store energy originating at the source and return it to the circuit at different points along the harmonic cycle. Equation E.9 can be rewritten in terms of the *impedance, $Z(\omega)$,*

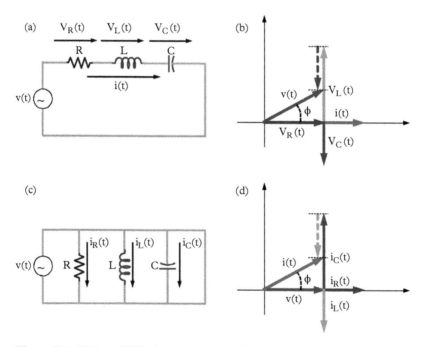

Figure E.2 *(a) Series RLC circuit with harmonic voltage source and a common current i(t) running through each circuit element. (b) Voltage phasor diagram showing the phase 'lead' for the capacitive voltage term, the phase 'lag' for the inductive voltage term, and the net phase difference between the driving voltage and the resistive voltage drop. Panels (c) and (d) show similar diagrams for current phasors in RLC parallel circuit.*

$$V(t) = Z(\omega)i_0 e^{-i\omega t} \tag{E.13}$$

$$Z(\omega) = R + (\omega L)e^{-i(\pi/2)} + \left(\frac{1}{\omega C}\right)e^{i(\pi/2)} \tag{E.14}$$

$$= R - i\omega L + \left(\frac{i}{\omega C}\right) \tag{E.15}$$

$$= R + iX_L(\omega) + iX_C(\omega) \tag{E.16}$$

where X_L and X_C are the inductive and capacitive reactances in phasor form. We can take this analysis further by considering the RLC circuit in series and in parallel. Figure E.2 shows circuit and phasor diagrams for the two cases. In the series circuit, the property common to all lumped elements is the current $i(t)$ so the relevant phasor quantities are the voltages across the resistive, inductive, and capacitive elements. For the parallel circuit, the common property is the voltage $V(t)$ and the relevant phasor diagram is expressed in terms of the individual currents passing through each circuit element.

Appendix F
Properties of the Laguerre Functions

The Laguerre polynomials and the 'associated' Laguerre polynomials are solutions to Laguerre's differential equation and are closely related to the radial solutions of the Schrödinger equation for the hydrogen atom.

F.1   Generating function and recursion relations

The generating function for the Laguerre polynomials is given by

$$g(x, z) = \frac{e^{-xz/(1-z)}}{1-z} = \sum_{n=0}^{\infty} L_n(x)z^n \qquad |z| \le 1 \tag{F.1}$$

As we did with Legendre and Hermite polynomials, we find recurrence relations between adjacent members of the Laguerre polynomials by differentiating the generating function with respect to z and x:

$$(1 - z^2)\frac{\partial g(x, z)}{\partial z} = (1 - x - z)g(x, z) \tag{F.2}$$

$$(z - 1)\frac{\partial g(x, z)}{\partial x} = zg(x, z) \tag{F.3}$$

Substituting Equation F.1 into Equation F.2, adjusting the summation indices, and equating terms with equal powers of z^n, results in the first recurrence relation:

$$L_{n+1}(x) = \left(\frac{2n + 1 - x}{n + 1}\right)L_n - nL_{n-1} \tag{F.4}$$

It is often useful to have a recurrence relation involving the derivatives of the polynomials, and we can obtain one by using Equations F.2 and F.3 to derive a third relation,

$$x\frac{\partial g(x, z)}{\partial x} = z\frac{\partial g(x, z)}{\partial z} - z\frac{\partial[zg(x, z)]}{\partial z} \tag{F.5}$$

Making the appropriate substitutions from Equation F.1, adjusting summation indices, and equating terms with like powers of z^n results in

$$xL'_n(x) = nL_n(x) - nL_{n-1}(x) \tag{F.6}$$

which is the second recursion relation for Laguerre polynomials.

F.2 Orthogonality and normalisation

The Laguerre polynomials themselves do not form an orthogonal set of functions, but they can be made orthogonal by joining a factor e^{-x} to them. Thus:

$$\int_0^\infty e^{-x} L_m(x) L_n(x) = \delta_{mn} \tag{F.7}$$

where the polynomials defined by the generating function, Equation F.1 are already normalised. As in the case of Hermite polynomials, we can identify a closely related Laguerre function $\varphi_n(x)$ that has the orthonormal properties of a complete set of functions involving the Laguerre polynomials,

$$\varphi_n(x) = e^{-x/2} L_n(x) \tag{F.8}$$

The ordinary differential equation that $\varphi_n(x)$ satisfies, however, is not the physically relevant one with which we associate the Schrödinger radial equation for the hydrogen atom. To find the solutions for *that* equation we need to examine the 'associated' Laguerre polynomials.

F.3 Associated Laguerre polynomials

The associated Laguerre polynomials are defined in terms of the Laguerre polynomials by,

$$L_n^k(x) = (-1)^k \frac{d^k}{dx^k} L_{n+k}(x) \tag{F.9}$$

and given the Rodrigues form for the Laguerre polynomials,

$$L_n(x) = \frac{e^x}{n!} \frac{d^n}{dx^n} (x^n e^{-x}) \tag{F.10}$$

from which, after writing the Taylor series expansion of the exponential terms, we can write a series representation for the 'normal' Laguerre polynomials with integral n as

$$L_n(x) = \sum_{m=0}^n \frac{(-1)^m n! x^m}{(n-m)! m! m!} \tag{F.11}$$

We can obtain a similar series representation for the associated Laguerre polynomials by substituting Equation F.11 into the definition, Equation F.9:

$$L_n^k(x) = \sum_{m=0}^{n} (-1)^m \frac{(n+k)!}{(n-m)!(k+m)!m!} x^m \qquad k > -1 \tag{F.12}$$

The first few associated Laguerre polynomials are, therefore,

$$L_0^k(x) = 1$$

$$L_1^k(x) = -x + k + 1$$

$$L_2^k(x) = \frac{x^2}{2} - (k+2)x + \frac{(k+2)(k+1)}{2}$$

F.3.1   Generating function, recurrence relations, and orthonormality

The generating function for the associated Laguerre polynomials is given by

$$\frac{e^{-xz/(1-z)}}{(1-z)^{k+1}} = \sum_{n=0}^{\infty} L_n^k(x) z^n \qquad |z| < 1 \tag{F.13}$$

from which recurrence relations can be derived by differentiation and equating of terms with like powers in z in the usual way. The resulting recurrence relation is

$$(n+1)L_{n+1}^k(x) = (2n+k+1-x)L_n^k(x) - (n+k)L_{n-1}^k(x) \tag{F.14}$$

The associated Laguerre polynomials are solutions to the associated Laguerre ordinary differential equation, but this equation is not 'self-adjoint' or hermitian, and therefore cannot function as a legitimate quantum mechanical operator. However, it can be cast into a form suitable for quantum mechanics by multiplying the associated Laguerre polynomials by a factor, $e^{-x/2} x^{k/2} L_n^k(x)$. This factor transforms the polynomials into the *associated Laguerre functions*:

$$\varphi_n^k(x) = e^{-x/2} x^{k/2} L_n^k(x) \tag{F.15}$$

These associated Laguerre functions have the desired orthonormality,

$$\int_0^{\infty} e^{-x} x^k L_n^k(x) L_m^k(x)\, dx = \frac{(n+k)!}{n!} \delta_{mn} \tag{F.16}$$

and they satisfy the 'quantum mechanically correct' ordinary differential equation,

$$x \frac{d^2 \varphi_n^k(x)}{dx^2} + \frac{d\varphi_n^k(x)}{dx} + \left(-\frac{x}{4} + \frac{2n+k+1}{2} - \frac{k^2}{4x} \right) \varphi_n^k(x) = 0 \tag{F.17}$$

This equation is 'correct' in the sense that the operator integrals corresponding to average values of observable variables, and the eigenvalues themselves, are guaranteed to be real. But in fact, Equation F.17 is still not the ordinary differential equation corresponding to the form of the radial Schrödinger equation for the hydrogen atom. The problem is that the Schrödinger equation does not exhibit a term with a first-derivative operator. However, this term can be eliminated by multiplying the Laguerre function by a factor $x^{1/2}$:

$$\Phi_n^k(x) = e^{-x/2} x^{(k+1)/2} L_n^k(x) \tag{F.18}$$

This family of modified Laguerre functions is a solution to the ordinary differential equation,

$$\frac{d^2 \Phi_n^k(x)}{dx^2} + \left(-\frac{1}{4} + \frac{2n+k+1}{2x} - \frac{k^2-1}{4x^2} \right) \Phi_n^k(x) = 0 \tag{F.19}$$

and has a normalisation:

$$\int_0^\infty e^{-x} x^{k+1} \left[L_n^k(x) \right]^2 dx = \frac{(n+k)!}{n!} (2n+k+1) \tag{F.20}$$

Notice that in Equation F.19 there is no term in the first derivative of x, and therefore this second-order differential equation and its family of modified Laguerre functions can serve as a prototype for the radial Schrödinger equation of the one-electron atom.

Appendix G
Properties of the Legendre Functions

The Legendre functions are a family of solutions to the Legendre differential equation. This equation determines the angular behaviour of many physical problems including the scalar Helmholtz wave equation in optics, Maxwell's equations in classical electro-dynamics, and the Schrödinger wave equation in quantum mechanics. We discuss here the properties of these functions common to all these physical situations.

The Legendre equation itself is

$$\left(1 - x^2\right) P_n''(x) - 2x P_n'(x) + n(n+1)P_n(x) = 0 \tag{G.1}$$

where $n = 0, 1, 2, \ldots$. In most physics problems $x = \cos\theta$, and the Legendre equation takes the form

$$\frac{1}{\sin\theta} \frac{d}{d\theta}\left(\sin\theta \frac{dP_n(\cos\theta)}{d\theta}\right) + n(n+1)P_n(\cos\theta) = 0 \tag{G.2}$$

or more explicitly, carrying out the derivative,

$$\frac{d^2 P_n(\cos\theta)}{d\theta} + \cot\theta \frac{dP_n(\cos\theta)}{d\theta} + n(n+1)P_n(\cos\theta) = 0 \tag{G.3}$$

where the derivative is with respect to the polar coordinate θ rather than x. We have already encountered this latter form in Equations 2.101, 2.102 and 2.103 when studying the angular solutions of Laplace's equation.

G.1 Generating function

The Legendre polynomials can always be obtained from a *generating function*:

$$g(t,x) = (1 - 2xt + t^2)^{-1/2} = \sum_{n=0}^{\infty} P_n(x)t^n, \qquad |t| < 1 \tag{G.4}$$

The left-hand side of this equation can be cast into a power series in t^n so that the left-hand side and right-hand sides, also a power series in t^n, can be equated term by term:

$$(1 - 2xt + t^2)^{-1/2} = \sum_{n=0}^{\infty} \sum_{k=0}^{[n/2]} (-1)^k \frac{(2n - 2k)!}{2^{2n-2k} n! k! (n - k)!} (2x)^{(n-2k)} t^n = \sum_{n=0}^{\infty} P_n(x) t^n \qquad (G.5)$$

where the upper limit on the sum, $[n/2]$, means $n/2$ when n is even and $(n - 1)/2$ when n is odd. Therefore, we can obtain an expression for each $P_n(x)$ in terms of a sum over the index k:

$$P_n(x) = \sum_{k=0}^{[n/2]} (-1)^k \frac{(2n - 2k)!}{2^n k! (n - k)! (n - 2k)!} x^{n-2k} \qquad (G.6)$$

Using the Equation G.6 function we find, for example,

$$P_0(x) = 1 \qquad P_1(x) = x \qquad P_2(x) = \frac{1}{2} (3x^2 - 1) \qquad P_3(x) = \frac{5}{2} (x^3 - 3x)$$

G.2   Recurrence relations

Families of functions that are solutions to a differential equation often exhibit *recurrence relations* that express how a given solution $P_n(x)$ is related in some simple way to its neighbours, $P_{n+1}(x)$ and $P_{n-1}(x)$. These recurrence relations are often another, simpler way to spawn all the needed members of the family once one member is known. In the case of the Legendre functions we can find a recurrence relation by starting from the generating function, Equation G.4, and differentiating it:

$$\frac{\partial g(t, x)}{\partial t} = \frac{x - t}{(1 - 2xt + t^2)^{3/2}} = \sum_{n=0}^{\infty} n P_n(x) t^{n-1} \qquad (G.7)$$

Using Equation G.4 we can write this last expression as the sum of four terms, each of which is a power series in t:

$$(1 - 2xt + t^2) \sum_{n=0}^{\infty} n P_n(x) t^{n-1} + (t - x) \sum_{n=0}^{\infty} P_n(x) t^n = 0 \qquad (G.8)$$

or

$$\sum_{n=0}^{\infty} n P_n(x) t^{n-1} - \sum_{n=0}^{\infty} x(2n + 1) P_n(x) t^n + \sum_{n=0}^{\infty} (n + 1) P_n(x) t^{n+1} = 0 \qquad (G.9)$$

Now we seek to rewrite the various terms so that they are all sums of a power series in t^n. We can do this by adjusting the indices n. Thus,

$$\sum_{n=0}^{\infty} nP_n(x)t^{n-1} \quad \text{is equivalent to} \quad \sum_{n=0}^{\infty} (n+1)P_{n+1}t^n \qquad (G.10)$$

and

$$\sum_{n=0}^{\infty} (n+1)P_n(x)t^{n+1} \quad \text{is equivalent to} \quad \sum_{n=0}^{\infty} nP_{n-1}(x)t^n \qquad (G.11)$$

and therefore we can write

$$(2n+1)xP_n(x) = (n+1)P_{n+1}(x) + nP_{n-1}(x) \qquad (G.12)$$

Equation G.12 is the recurrence relation for Legendre polynomials. It is a prescription for expressing a given $P_n(x)$ in terms of the adjacent polynomials in the series, $P_{n-1}(x)$ and $P_{n+1}(x)$. For example, it is easy to remember the first two Legendre polynomials, $P_0(x) = 1$ and $P_1 = x$. Using Equation G.12 we find immediately $P_2 = 1/2(3x^2-1)$. Then, with $P_1(x)$ and $P_2(x)$ in hand, we find $P_3(x) = 1/2(5x^3 - 3x)$. Therefore, in principle, any member of the series may be found from the recurrence relation.

G.3 Parity

It appears by inspection that even-index Legendre polynomials are even and odd-index polynomials are odd. In general, the parity of the polynomials can be demonstrated by recourse to the generating function, Equation G.4:

$$g(t, x) = g(-t, -x)$$

$$\sum_{n=0}^{\infty} P_n(x)t^n = \sum_{n=0}^{\infty} P_n(-x)(-t)^n = \sum_{n=0}^{\infty} P_n(-x)(-1)^n t^n$$

Equating equal powers of t^n we have

$$P_n(x) = (-1)^n P_n(-x) \qquad (G.13)$$

which shows that, in general, even-index Legendre polynomials are even and odd-index polynomials are odd in the $x \to -x$ parity operation.

G.4 Orthogonality and normalisation

The Legendre equation, Equation G.1, can be regarded as an operator-eigenfunction-eigenvalue equation, $\hat{O}\psi - \lambda\psi = 0$, where $\hat{O}$ the operator is

$$\hat{O} = \frac{d}{dx}(1-x^2)\frac{d}{dx} \tag{G.14}$$

The eigenfunctions ψ are the Legendre polynomials $P_n(x)$ and the eigenvalues $\lambda = n(n+1)$. Viewed in this way, it can be shown that the 'operator' is hermitian, and therefore the eigenfunctions are orthogonal and the eigenvalues real. Therefore, the Legendre polynomials are orthogonal:

$$\int_{-1}^{1} P_n(x)P_m(x)\,dx = 0 \qquad m \neq n \tag{G.15}$$

and in fact they form a complete set, spanning the space of x from -1 to +1. The normalisation factor comes from the determination of the integral when $m = n$. We start with the generating function, square it, and write

$$(1 - 2tx + t^2)^{-1} = \left[\sum_{n=0}^{\infty} P_n(x)t^n\right]^2 \tag{G.16}$$

Then we integrate both sides and note that all the $P_n P_m$ cross terms will vanish by virtue of Equation G.15:

$$\int_{-1}^{1} \frac{dx}{1 - 2tx + t^2} = \sum_{n=0}^{\infty} \int_{-1}^{1} [P_n(x)]^2\,dx\, t^{2n} \tag{G.17}$$

The integral on the left can be readily evaluated by making a change of variable, $y = 1 + 2t + t^2$ and $dy = -2t\,dx$. Making the substitution into Equation G.17, together with the change in limits $x = -1 \rightarrow y = (1 + t)^2$ and $x = 1 \rightarrow y = (1 - t)^2$, we have

$$\int_{-1}^{1} \frac{dx}{1 - 2tx + t^2} = \frac{1}{t}\ln\left(\frac{1+t}{1-t}\right) = \sum_{n=0}^{\infty} \int_{-1}^{1} [P_n(x)]^2\,dx\, t^{2n} \tag{G.18}$$

Now, the ln expression can be expanded in a power series,

$$\frac{1}{t}\ln\left(\frac{1+t}{1-t}\right) = 2\sum_{n=0}^{\infty} \frac{t^{2n}}{2n+1} \tag{G.19}$$

and therefore,

$$2 \sum_{n=0}^{\infty} \frac{t^{2n}}{2n+1} = \sum_{n=0}^{\infty} \int_{-1}^{1} [P_n(x)]^2 \, dx \, t^{2n} \tag{G.20}$$

Finally, equating each term in the power series t^{2n} on each side of Equation G.20, we have

$$\int_{-1}^{1} [P_n(x)]^2 \, dx = \frac{2}{2n+1} \tag{G.21}$$

which is the normalisation condition.

Using the conventional expression that summarises orthonormality, we write

$$\int_{-1}^{1} P_m(x) P_n(x) \, dx = \frac{2\delta_{mn}}{2n+1} \tag{G.22}$$

Appendix H
Properties of the Hermite Polynomials

The Hermite polynomials constitute the essential part of the solutions to the Schrödinger equation for the one-dimensional harmonic oscillator. Their properties are the starting point for interpreting vibrational motion and dipole transition selection rules in diatomic molecules. The Hermite polynomials can be defined from the 'Rodrigues representation':

$$H_n(x) = (-1)^n \left(\exp x^2 \right) \frac{d^n}{dx^n} \left[\exp \left(-x^2 \right) \right] \tag{H.1}$$

or by use of the generating function.

H.1 Generating function and recurrence relations

The generating function is given by

$$g(x, t) = e^{-t^2 + 2tx} = \sum_{n=0}^{\infty} H_n(x) \frac{t^n}{n!} \tag{H.2}$$

from which we can obtain the recurrence relations similarly to the procedure followed for the Legendre polynomials in Appendix G. First we take the partial derivative of $g(x, t)$ with respect to t,

$$\frac{\partial g(x, t)}{\partial t} = 2(x - t)e^{-t^2 + 2tx} = \sum_{n=0}^{\infty} H_n(x) \frac{nt^n}{n!} = \sum_{n=0}^{\infty} H_n(x) \frac{t^{n-1}}{(n-1)!} \tag{H.3}$$

From which we find, after adjusting the summation indices,

$$-2n \sum_{n=0}^{\infty} H_{n-1}(x) \frac{t^n}{n!} + 2x \sum_{n=0}^{\infty} H_n(x) \frac{t^n}{n!} - \sum_{n=0}^{\infty} H_{n+1}(x) \frac{t^n}{n!} = 0 \tag{H.4}$$

Terms of equal powers of t^n must be equal and the recurrence relation is, therefore,

$$H_{n+1}(x) = 2xH_n(x) - 2nH_{n-1}(x) \tag{H.5}$$

We can also get a recurrence relation involving the derivative of the Hermite polynomials by taking the partial derivative of the generating function with respect to x:

Table H.1 *Hermite polynomials.*

Index n	Function	Polynomial
0	$H_0(x)$	1
1	$H_1(x)$	$2x$
2	$H_2(x)$	$4x^2 - 2$
3	$H_3(x)$	$8x^3 - 12x$
4	$H_4(x)$	$16x^4 - 48x^2 + 12$

$$\frac{\partial g(x, t)}{\partial x} = 2te^{-t^2+2tx} = 2t\sum_{n=0}^{\infty} H_n(x)\frac{t^n}{n!} = \sum_{n=0}^{\infty} H_n'(x)\frac{t^n}{n!} \tag{H.6}$$

Then

$$2\sum_{n=0}^{\infty} H_n(x)\frac{t^{n+1}}{n!} = 2n\sum_{n=0}^{\infty} H_{n-1}(x)\frac{t^n}{n!} = \sum_{n=0}^{\infty} H_n'(x)\frac{t^n}{n!} \tag{H.7}$$

and

$$2nH_{n-1}(x) = H_n'(x) \tag{H.8}$$

We can get the first two members of the Hermite polynomials by expanding the generating function exponential in a Taylor series,

$$e^{-t^2+2tx} = \sum_{n=0}^{\infty} \frac{\left(2tx - t^2\right)^n}{n!} = 1 + (2tx - t^2) + \dots \tag{H.9}$$

that results in $H_0(x) = 1$ and $H_1(x) = 2x$. Then from these, and Equation H.5, all the polynomials up to arbitrary n can be calculated. The first few Hermite polynomials are shown in Table H.1.

H.2   Orthogonality and normalisation

H.2.1   Orthogonality

The Hermite polynomials themselves are not orthogonal in the sense that

$$\int_{-\infty}^{\infty} H_n(x)H_m(x)\, dx \neq 0 \qquad m \neq n \tag{H.10}$$

but a closely related integral is orthogonal,

$$\int_{-\infty}^{\infty} H_n(x)H_m(x)\exp\left(-x^2\right) dx = 0 \qquad m \neq n \tag{H.11}$$

In fact, we can define a new orthogonal function,

$$\varphi_n(x) = \exp(-x^2/2)H_n(x) \tag{H.12}$$

Now using the two recurrence relations, we can obtain a second order differential equation in the Hermite polynomials of index n in the following way. First, by substituting Equation H.8 into Equation H.5 we eliminate the $n-1$ index,

$$H_{n+1}(x) = 2xH_n(x) - H_n'(x) \tag{H.13}$$

Then differentiate the result,

$$H_{n+1}'(x) = 2H_n(x) + 2xH_n' - H_n''(x) \tag{H.14}$$

Use Equation H.8 again to eliminate the term with the $n+1$ index,

$$H_{n+1}'(x) = 2(n+1)H_n(x) \tag{H.15}$$

Now regroup the terms in n,

$$H_n''(x) - 2xH_n'(x) + 2nH_n(x) = 0 \tag{H.16}$$

Differentiating Equation H.12 and substituting into Equation H.16 results in

$$\varphi_n''(x) + (2n + 1 - x^2)\varphi_n(x) = 0 \tag{H.17}$$

which is the equation for the one-dimensional quantum mechanical oscillator. Therefore, we see that the physically significant entities are not the Hermite polynomials *per se* but the orthogonal functions, $\varphi_n(x) = H_n(x)\exp(-x^2/2)$.

H.2.2   Normalisation
We find the normalisation of the Hermite functions, Equation H.12, by multiplying the generating function for $H_m(x)$ with that for $H_n(x)$, similar to the approach we take with the squared generating function of the Legendre polynomials, Equation G.16,

$$\exp(-s^2 + 2sx)\exp(-t^2 + 2tx) = \sum_{m,n=0}^{\infty} H_m(x)H_n(x)\frac{s^m t^n}{m!n!} \tag{H.18}$$

Then multiply both sides by $\exp(-x^2)$ and integrate over the interval from $-\infty$ to ∞. Cross terms in the $\exp(-x^2)H_m(x)H_n(x)$ products drop out of the integral due to orthogonality:

$$\sum_{n=0}^{\infty} \frac{(st)^n}{n!n!} \int_{-\infty}^{\infty} \exp(-x^2) \, [H_n(x)]^2 \, dx = \int_{-\infty}^{\infty} \exp-[(x-s-t)^2] \exp(2st) \, dx \qquad \text{(H.19)}$$

The first exponential term under the integral on the right-hand side is just the error function, and the integral over it is known to be $\sqrt{\pi}$. The remaining term, $\exp(2st)$, can be expanded in a Taylor series and we have, therefore,

$$\sum_{n=0}^{\infty} \frac{(st)^n}{n!n!} \int_{-\infty}^{\infty} \exp(-x^2) \, [H_n(x)]^2 \, dx = \sqrt{\pi} \sum_{n=0}^{\infty} \frac{2^n (st)^n}{n!} \qquad \text{(H.20)}$$

Now in the usual way we equate equal powers of $(st)^n$ that results in

$$\int_{-\infty}^{\infty} \exp(-x^2) \, [H_n(x)]^2 \, dx = 2^n \sqrt{\pi} n! \qquad \text{(H.21)}$$

Equation H.21 is the normalisation expression for the Hermite functions.

Index

AUTHORS

John Weiner graduated from Mrs. Warnock's kindergarten on Hathaway Lane in Havertown, Pennsylvania, with a dual major in finger painting and rhythm sticks, in 1949. He was subsequently promoted from Mrs. Warnock's to the adjacent Oakmont Elementary School. In the 5th and 6th grades he played right tackle, left end, and left halfback (T-formation) on the football team, graduating in June of 1955. He advanced to the Haverford Junior High School in September, but he was too small to play on the football team as a 7th grader. Instead he joined the marching band (trombone and baritone); and because the trombones always marched in the first rank (to leave enough space for their slides) and right behind the majorettes, he had occasion to fall in love with the head drum majorette, marching right in front of him. His love was unrequited (she was a 9th grader and the most popular girl in the high school). Despite this disappointment he managed to pass the 7th, 8th, and 9th grades. Regretfully leaving his majorette heart-throb at Haverford High, he and his family in the summer of 1958 moved to Chambersburg, Pennsylvania, where he enrolled in the Chambersburg Area Senior High School. He finally made the junior varsity (JV) football squad but was still too small for the varsity. Nevertheless during the JV game between Chambersburg and Scotland School he caught a 40 yard pass from quarterback Kirby Keller and ran for the only touchdown that the team was to score that day. It was a unique moment of unalloyed happiness. In the summer of 1960 his family moved again to Bethesda, Maryland, where he graduated from high school in June of 1961. From 1961 to 1964 he attended Penn State in the town of State College, leaving in June of 1964 with a BS degree in Chemistry (at Penn State playing football was out of the question). From 1964 to 1970 he attended graduate school at the University of Chicago, earning a PhD in Chemical Physics in June of 1970. After a post-doctoral interlude at Yale he got his first 'real' job as an assistant professor at Dartmouth College in September of 1973. In 1978 he was invited to the Université de Paris for a year, and while there accepted an offer from the University of Maryland, College Park.

At Maryland Weiner established a research group studying the collisional process, associative ionisation. Sodium atoms were a convenient experimental choice for these studies, and he became interested in 'laser-induced' collisions – the idea of colliding atoms or ions absorbing a light quantum to produce chemical binding. Coincidently, at the National Institute of Standards and Technology (NIST), W. D. Phillips' group was learning how to cool a beam of sodium atoms to submilliKelvin temperatures. Using cold atoms to study extremely slow associative ionisation collisions was an obvious consequence, and Weiner spent the next ten years focused on various aspects of 'ultracold collisions'. In 1997 he joined the faculty at the Université Paul Sabatier in Toulouse, France, where his interests shifted to how nanoscale structures might be used to manipulate cooled and trapped atom clouds and condensed quantum gases. In 2006 he retired from his post in Toulouse becoming a visiting researcher at the Instituto de Física de S ao Carlos (IFSC) in the Universidade de S ao Paulo, Brazil. In the years 2009-2010 he was a visiting fellow at the Center for Nanoscale Science and Technology (CNST) at NIST, Gaithersburg before returning to the IFSC for another and final year. Weiner is currently living in Paris, France, where he is tolerated by Samba and Annick, his cat and wife, respectively. He reads a lot, attempts to think clearly about some difficult subjects in the morning, and tries not to make stupid mistakes for the rest of the day. The success or failure of this latter effort will be judged by readers of this book.

Frederico Nunes was born in Recife, Pernambuco State, Brazil, in 1947. He obtained an undergraduate degree in electrical engineering from the School of Engineering at the Federal University of Pernambuco (UFPE) in 1971. He continued his advanced education by enrolling in the Physics and Pure Mathematics curriculum at the Institute of Professor Gleb Wataghin, State University of Campinas (UNICAMP), obtaining his MS and PhD degrees in 1974 and 1977, respectively. Since then he has been on the faculty at several Brazilian universities and since 1997, he has been an associate professor at UFPE in Recife. He has received the Marechal Trompowski award for excellence in teaching at UFPE and has made many research contributions in the area of optics, semiconductors, photonics, and nanoplasmonics. He has authored many scientific publications and holds a number of patents in the technologies associated with these scientific disciplines.

John Weiner   *Frederico Nunes*

All Oakmont Elementary School right tackle,
1954. 'Always fight for Brown and White
and march to vic-tor-y'.